DOPAMINE–GLUTAMATE INTERACTIONS in the BASAL GANGLIA

FRONTIERS IN NEUROSCIENCE

Series Editors
Sidney A. Simon, Ph.D.
Miguel A.L. Nicolelis, M.D., Ph.D.

Published Titles

Apoptosis in Neurobiology
Yusuf A. Hannun, M.D., Professor of Biomedical Research and Chairman, Department
of Biochemistry and Molecular Biology, Medical University of South Carolina, Charleston,
South Carolina
Rose-Mary Boustany, M.D., tenured Associate Professor of Pediatrics and Neurobiology, Duke
University Medical Center, Durham, North Carolina

Neural Prostheses for Restoration of Sensory and Motor Function
John K. Chapin, Ph.D., Professor of Physiology and Pharmacology, State University
of New York Health Science Center, Brooklyn, New York
Karen A. Moxon, Ph.D., Assistant Professor, School of Biomedical Engineering, Science,
and Health Systems, Drexel University, Philadelphia, Pennsylvania

Computational Neuroscience: Realistic Modeling for Experimentalists
Eric DeSchutter, M.D., Ph.D., Professor, Department of Medicine, University of Antwerp,
Antwerp, Belgium

Methods in Pain Research
Lawrence Kruger, Ph.D., Professor of Neurobiology (Emeritus), UCLA School of Medicine and
Brain Research Institute, Los Angeles, California

Motor Neurobiology of the Spinal Cord
Timothy C. Cope, Ph.D., Professor of Physiology, Wright State University, Dayton, Ohio

Nicotinic Receptors in the Nervous System
Edward D. Levin, Ph.D., Associate Professor, Department of Psychiatry and Pharmacology and
Molecular Cancer Biology and Department of Psychiatry and Behavioral Sciences,
Duke University School of Medicine, Durham, North Carolina

Methods in Genomic Neuroscience
Helmin R. Chin, Ph.D., Genetics Research Branch, NIMH, NIH, Bethesda, Maryland
Steven O. Moldin, Ph.D., University of Southern California, Washington, D.C.

Methods in Chemosensory Research
Sidney A. Simon, Ph.D., Professor of Neurobiology, Biomedical Engineering,
and Anesthesiology, Duke University, Durham, North Carolina
Miguel A.L. Nicolelis, M.D., Ph.D., Professor of Neurobiology and Biomedical Engineering,
Duke University, Durham, North Carolina

The Somatosensory System: Deciphering the Brain's Own Body Image
Randall J. Nelson, Ph.D., Professor of Anatomy and Neurobiology,
University of Tennessee Health Sciences Center, Memphis, Tennessee

The Superior Colliculus: New Approaches for Studying Sensorimotor Integration
William C. Hall, Ph.D., Department of Neuroscience, Duke University, Durham, North Carolina
Adonis Moschovakis, Ph.D., Department of Basic Sciences, University of Crete, Heraklion, Greece

New Concepts in Cerebral Ischemia
Rick C. S. Lin, Ph.D., Professor of Anatomy, University of Mississippi Medical Center,
  Jackson, Mississippi

DNA Arrays: Technologies and Experimental Strategies
Elena Grigorenko, Ph.D., Technology Development Group, Millennium Pharmaceuticals,
  Cambridge, Massachusetts

Methods for Alcohol-Related Neuroscience Research
Yuan Liu, Ph.D., National Institute of Neurological Disorders and Stroke,
  National Institutes of Health, Bethesda, Maryland
David M. Lovinger, Ph.D., Laboratory of Integrative Neuroscience, NIAAA,
  Nashville, Tennessee

Primate Audition: Behavior and Neurobiology
Asif A. Ghazanfar, Ph.D., Princeton University, Princeton, New Jersey

Methods in Drug Abuse Research: Cellular and Circuit Level Analyses
Barry D. Waterhouse, Ph.D., MCP-Hahnemann University, Philadelphia, Pennsylvania

Functional and Neural Mechanisms of Interval Timing
Warren H. Meck, Ph.D., Professor of Psychology, Duke University, Durham, North Carolina

Biomedical Imaging in Experimental Neuroscience
Nick Van Bruggen, Ph.D., Department of Neuroscience Genentech, Inc.
Timothy P.L. Roberts, Ph.D., Associate Professor, University of Toronto, Canada

The Primate Visual System
John H. Kaas, Department of Psychology, Vanderbilt University, Nashville, Tennessee
Christine Collins, Department of Psychology, Vanderbilt University, Nashville, Tennessee

Neurosteroid Effects in the Central Nervous System
Sheryl S. Smith, Ph.D., Department of Physiology, SUNY Health Science Center,
  Brooklyn, New York

Modern Neurosurgery: Clinical Translation of Neuroscience Advances
Dennis A. Turner, Department of Surgery, Division of Neurosurgery,
  Duke University Medical Center, Durham, North Carolina

Sleep: Circuits and Functions
Pierre-Hervé Luppi, Université Claude Bernard, Lyon, France

Methods in Insect Sensory Neuroscience
Thomas A. Christensen, Arizona Research Laboratories, Division of Neurobiology,
  University of Arizona, Tuscon, Arizona

Motor Cortex in Voluntary Movements
Alexa Riehle, INCM-CNRS, Marseille, France
Eilon Vaadia, The Hebrew University, Jerusalem, Israel

Neural Plasticity in Adult Somatic Sensory-Motor Systems
Ford F. Ebner, Vanderbilt University, Nashville, Tennessee

Advances in Vagal Afferent Neurobiology
Bradley J. Undem, Johns Hopkins Asthma Center, Baltimore, Maryland
Daniel Weinreich, University of Maryland, Baltimore, Maryland

The Dynamic Synapse: Molecular Methods in Ionotropic Receptor Biology
Josef T. Kittler, University College, London, England
Stephen J. Moss, University College, London, England

Animal Models of Cognitive Impairment
Edward D. Levin, Duke University Medical Center, Durham, North Carolina
Jerry J. Buccafusco, Medical College of Georgia, Augusta, Georgia

The Role of the Nucleus of the Solitary Tract in Gustatory Processing
Robert M. Bradley, University of Michigan, Ann Arbor, Michigan

Brain Aging: Models, Methods, and Mechanisms
David R. Riddle, Wake Forest University, Winston-Salem, North Carolina

Neural Plasticity and Memory: From Genes to Brain Imaging
Frederico Bermudez-Rattoni, National University of Mexico, Mexico City, Mexico

Serotonin Receptors in Neurobiology
Amitabha Chattopadhyay, Center for Cellular and Molecular Biology, Hyderabad, India

TRP Ion Channel Function in Sensory Transduction and Cellular Signaling Cascades
Wolfgang B. Liedtke, M.D., Ph.D., Duke University Medical Center, Durham, North Carolina
Stefan Heller, Ph.D., Stanford University School of Medicine, Stanford, California

Methods for Neural Ensemble Recordings, Second Edition
Miguel A.L. Nicolelis, M.D., Ph.D., Professor of Neurobiology and Biomedical Engineering,
    Duke University Medical Center, Durham, North Carolina

Biology of the NMDA Receptor
Antonius M. VanDongen, Duke University Medical Center, Durham, North Carolina

Methods of Behavioral Analysis in Neuroscience
Jerry J. Buccafusco, Ph.D., Alzheimer's Research Center, Professor of Pharmacology and Toxicology,
    Professor of Psychiatry and Health Behavior, Medical College of Georgia,
    Augusta, Georgia

***In Vivo* Optical Imaging of Brain Function, Second Edition**
Ron Frostig, Ph.D., Professor, Department of Neurobiology, University of California,
    Irvine, California

Fat Detection: Taste, Texture, and Post Ingestive Effects
Jean-Pierre Montmayeur, Ph.D., Centre National de la Recherche Scientifique, Dijon, France
Johannes le Coutre, Ph.D., Nestlé Research Center, Lausanne, Switzerland

The Neurobiology of Olfaction
Anna Menini, Ph.D., Neurobiology Sector International School for Advanced Studies, (S.I.S.S.A.),
    Trieste, Italy

Neuroproteomics
Oscar Alzate, Ph.D., Department of Cell and Developmental Biology,
    University of North Carolina, Chapel Hill, North Carolina

Translational Pain Research: From Mouse to Man
Lawrence Kruger, Ph.D., Department of Neurobiology, UCLA School of Medicine, Los Angeles,
    California
Alan R. Light, Ph.D., Department of Anesthesiology, University of Utah, Salt Lake City, Utah

Advances in the Neuroscience of Addiction
Cynthia M. Kuhn, Duke University Medical Center, Durham, North Carolina
George F. Koob, The Scripps Research Institute, La Jolla, California

DOPAMINE–GLUTAMATE INTERACTIONS in the BASAL GANGLIA

Edited by
Susan Jones
University of Cambridge
England

CRC Press is an imprint of the
Taylor & Francis Group, an **informa** business

CRC Press
Taylor & Francis Group
6000 Broken Sound Parkway NW, Suite 300
Boca Raton, FL 33487-2742

First issued in paperback 2019

© 2012 by Taylor & Francis Group, LLC
CRC Press is an imprint of Taylor & Francis Group, an Informa business

No claim to original U.S. Government works

ISBN-13: 978-1-4200-8879-3 (hbk)
ISBN-13: 978-0-367-38197-4 (pbk)

Library of Congress Cataloging-in-Publication Data

Dopamine-glutamate interactions in the basal ganglia / editor, Susan Jones.
   p. ; cm. -- (Frontiers in neuroscience)
  Includes bibliographical references and index.
  ISBN 978-1-4200-8879-3 (hardback : alk. paper)
  I. Jones, Susan, Ph. D. II. Series: Frontiers in neuroscience.
  [DNLM: 1. Basal Ganglia--physiology. 2. Basal Ganglia Diseases--physiopathology. 3. Dopamine--physiology. 4. Receptors, Glutamate. WL 307]

616.83--dc23   2011039590

Visit the Taylor & Francis Web site at
http://www.taylorandfrancis.com

and the CRC Press Web site at
http://www.crcpress.com

I dedicate this book to the teachers and mentors who inspired my interest in neuroscience: my lecturers in the Department of Pharmacology, King's College London (1987–1991); my PhD supervisor, Professor David A. Brown; and my postdoctoral advisors, Dr. Jerrel L. Yakel and Dr. Julie A. Kauer.

Contents

Series Preface

The Frontiers in Neuroscience Series presents the insights of experts on emerging fields and theoretical concepts that are, or will be, at the vanguard of neuroscience.

The books cover new and exciting multidisciplinary areas of brain research and describe breakthroughs in fields like visual, gustatory, auditory, and olfactory neuroscience as well as aging and biomedical imaging. Recent books cover the rapidly evolving fields of multisensory processing, glia, and depression, and different aspects of reward.

Each book is edited by experts and consists of chapters written by leaders in a particular field. The books are richly illustrated and contain comprehensive bibliographies. The chapters provide substantial background material relevant to the particular subject.

The goal is for these books to be the references every neuroscientist uses in order to acquaint themselves with new information and methodologies in brain research. We view our task as series editors to produce outstanding products that contribute to the broad field of neuroscience. Now that the chapters are available online, the effort put in by us, the publisher, and the book editors hopefully will contribute to the further development of brain research. To the extent that you learn from these books we will have succeeded.

Sidney A. Simon
Miguel A.L. Nicolelis
Duke University

Preface

Between 2002 and 2006, I gave lectures on the basal ganglia to Neuroscience students at the University of Cambridge. Although fascinating, I found the basal ganglia unsatisfying as a lecture topic; the questions asked by the students echoed my own. Can we separate the roles of the basal ganglia in movement control versus motivational processing versus learning? Is synaptic plasticity in the basal ganglia fundamentally different from other brain regions? At the network level, are cortical and thalamic inputs similarly influential over basal ganglia output? Does the "direct/indirect" model of basal ganglia function have validity? To give a few examples.

In the early noughties, it felt to me that we were a long way from answering these questions. However, in the past few years enormous progress has been made in basal ganglia research. Elegant genetic and optical methods have been used to probe function at the level of individual identified cells and synapses. Detailed circuit analysis, applying improved labeling, imaging, and recording tools, has shed light on the functional connectivity between and within the basal ganglia nuclei. Sophisticated approaches have been applied to learning and behavioral studies in awake animals. Against a background of historical work, we present some of the recent advances made using these new approaches in the context of the focus of this book: interactions between glutamatergic and dopaminergic systems in the basal ganglia.

In our definition of the basal ganglia we include the input nuclei of the neostriatum: caudate, putamen, nucleus accumbens—where the medium spiny neurons (MSN) are located; the output nuclei—globus pallidus internal segment (GPi) in primates and the equivalent substantia nigra pars reticulata (SNr) or entopeduncular nucleus (EP) in rodents; associated nuclei, including external globus pallidus (GPe) or the equivalent globus pallidus (GP) in rodents; the subthalamic nucleus (STN), and the dopamine (DA) neurons of the substantia nigra pars compacta (SNc) and ventral tegmental area (VTA). A simple relationship of these nuclei, based on the model of direct projecting versus indirect projecting MSNs, is schematized in Figure P.1. More detailed explanations of the basal ganglia circuitry are provided in chapters throughout the book, notably in Chapters 1 and 4 and, in a reversal of viewpoint, Chapter 8, where the circuitry is considered in the context of the substantia nigra rather than in the context of the striatum.

Within the focus of this book, we try to convey how the interplay between glutamate and dopamine, particularly at the level of MSNs in the striatum and at the level of output from the substantia nigra, contributes to the fine-tuning of analogue inhibitory signaling through the basal ganglia. In Chapters 1 through 5, we present an overview of what is known about the metabotropic and ionotropic glutamate receptors present in basal ganglia nuclei, their interactions with dopamine, and the functional and dysfunctional consequences of these interactions. For example, Kari Johnson and Jeffrey Conn (Chapter 1) provide a comprehensive overview of the functional roles of metabotropic glutamate receptors (mGluRs) in all nuclei of the basal ganglia, highlighting interactions with dopamine. It becomes clear that mGluRs are

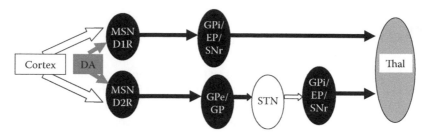

FIGURE P.1 Schematic representation of basal ganglia nuclei.

likely to be casualties during dopamine depletion, such as occurs in Parkinson's disease (PD). Indeed, antagonism of group I mGluRs in the striatum or agonist activity at group III mGluRs at striatopallidal synapses can attenuate motor signs in animal models of PD.

In the past five years, great strides have been made in unraveling the mechanisms of synaptic plasticity and modulation in the striatum. This work is reviewed in Chapters 3 through 6. The complexity of molecular interactions between dopamine receptors and NMDA glutamate receptors is described by Alasdair Gibb and Huaxia Tong in Chapter 3, then Danny Winder and colleagues provide a detailed consideration of the dendritic spines on striatal MSNs, the sites for convergence of glutamatergic, dopaminergic and neuromodulatory afferents: the forms of plasticity at these synapses, and the precise roles of dopamine and other neuromodulators (Chapter 4). The authors ponder the challenges of inducing reliable long-term potentiation (LTP) in the striatum, compared with, for example, the hippocampus, outlining the complexities established by cell-specific, activity-specific, and neuromodulator-specific requirements. Extraordinary progress has been made since the development of bacterial artificial chromosome (BAC) transgenic mice, expressing fluorescent proteins under the control of cell-specific promoters, which has allowed investigators to distinguish different cell types in the striatum: the specific contributions from these studies are beautifully summarized by D. James Surmeier and colleagues in Chapter 5.

To circumvent the danger of implying that glutamate and dopamine are the only important neurotransmitters in the basal ganglia, Chapter 6 is largely dedicated to acetylcholine in the striatum, released by tonically active striatal cholinergic interneurons. Sarah Threlfell and Stephanie Cragg review their elegant work on the modulation of dopamine release in the striatum by both nicotinic and muscarinic acetylcholine receptors, revealing that cholinergic modulation of striatal dopamine depends on the firing patterns of dopaminergic inputs. Then, in Chapter 8, Jennifer Brown focuses on GABA–dopamine interactions in the synaptic control of the substantia nigra. She highlights the importance of considering not only extrinsic inputs to the SN, but also of the local circuitry.

A key determinant of basal ganglia output is cortical control of the striatum, and this is explored in Chapter 7. Stéphane Charpier and colleagues provide close inspection of the properties of corticostriatal projection neurons and of MSNs and consider how this determines information processing in corticostriatal networks. They explain how, in the massively convergent cortical innervation of MSNs, spatially and temporally correlated inputs are integrated; they illustrate this with seven figures of data.

Does the interplay between glutamate and dopamine have anything to do with learning and behavior in a living animal? Fascinating research presented by Henry Yin and Rui Costa in Chapter 9 indicates that it does. While considering the challenges of inferring basal ganglia function from observation of simple behaviors in the laboratory, they go on to document progress in understanding the role of distinct striatal networks in specific aspects of instrumental learning and behavior and the contribution of dopamine to these functions.

Ultimately, we hope that our research efforts will translate into the development of preventative or better palliative treatments for patients with disorders that are seated in basal ganglia dysfunction. In Chapter 10, Miriam Hickey and Carlos Cepeda present mechanistic evidence for glutamate–dopamine dysfunction at the level of corticostriatal synapses in Huntington's disease (HD), PD, dyskinesia, and dystonia. While a pivotal role for dopamine dysfunction in PD is well established, the authors summarize a large body of evidence for dysregulation of striatal glutamate–dopamine interactions in HD. They document changes in these neurotransmitter systems in transgenic models of HD and, importantly, highlight therapies used by HD patients that modulate dopamine and glutamate function.

After working on this book, I have renewed hope that explaining the basal ganglia to students will be a less daunting task in the future. I hope that you find the following chapters as enlightening and fascinating as I did.

Editor

Susan Jones is a lecturer in the Department of Physiology, Development and Neuroscience (PDN) at the University of Cambridge, Cambridge, United Kingdom. Her research primarily focuses on glutamate receptors in midbrain dopaminergic circuits, with a particular interest in how glutamatergic function may be compromised in diseases that affect basal ganglia function.

Contributors

Shona L.C. Brothwell
College of Medical and Dental Sciences
University of Birmingham
Birmingham, United Kingdom

Jennifer Brown
Janelia Farm Research Campus
Howard Hughes Medical Institute
Ashburn, Virginia

and

Department of Physiology, Development
   and Neuroscience
University of Cambridge
Cambridge, United Kingdom

Carlos Cepeda
Intellectual and Developmental
   Disabilities Research Center
David Geffen School of Medicine
University of California, Los Angeles
Los Angeles, California

C. Savio Chan
Department of Physiology
Feinberg School of Medicine
Northwestern University
Chicago, Illinois

Stéphane Charpier
Centre de Recherche de l'Institut du
   Cerveau et de la Moelle épinière
Université Pierre et Marie Curie
Paris, France

Roger J. Colbran
Department of Molecular Physiology
   and Biophysics
Center for Molecular Neuroscience
and

Kennedy Center for Research on
   Human Development
Vanderbilt University School of Medicine
Nashville, Tennessee

P. Jeffrey Conn
Department of Pharmacology
and
Vanderbilt Center for Neuroscience
   Drug Discovery
Vanderbilt University Medical Center
Nashville, Tennessee

Rui M. Costa
Champalimaud Neuroscience Program
Instituto Gulbenkian de Ciência
Oeiras, Portugal

Stephanie J. Cragg
Department of Physiology, Anatomy
   and Genetics
and
Oxford Parkinson's Disease Centre
University of Oxford
Oxford, United Kingdom

Ariel Y. Deutch
Department of Psychiatry
and
Department of Pharmacology
Center for Molecular Neuroscience
Kennedy Center for Research
   on Human Development
Vanderbilt University School of Medicine
Nashville, Tennessee

Tracy S. Gertler
Department of Physiology
Feinberg School of Medicine
Northwestern University
Chicago, Illinois

Alasdair J. Gibb
Department of Neuroscience,
  Physiology and Pharmacology
University College London
London, United Kingdom

James Hallett
Department of Physiology, Development
  and Neuroscience
University of Cambridge
Cambridge, United Kingdom

Miriam A. Hickey
Department of Neurology
  and Neurobiology
Reed Neurological Research Center
University of California, Los Angeles
Los Angeles, California

Isabel Huang-Doran
University of Cambridge Metabolic
  Research Laboratories
Addenbrooke's Hospital
Institute of Metabolic Science
Cambridge, United Kingdom

Kari A. Johnson
Department of Pharmacology
and
Vanderbilt Center for Neuroscience
  Drug Discovery
Vanderbilt University Medical Center
Nashville, Tennessee

Susan Jones
Department of Physiology, Development
  and Neuroscience
University of Cambridge
Cambridge, United Kingdom

Jason R. Klug
Neuroscience Training Program
Vanderbilt University School of Medicine
Nashville, Tennessee

Séverine Mahon
Centre de Recherche de l'Institut du
  Cerveau et de la Moelle épinière
Université Pierre et Marie Curie
Paris, France

Morgane Pidoux
Centre de Recherche de l'Institut du
  Cerveau et de la Moelle épinière
Université Pierre et Marie Curie
Paris, France

Weixing Shen
Department of Physiology
Feinberg School of Medicine
Northwestern University
Chicago, Illinois

D. James Surmeier
Department of Physiology
Feinberg School of Medicine
Northwestern University
Chicago, Illinois

Sarah Threlfell
Department of Physiology, Anatomy
  and Genetics
and
Oxford Parkinson's Disease Centre
University of Oxford
Oxford, United Kingdom

Huaxia Tong
Department of Biology
University of Leicester
Leicester, United Kingdom

Danny G. Winder
Department of Molecular Physiology
    and Biophysics
Center for Molecular Neuroscience
and
Department of Neurobiology
Kennedy Center for Research
    on Human Development
Vanderbilt University School of Medicine
Nashville, Tennessee

Henry H. Yin
Department of Psychology
    and Neuroscience
and
Department of Neurobiology
and
Center for Cognitive Neuroscience
Duke University
Durham, North Carolina

1 Metabotropic Glutamate Receptor–Dopamine Interactions in the Basal Ganglia Motor Circuit

Kari A. Johnson and P. Jeffrey Conn

CONTENTS

1.1   INTRODUCTION

The basal ganglia are a group of interconnected subcortical nuclei that play critical roles in motor activity. Alterations in normal basal ganglia neurotransmission in disease states such as Parkinson's disease (PD) cause dramatic impairments of normal motor function. Metabotropic glutamate receptors (mGluRs) are an important class of G protein-coupled receptors (GPCRs) that modulate both excitatory and inhibitory transmission through the basal ganglia motor circuit. In recent years, mGluRs have been implicated as exciting new targets for the treatment of both motor symptoms and neurodegeneration in PD. Because the primary pathology underlying PD is the selective loss of dopamine neurons of the substantia nigra pars compacta (SNc), the significance of dopaminergic modulation of mGluR function in the basal ganglia has become increasingly appreciated. Importantly, changes in mGluR function in the absence of dopamine could impact the therapeutic potential of drugs targeting these receptors. In addition, aberrant mGluR function in the absence of dopamine could contribute to the changes in neurotransmission underlying the pathogenesis of PD symptoms. This chapter explores the physiological roles of mGluRs in the basal ganglia motor circuit, the changes in mGluR function in the absence of dopaminergic modulation, the regulation of dopaminergic transmission by mGluRs, and the significance of mGluR–dopamine interactions to PD pathophysiology and therapeutics.

1.2   BASAL GANGLIA

The flow of information through the basal ganglia involves two major pathways that have opposing effects on basal ganglia output (Wichmann and DeLong 1996). The primary input nucleus of the basal ganglia is the striatum, which receives excitatory projections from cortical regions including the primary motor cortex, as well as the thalamus (Figure 1.1). The striatal neuron population primarily consists of medium spiny neurons (MSNs), which are projection neurons that use γ-aminobutyric acid (GABA) as their neurotransmitter. These projection neurons relay information to the output nuclei through direct projections as well as indirectly through the globus pallidus external segment (GPe; GP in rodents) and the subthalamic nucleus (STN). The substantia nigra pars reticulata (SNr) and the internal globus pallidus (GPi, entopeduncular nucleus [EPN] in rodents) are the major output nuclei of the basal ganglia. The direct pathway exerts inhibitory control over the activity of these nuclei, whereas activity in the indirect pathway disinhibits excitatory STN neurons, providing increased excitation of the SNr and GPi. The output nuclei in turn provide information to the cortex via inhibitory projections to the thalamus. The balance of inhibitory control exerted by the direct pathway and the excitatory control provided by the indirect pathway are thought to be crucial for maintaining appropriate control of motor activity. Although this is a highly simplified model of basal ganglia connectivity, it provides a valuable model for understanding the neuromodulatory roles of mGluRs and dopamine, as well as the changes in neurotransmission that are thought to underlie motor symptoms of PD such as tremor, bradykinesia, and muscle rigidity.

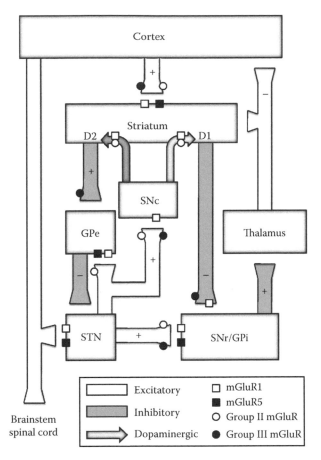

FIGURE 1.1 Localization of metabotropic glutamate receptors in the basal ganglia motor circuit. As discussed in the introduction, the basal ganglia motor circuit consists of two pathways, the direct pathway and the indirect pathway, which relay information from the striatum to the output nuclei of the basal ganglia (SNr and GPi). Dopaminergic input from the SNc to the striatum increases the activity of the direct pathway by activating D1 receptors on MSNs in the striatum, whereas dopamine acts on D2 receptors on MSNs to decrease activity through the indirect pathway. In PD, the degeneration of dopaminergic SNc neurons leads to increased activity of the indirect pathway and decreased activity of the direct pathway, resulting in increased inhibition of thalamocortical transmission. Symbols indicate synapses with increased (+) or decreased (−) activity in PD. mGluRs are localized both presynaptically and postsynaptically at several key synapses in the basal ganglia motor circuit. In contrast to dopamine, the overall effect of group I mGluR activation is to increase activity through the indirect pathway. Conversely, activation of group II and group III mGluRs tends to mimic the effect of striatal D2 receptor activation by reducing transmission through the indirect pathway. As discussed in the main text, dopamine modulates the function of various mGluR subtypes in several basal ganglia nuclei. Abbreviations: GPe, globus pallidus external segment (globus pallidus in rodents); STN, subthalamic nucleus; SNr, substantia nigra pars reticulata; GPi, globus pallidus internal segment (EPN in rodents); SNc, substantia nigra pars compacta; mGluR, metabotropic glutamate receptor. (Adapted from Wichmann, T. and DeLong, M.R., *Curr. Opin. Neurobiol.*, 6, 751, 1996; Conn, P. et al., *Nat. Rev. Neurosci.*, 6, 787, 2005.)

Dopaminergic neurons of the SNc primarily project to the striatum, but also modulate the activity of other key nuclei of the basal ganglia (Smith and Kieval 2000). Simplistically, MSNs that project directly to the output nuclei primarily express D1 receptors, whereas MSNs that give rise to the indirect pathway primarily express D2 receptors, allowing nigrostriatal dopamine to increase activity through the direct pathway and decrease activity of the indirect pathway. In PD, the loss of dopaminergic modulation of striatal MSNs therefore causes increased activity of the indirect pathway relative to the direct pathway (Figure 1.1). The increased activity of the indirect pathway causes disinhibition of the STN, resulting in increased burst firing of STN neurons and an increase in synaptic excitation of projection neurons in the output nuclei (Bergman et al. 1994; Vila et al. 2000; Ni et al. 2001a,b; Liu et al. 2002b). This pathological increase in activity and concurrent changes in firing patterns in the indirect pathway are thought to contribute to the motor symptoms observed in PD patients.

1.3 METABOTROPIC GLUTAMATE RECEPTORS IN THE BASAL GANGLIA

Current therapeutic strategies for treating the motor symptoms of PD primarily rely on dopamine replacement with drugs such as levodopa (L-DOPA; L-3,4-dihydroxy-phenylalanine) (Chen and Swope 2007). Unfortunately, long-term use of these drugs eventually results in loss of efficacy and severe adverse effects such as dyskinesias. These drawbacks have led to extensive efforts to identify alternative treatment options for PD, including highly successful surgical interventions such as deep brain stimulation of the STN or GPi (Walter and Vitek 2004; Wichmann and Delong 2006).

Recent studies have shown that mGluRs modulate synaptic transmission at every major synapse in the basal ganglia motor circuit (Conn et al. 2005), suggesting that these GPCRs are important for normal basal ganglia function, and may provide therapeutic targets for PD. There are currently eight known subtypes of mGluRs, which are classified into three groups according to sequence homology, G protein-coupling profiles, and pharmacological sensitivity (Conn and Pin 1997; Pin and Acher 2002). Group I mGluRs (mGluR1 and mGluR5) primarily couple to Gq and have frequently been found to modulate neurotransmission postsynaptically. Group II mGluRs (mGluR2 and mGluR3) and group III mGluRs (mGluR4, mGluR6, mGluR7, and mGluR8) primarily couple to Gi/o and often modulate neurotransmission by presynaptic inhibition of neurotransmitter release. With the exception of mGluR6, the expression of all subtypes of mGluRs has been detected in various basal ganglia nuclei (Figure 1.1), and recent studies have identified many physiological roles for these receptors. Because drugs targeting mGluRs can modulate neurotransmission at key basal ganglia synapses that are overactive in PD, many mGluR ligands have been evaluated in animal models of PD, and several subtypes have been identified as exciting therapeutic targets. Finally, the importance of dopamine in the basal ganglia and the relevance of mGluR function to PD have led to the evaluation of interactions between mGluRs and dopamine in the basal ganglia, revealing many ways in which dopaminergic signaling modulates the activity of mGluRs.

1.4 MODULATION OF BASAL GANGLIA NEUROTRANSMISSION BY GROUP I mGluRs

1.4.1 STRIATUM

In the striatum, both mGluR1 and mGluR5 are expressed by all cell types, which include MSNs, cholinergic interneurons, fast-spiking parvalbumin-containing GABA interneurons, and burst firing somatostatin-positive GABA interneurons (Tallaksen-Greene et al. 1998; Bell et al. 2002). Group I mGluRs can influence MSNs both directly and by modulating the activity of interneurons in the striatum, and mGluR1 and mGluR5 often have distinct roles despite being expressed in the same neurons and coupling to the same signal transduction cascades (reviewed in Bonsi et al. 2008).

Activation of group I mGluRs by the nonselective agonist 3,5-dihydroxyphenyl-glycine (DHPG) potentiates N-methyl-D-aspartate (NMDA) receptor-mediated currents in MSNs through a mechanism that is dependent on protein kinase C (PKC) activation (Pisani et al. 1997b, 2001b). This effect is present in mGluR1 knockout mice but absent in mGluR5 knockout mice, suggesting that mGluR5 exclusively mediates the enhancement of NMDA receptor currents. In cholinergic interneurons, activation of mGluR1 or mGluR5 raises intracellular calcium levels and induces membrane depolarization by inhibiting potassium conductance (Takeshita et al. 1996; Calabresi et al. 1999; Pisani et al. 2001a; Bonsi et al. 2005), resulting in an increase in acetylcholine release (Marti et al. 2001). Acetylcholine can then activate M1 muscarinic receptors on MSNs, which can also potentiate NMDA receptor currents (Calabresi et al. 1998), indicating that group I mGluRs can enhance NMDA receptor activity both directly and indirectly. Fast-spiking parvalbumin-containing interneurons also express mGluR1 and mGluR5, and activation of mGluR1 mediates direct excitation of these neurons (Bonsi et al. 2007b).

mGluR1 is also present on dopaminergic terminals in the striatum (Zhang and Sulzer 2003). Activation of these receptors by glutamate spillover from corticostriatal terminals inhibits striatal dopamine release. Conversely, extrasynaptic dopamine activates dopamine receptors on glutamatergic terminals in the striatum, and activation of these receptors inhibits glutamate release (Cepeda et al. 2001; Wang and Pickel 2002). Corticostriatal transmission increases in response to nigrostriatal lesions, raising the possibility that following nigrostriatal damage, increased striatal glutamate could decrease dopamine release from the remaining functional nigrostriatal terminals and worsen the effects of nigrostriatal lesion on basal ganglia neurotransmission.

mGluR1 activation directly mediates glutamatergic transmission in the SNc, resulting in complex effects on membrane properties. Although glutamate receptors are typically thought to mediate the excitation of neurons, brief activation of mGluR1 hyperpolarizes SNc neurons (Fiorillo and Williams 1998). This effect involves increased potassium conductance and depends on intracellular calcium stores. Prolonged activation of mGluR1 depolarizes SNc neurons and activates burst firing (Mercuri et al. 1992, 1993).

Interestingly, group I mGluRs cooperate with dopamine receptors to participate in the induction of long-term synaptic plasticity at glutamatergic corticostriatal synapses (reviewed in Gubellini et al. 2004). Plasticity at this synapse is thought to be important for motor learning and habit formation, and is likely disrupted in pathological states such as PD (Calabresi et al. 2009). Induction of long-term depression (LTD) at corticostriatal synapses by high-frequency stimulation of cortical afferents requires mGluR1 as well as D1-like and D2-like dopamine receptor activation, dopamine- and cAMP-regulated phosphoprotein (DARPP-32) phosphorylation, and mGluR-mediated retrograde endocannabinoid signaling (Calabresi et al. 1992a,b, 1994, 2000, 2007; Choi and Lovinger 1997; Gubellini et al. 2001; Sung et al. 2001; Gerdeman et al. 2002; Sergeeva et al. 2007). High-frequency stimulation fails to induce corticostriatal LTD in slices obtained from 6-hydroxydopamine-lesioned rats (Calabresi et al. 1992b), highlighting the critical role of dopamine receptor coactivation in this form of mGluR-dependent synaptic plasticity and the potential relevance of impaired corticostriatal plasticity to PD.

Under certain conditions, high-frequency stimulation can also induce NMDA receptor-dependent long-term potentiation (LTP) of corticostriatal transmission (Calabresi et al. 1992c; Partridge et al. 2000). While there is evidence that only mGluR1 activation is necessary for LTD induction, both mGluR1 and mGluR5 are involved in the induction of LTP (Gubellini et al. 2003). In addition, D1-like receptor activation is necessary for LTP induction, whereas D2-like receptors negatively modulate LTP (Calabresi et al. 2000, 2007). These findings highlight the importance of mGluR-dopamine interactions to multiple forms of synaptic plasticity.

Activation of group I mGluRs also modulates basal and dopamine-induced gene expression in MSNs (reviewed in Mao et al. 2008). Coactivation of mGluR5 and NMDA receptors leads to calcium-dependent downstream activation of MAP kinase cascades and phosphorylation of transcription factors such as ELK1 and cyclic AMP (cAMP) responsive element-binding protein (CREB) (Choe and McGinty 2001; Choe and Wang 2001a,b; Yang et al. 2004; Mao et al. 2005; Voulalas et al. 2005). This signal transduction pathway leads to transcription of the immediate early gene c-Fos as well as prodynorphin and proenkephalin in MSNs (Mao and Wang 2001; Mao et al. 2002; Parelkar and Wang 2003; Yang et al. 2004). The efficiency of signaling downstream of group I mGluRs is improved by the formation of protein complexes including Homer proteins (Mao et al. 2005). Interestingly, group I mGluR blockade impairs the ability of amphetamine to stimulate CREB phosphorylation or induce c-Fos or prodynorphin expression (Choe et al. 2002), indicating considerable cross-talk between group I mGluRs and dopamine receptors in the control of inducible gene expression in MSNs. Amphetamine or D1 agonist-induced c-Fos and prodynorphin expression are also attenuated in mGluR1 knockout mice (Mao et al. 2001, 2002), supporting the hypothesis that coactivation of group I mGluRs and dopamine receptors plays an important role in mediating dopamine receptor-induced gene expression in striatal neurons.

Coactivation of dopamine receptors and group I mGluRs can have opposing effects on downstream signaling events. In MSNs that give rise to the indirect pathway, mGluR5 and A_{2A} adenosine receptors physically interact and promote downstream events such as activation of the MAP kinase cascade, whereas D2 receptors have an inhibitory effect on the same downstream effectors (Ferre et al. 2002;

Nishi et al. 2003). In D1-expressing neurons, group I mGluRs can also oppose the effects of dopamine by modulating the activity of DARPP-32. D1 activation leads to the protein kinase A (PKA)-dependent phosphorylation of DARPP-32 at Thr34, which in turn inhibits the activity of protein phosphatase 1 (PP1) and results in amplification of D1-mediated downstream signaling (Svenningsson et al. 2004). Conversely, activation of group I mGluRs leads to phosphorylation of DARPP-32 on Thr75 and serine 137 by casein kinase 1 and cyclin-dependent kinase 5 (Liu et al. 2001, 2002a; Nishi et al. 2005). This form of DARPP-32 negatively regulates PKA activity, thereby reducing the downstream effects of D1 receptor activation. Thus, for some signaling pathways, the overall effects of group I mGluR activation in MSNs tend to oppose the effects of dopamine receptor activation.

1.4.2 DOPAMINERGIC MODULATION OF STRIATAL GROUP I mGluRs

Dopamine depletion and nigrostriatal lesion have been shown to regulate the expression of group I mGluRs in the striatum. In reserpinized rats, a significant increase in mGluR5 mRNA and receptor density has been found in the striatum (Ismayilova et al. 2006). Similarly, positron emission tomography (PET) studies in 6-hydroxy-dopamine-lesioned rats and 1-methyl-4-phenyl-1,2,3,6-tetrahydropyridine (MPTP)-treated primates have demonstrated an increase in striatal mGluR5 PET tracer binding when compared with tracer binding in normal animals (Pellegrino et al. 2007; Sanchez-Pernaute et al. 2008). In addition, expression levels of mGluR1a are reduced in MPTP-treated mice, and mGluR1a and mGluR5 undergo complex changes in trafficking resulting in abnormal ultrastructural localization in response to nigrostriatal lesion (Kuwajima et al. 2007). Taken together, these findings indicate that dopaminergic tone regulates both the expression levels and trafficking of group I mGluRs. It is possible that increases in striatal group I mGluR expression could contribute to the pathogenesis of PD symptoms by increasing the activity of the indirect pathway.

1.4.3 GLOBUS PALLIDUS (EXTERNAL)

Both group I mGluR subtypes are expressed in the rodent GP and the primate GPe (Testa et al. 1994, 1998; Hanson and Smith 1999; Rouse et al. 2000; Poisik et al. 2003; Kuwajima et al. 2007), and several physiological consequences have been identified. Activation of group I mGluRs by the nonselective agonist DHPG inhibits calcium currents in GP neurons, most likely by inhibiting calcium influx through N- and P-type calcium channels (Stefani et al. 1998). In addition, DHPG directly depolarizes GP neurons in rat brain slices, and this effect is blocked by the mGluR1-selective antagonist LY367385, but not the mGluR5-selective antagonist 2-methyl-6-(phenylethynyl)-pyridine (MPEP) (Poisik et al. 2003). Interestingly, repetitive activation of the internal capsule results in a long-lasting excitation of GP neurons that is mediated by mGluR1 (Kaneda et al. 2007). In addition, *in vivo* electrophysiology recordings in nonhuman primates have demonstrated that mGluR1 blockade decreases firing rates in GPe neurons (Kaneda et al. 2005). Together, these findings suggest that mGluR1 plays a role in mediating glutamatergic excitation of GPe neurons. In rat brain slices, mGluR5 blockade potentiates the DHPG-induced membrane depolarization (Poisik et al. 2003). Further, repeated application of DHPG results in desensitization of mGluR1-mediated depolarization, but blockade of mGluR5 during repeated DHPG application prevents mGluR1 desensitization. These findings suggest that mGluR5 activation causes cross-desensitization of mGluR1 in GP neurons and highlight the fact that mGluR1 and mGluR5 can perform distinct functions in the same population of neurons. Pharmacological inhibition of PKC activity mimics the effect of mGluR5 blockade, so it is likely that mGluR5-induced PKC activity mediates the desensitization of mGluR1 in GP neurons.

The physiological effects of mGluR1 and mGluR5 in GP neurons are significantly altered in dopamine-depleted brain slices, suggesting that dopaminergic signaling plays a critical role in the normal function of GP group I mGluRs (Poisik et al. 2007). Following reserpine treatment, mGluR1 activation no longer induces depolarization of GP neurons, whereas mGluR5 gains the ability to depolarize neurons. Activation of D1-like and D2-like receptors by exogenous dopamine application or selective agonists of these receptors partially restores the ability of mGluR1, while simultaneously reversing the ability of mGluR5, to depolarize GP neurons. Blockade of PKA activity mimics the ability of dopamine to restore the normal function of the group I mGluRs in the GP of reserpinized animals, suggesting that dopamine receptor-mediated inhibition of PKA may play a role in determining the normal functions of mGluR1 and mGluR5 in GP neurons. In addition, mGluR1 is downregulated in the GPe of MPTP-treated monkeys, whereas mGluR5 protein levels are not altered, suggesting that dopaminergic tone can differentially influence group I mGluR expression levels in this nucleus (Kaneda et al. 2005).

1.4.4 Subthalamic Nucleus

Both mGluR1 and mGluR5 are expressed in STN neurons (Testa et al. 1994; Awad et al. 2000). Similar to the effects of group I mGluRs in the rodent GP, these receptors have distinct roles in STN neurons that are regulated by dopamine receptor signaling (Valenti et al. 2002). Activation of group I mGluRs by DHPG increases intracellular calcium levels, and both mGluR1 and mGluR5 contribute to this effect (Marino et al. 2002). DHPG also directly depolarizes STN neurons and increases firing frequency and bursting activity (Beurrier et al. 1999; Awad et al. 2000), but in contrast to GP neurons, this is mediated by mGluR5, whereas mGluR1 activation reduces evoked excitatory transmission in the STN by a presynaptic mechanism (Awad-Granko and Conn 2001). Interestingly, prolonged treatment of rats with haloperidol prior to brain slice preparation produces a significant change in the roles of group I mGluRs in STN neurons; mGluR1 contributes to STN depolarization in haloperidol-treated animals. This finding provides another intriguing example of regulation of group I mGluR function by intact dopamine signaling.

1.4.5 Substantia Nigra Pars Reticulata (and Internal Globus Pallidus)

Finally, group I mGluRs are also expressed in the output nuclei of rodents and primates (Testa et al. 1994, 1998; Marino et al. 2001; Wittmann et al. 2001a; Messenger et al. 2002; Kaneda et al. 2005), suggesting that they may play an important role in modulating basal ganglia output. Activation of group I mGluRs by DHPG reduces both excitatory and inhibitory transmission in rat SNr GABAergic projection neurons (Marino et al. 2001; Wittmann et al. 2001a). The ability of DHPG to inhibit excitatory transmission is presynaptically mediated and involves mGluR1 but not mGluR5 activation (Wittmann et al. 2001a). While most group I mGluR immunoreactivity in the rat SNr is detected postsynaptically, some group I mGluR labeling has also been detected in pre-terminal axons at both asymmetric and symmetric SNr synapses (Marino et al. 2001; Wittmann et al. 2001a), suggesting that group I mGluR agonists may reduce both excitatory and inhibitory transmission by activation of presynaptic receptors. Activation of group I mGluRs by DHPG also directly excites SNr neurons. Application of DHPG to rat midbrain slices depolarizes SNr neurons, and this effect is exclusively mediated via mGluR1 (Marino et al. 2001, 2002; Valenti et al. 2002). Interestingly, *in vivo* electrophysiological recordings from GPi neurons in monkeys have demonstrated that the mGluR1-selective antagonist LY367385 reduces the firing rate of GPi neurons, providing direct evidence that mGluR1 plays a role in mediating glutamatergic excitation of neurons in the basal ganglia output nuclei (Kaneda et al. 2005).

mGluR1 expression levels are reduced in the GPi of MPTP-treated monkeys, and the effect of LY367385 on the firing rate of individual GPi neurons is reduced (Kaneda et al. 2005). Disruption of dopaminergic signaling also disrupts the normal function of group I mGluRs in the rat SNr; while mGluR1 exclusively mediates DHPG-induced depolarization of SNr neurons in normal animals, mGluR5 gains the ability to mediate depolarization after prolonged haloperidol treatment (Marino et al. 2002), similar to the effect observed in the STN, providing further evidence that dopamine receptor signaling is important for the segregation of physiological roles of group I mGluRs in the basal ganglia.

1.5   GROUP I mGluR ANTAGONISTS IN ANIMAL MODELS OF PD

Because the overall effects of group I mGluR activation lead to increased transmission through the indirect pathway and direct excitation of nuclei that are overactive in the parkinsonian brain, antagonists of these receptors may be targets for treating the motor symptoms of PD. In agreement with this hypothesis, several studies have demonstrated that systemic administration of negative allosteric modulators of mGluR5 such as MPEP and 3-((2-methyl-1,3-thiazol-4-yl)ethynyl)pyridine (MTEP) yields antiparkinsonian effects in animal models of PD (Spooren et al. 2000; Ossowska et al. 2001, 2005; Breysse et al. 2002, 2003; Coccurello et al. 2004; Turle-Lorenzo et al. 2005; De Leonibus et al. 2009). Interestingly, combined blockade of mGluR5 and A_{2A} adenosine receptors produces very robust antiparkinsonian effects (Coccurello et al. 2004; Kachroo et al. 2005), possibly due to the interaction of these receptors in the striatal neurons that give rise to the indirect pathway.

Several mechanisms could mediate the antiparkinsonian effects of mGluR5 antagonism. Direct-site infusion of group I mGluR agonists into the rat striatum increases activity of the indirect pathway while reducing motor activity (Kearney et al. 1997, 1998), providing evidence that mGluR5 antagonists may reverse PD-like motor impairments by reducing striatopallidal transmission. mGluR5 antagonists may also increase excitatory transmission in the GPe by relieving mGluR1 desensitization (Poisik et al. 2003), although dopamine depletion may diminish this effect (Poisik et al. 2007). Because mGluR5 activation depolarizes STN neurons and promotes burst firing (Beurrier et al. 1999; Awad et al. 2000), reducing mGluR5 activity in the STN represents another putative mechanism of action of mGluR5 antagonists. Consistent with this prediction, infusion of MPEP into the STN reverses motor asymmetries caused by unilateral 6-hydroxydopamine lesion (Phillips et al. 2006). Finally, blockade of mGluR1 or mGluR5 in striatal cholinergic interneurons may reduce the release of acetylcholine, which would also be predicted to have antiparkinsonian effects (Pisani et al. 2003).

Although the mechanisms underlying the generation of L-DOPA-induced dyskinesias (LIDs) are not clear, recent studies have associated the upregulation of mGluR5 in the primate striatal complex with the development of LIDs, suggesting that blockade of mGluR5 may prevent the development of LIDs or reduce their severity (Samadi et al. 2007). In support of this hypothesis, mGluR5 blockade reduces dyskinetic behaviors in rat models of LID (Dekundy et al. 2006; Mela et al. 2007; Levandis et al. 2008). Excitingly, negative allosteric modulators of mGluR5 are currently being investigated in humans for the treatment of LIDs, and the results of these studies may lead to improvements in the treatment of PD by combining dopamine-replacement therapies with mGluR5 antagonists to alleviate the adverse effects of L-DOPA. Further studies aimed at discovering the mechanism responsible for mGluR5 antagonist-mediated attenuation of LIDs will likely increase our understanding of interactions between group I mGluRs and dopaminergic transmission in the basal ganglia.

1.6   GROUP I mGluR–MEDIATED PROTECTION AGAINST NIGROSTRIATAL DEGENERATION

The possibility that antagonists of group I mGluRs can reduce activity through the indirect pathway suggests that blockade of group I mGluRs may provide protection against progressive nigrostriatal degeneration. An interesting finding in support of this hypothesis is that mGluR5 knockout mice are more resistant to MPTP-induced nigrostriatal toxicity than their wildtype counterparts (Battaglia et al. 2004), suggesting that endogenous activation of mGluR5 enhances MPTP-induced nigrostriatal degeneration. Furthermore, mGluR5 antagonists reduce nigrostriatal damage in MPTP-treated mice (Battaglia et al. 2004; Aguirre et al. 2005), methamphetamine-treated mice (Battaglia et al. 2002), and 6-hydroxydopamine-treated rats (Vernon et al. 2005, 2007; Armentero et al. 2006). Intranigral infusion of the mGluR5 antagonist MPEP protects against 6-hydroxydopamine-induced SNc degeneration in rats, suggesting that decreasing endogenous mGluR5 activation in the SNc is a potential neuroprotective mechanism. Antagonists of mGluR5 may also act in the STN by reducing mGluR5-mediated increases in neuronal excitability and burst firing. Because activation of mGluR1 has been shown to mediate direct excitation of SNc neurons (Mercuri et al. 1992, 1993), it is also possible that antagonists of mGluR1 could protect against excitotoxicity in the SNc; intranigral infusion of the mGluR1-selective antagonist LY367385 reduces the extent of 6-hydroxydopamine-induced nigrostriatal lesion in rats, providing support for this hypothesis. Antagonists of both of the group I mGluRs may therefore be useful for slowing the loss of dopaminergic neurons in PD.

1.7 MODULATION OF BASAL GANGLIA NEUROTRANSMISSION BY GROUP II mGluRs

1.7.1 STRIATUM

Group II mGluR expression has been detected in several basal ganglia nuclei (Testa et al. 1998; Bradley et al. 2000; Kahn et al. 2001; Pisani et al. 2002), and multiple physiological effects of these receptors have been identified, many of which are sensitive to alterations in dopaminergic transmission. In the rat striatum, activation of presynaptic group II mGluRs reversibly reduces glutamatergic corticostriatal transmission onto MSNs by inhibiting glutamate release (Lovinger and McCool 1995; Battaglia et al. 1997; Cozzi et al. 1997; Picconi et al. 2002). *In vivo* microdialysis studies have shown that local administration of group II mGluR antagonists in the striatum increases extracellular glutamate levels, suggesting that these receptors tonically inhibit glutamate release (Cozzi et al. 1997). Further studies have revealed that extracellular glutamate derived from cysteine-glutamate antiporter activity is the main source of glutamate responsible for the tonic activation of striatal group II mGluRs (Baker et al. 2002). In contrast to the reversible inhibition of excitatory transmission in rat brain slices, activation of group II mGluRs at the mouse corticostriatal synapse causes an LTD of excitatory transmission that persists after drug-washout (Kahn et al. 2001), although a more recent study reported a reversible depression of excitatory transmission at the corticostriatal synapse in mice (Martella et al. 2009).

The potency of group II agonists for inhibiting corticostriatal transmission is enhanced in brain slices obtained from 6-hydroxydopamine-lesioned rats, and this enhancement is associated with an increase in receptor density following dopamine denervation (Picconi et al. 2002); this upregulation could represent a mechanism for reducing excessive corticostriatal transmission resulting from nigrostriatal degeneration by increasing autoreceptor activity (Picconi et al. 2002). Chronic administration of L-DOPA restores the potency of group II agonists and the expression level of group II mGluRs to normal levels, indicating that L-DOPA treatment can reverse changes in mGluR function that are caused by dopaminergic denervation. In MPTP-treated monkeys, there is no change in group II mGluR receptor density in the striatum (Samadi et al. 2008); however, in MPTP-treated monkeys a combination of L-DOPA and the D2 receptor agonist cabergoline reduces specific binding of the group II mGluR antagonist [^{3}H]LY341495 in the neostriatum, whereas treatment with L-DOPA alone has no effect, suggesting an interaction between group II mGluRs and D2 receptors (Samadi et al. 2008). Recent studies evaluating the effects of group II mGluRs on corticostriatal transmission in genetic models of PD have also found alterations in group II mGluR function. In mice lacking the familial PD-linked genes *PINK1* or *Parkin*, the potency of the group II mGluR agonist LY379268 for reducing excitatory corticostriatal transmission is increased (Martella et al. 2009). This increase in potency is not reversed by acute L-DOPA administration, so a possible role of the dopaminergic system in mediating this change in group II mGluR function has not been delineated.

Electrophysiological recordings from striatal cholinergic interneurons have revealed that activation of group II mGluRs directly modulates the excitability of these cells (Pisani et al. 2002). In rat striatal slices, bath application of a group II mGluR agonist such as LY379268 or (2S,2′R,3′R)-2-(2′,3′-dicarboxycyclopropyl)glycine (DCG-IV) reduces both excitatory postsynaptic potentials (EPSPs) and inhibitory postsynaptic potentials (IPSPs) evoked by intrastriatal stimulation, and reduces calcium-dependent plateau potentials, possibly by modulating the activity of P-type calcium channels. In addition, group II mGluR agonists reduce electrically evoked acetylcholine release from these neurons, possibly by acting as heteroreceptors at cholinergic terminals. In situ hybridization studies demonstrate that mGluR2 mRNA can be detected in cholinergic interneurons, whereas mGluR3 mRNA is absent, suggesting that these effects are mediated exclusively by mGluR2 activation. In contrast to the effect of nigrostriatal lesion on group II mGluR function at the corticostriatal synapse, the potency of the group II mGluR-selective agonist LY354740 for depressing evoked acetylcholine release is reduced following 6-hydroxydopamine lesion (Marti et al. 2003). As a consequence, reduction in mGluR2-mediated inhibition of striatal acetylcholine release following nigrostriatal degeneration could contribute to pathological increases in striatal acetylcholine levels (Marti et al. 2003). Treatment of 6-hydroxydopamine-lesioned rats with L-DOPA restores the sensitivity of mGluR2, suggesting that the ability of mGluR2 to reduce acetylcholine release is indeed dependent upon dopaminergic transmission, and supporting the idea that L-DOPA treatment can reverse some changes in mGluR function that are caused by dopaminergic denervation.

A series of microdialysis studies have identified several mechanisms by which group II mGluRs regulate the release of dopamine in the striatum. Group II mGluR agonists decrease whilst group II/III mGluR antagonists increase extracellular dopamine levels (Hu et al. 1999). Decreasing extrasynaptic glutamate levels in the striatum by inhibiting glial cysteine–glutamate exchange increases extracellular dopamine levels, and this is reversed by the group II mGluR agonist (2R,4R)-4-aminopyrrolidine-2,4-dicarboxylate (APDC) (Baker et al. 2002). These data suggest that under normal conditions, striatal dopamine release is decreased by tonic activation of group II mGluRs on nigrostriatal terminals. In addition, activation of group II mGluRs at the STN-SNc synapse reduces excitatory transmission (Wigmore and Lacey 1998; Wang et al. 2005), which would also be predicted to reduce dopamine release in the striatum by reducing the excitatory drive onto SNc neurons. However, a microdialysis study testing the effect of systemic administration of the group II mGluR-selective agonist LY379268 failed to find any effect on extracellular dopamine levels in the striatum (Cartmell et al. 2000). Further studies are therefore needed to fully elucidate the role of group II mGluRs in regulating striatal dopamine release *in vivo*.

1.7.2   GPe AND STN

Group II mGluRs modulate excitatory transmission in the indirect pathway through actions in both the globus pallidus and the STN. The primary source of excitatory input to the globus pallidus is from the STN, and mGluR2 mRNA has been detected in STN neurons (Testa et al. 1994; Messenger et al. 2002), raising the possibility that mGluR2 may be expressed on STN terminals in the globus pallidus. In agreement with this prediction, immunolocalization studies have detected group II mGluR reactivity in glutamatergic pre-terminal axons in the rat globus pallidus (Poisik et al. 2005). Activation of these receptors by a group II mGluR agonist reduces the amplitude of evoked excitatory postsynaptic currents (EPSCs), and this effect is potentiated by the mGluR2-selective positive allosteric modulator LY487379. Conversely, group II mGluR agonists have no effect on inhibitory transmission in the globus pallidus. Although the primary source of glutamatergic input to the globus pallidus is from the STN, effects on glutamatergic inputs from the cortex, brainstem, and thalamus may also contribute to this effect.

In the STN of adult rats, activation of group II mGluRs also reduces the amplitude of EPSCs by reducing glutamate release from presynaptic terminals, via PKC activation (Shen and Johnson 2003), although this effect was not observed in young (15–18 day old) animals (Awad-Granko and Conn 2001). Potential interactions between dopamine and group II mGluRs have not been evaluated in the globus pallidus or STN, but may be the subject of future study.

1.7.3   SNr

Anatomical studies using specific antibodies have demonstrated the presence of group II mGluRs on presynaptic glutamatergic axon terminals in the SNr, which primarily receives glutamatergic inputs from the STN (Bradley et al. 2000), raising the possibility that these receptors modulate excitatory neurotransmission in the SNr. Consistent with this, brief application of group II mGluR-selective agonists such as LY354740 reversibly reduces the amplitude EPSCs recorded from SNr GABAergic neurons after stimulation of STN afferents (Bradley et al. 2000). LY354740 does not alter the frequency or amplitude of spontaneous miniature EPSCs recorded from these neurons, suggesting that the effect of LY354740 on excitatory transmission is mediated presynaptically. In addition, LY354740 reduces the frequency of EPSCs evoked by applying glutamate directly to the STN without affecting the amplitude, confirming that group II mGluR activation inhibits excitatory transmission arising from STN neurons.

Interestingly, the ability of LY354740 to inhibit excitatory neurotransmission in the SNr is substantially reduced in dopamine-depleted animals, suggesting that dopaminergic tone is required for this effect (Wittmann et al. 2002). Bath application of dopamine to reserpinized slices rescues the effect of LY354740, confirming that the reduced effect of LY354740 is due to a loss of dopamine rather than other catecholamines or serotonin. In contrast to changes in expression levels of group II mGluRs at corticostriatal synapses following dopamine denervation, the alteration in group II mGluR effects in SNr seems to involve an acute interaction between dopamine receptor and mGluR signaling. The effect of reserpine is mimicked by bath application of the dopamine receptor antagonist haloperidol as well as the D1 receptor-selective antagonist SCH23390, but not by the D2 receptor-selective antagonist sulpiride, suggesting that tonic stimulation of D1 receptors in the substantia nigra by ambient dopamine is required for group II mGluR-mediated inhibition of excitatory transmission. In agreement with this interpretation, pharmacological activation of D1 receptors in dopamine-depleted slices is able to rescue the effects of LY354740, whereas D2 receptor activation does not rescue the effects of LY354740.

A potential consequence of the reduction in group II mGluR efficacy following dopamine depletion is that loss of group II mGluR autoreceptor activity at the STN-SNr synapse could contribute to pathological hyperactivity by removing feedback inhibition of glutamate release, and could therefore contribute to the symptoms of PD (Wittmann et al. 2002). Increasing evidence suggests that modulation of neurotransmission in the SNr by somatodendritically released dopamine plays a crucial role in basal ganglia output, raising the possibility that the loss of dopaminergic modulation at both striatal and extrastriatal sites in the basal ganglia may contribute to the symptoms of PD. For example, local depletion of nigral dopamine by infusing tetrabenazine through a microdialysis probe causes an impairment of motor function in a rotarod performance test, despite the fact that striatal dopamine levels are unaffected (Andersson et al. 2006). Surprisingly, in the same study, local infusion of tetrabenazine into the striatum greatly reduced dopamine levels, but only produced a mild impairment of rotarod performance. These findings suggest that loss of local actions of dopamine in the basal ganglia output nuclei may play a more prominent role in the pathogenesis of PD symptoms than previously thought. If a loss of D1 activation by dopamine released by SNc neurons reduces the effectiveness of group II mGluR autoreceptor activity at the STN-SNr synapse, this dopamine–mGluR interaction could contribute to hyperactivity of this synapse and the generation of the motor symptoms characteristic of PD.

Group II mGluRs also regulate somatodendritic dopamine release from SNc neurons, and this effect could have a local influence on output nuclei function. In contrast to the decrease in dopamine release observed when group II mGluR agonists are applied to the striatum, addition of group II mGluR agonists to nigral brain slices significantly increases local dopamine release (Campusano et al. 2002). Interestingly, partial lesion of the SNc by 6-hydroxydopamine enhances the ability of a group II mGluR agonist to evoke nigral dopamine release.

1.8 GROUP II mGluR AGONISTS IN ANIMAL MODELS OF PD

Several of the physiological effects of group II mGluR activation suggest that agonists could counteract the pathological changes in basal ganglia neurotransmission that contribute to the motor symptoms of PD. In particular, inhibition of excitatory transmission at the STN-SNr synapse could reduce excessive STN-SNr transmission, and would therefore be predicted to have antiparkinsonian effects. Consistent with this hypothesis, intranigral or intracerebroventricular administration of group II mGluR agonists reverses reserpine-induced akinesia in rats (Dawson et al. 2000; Murray et al. 2002), and systemic administration of the group II agonist LY354740 reverses catalepsy and muscle rigidity induced by haloperidol (Konieczny et al. 1998; Bradley et al. 2000). The antiparkinsonian effect of intranigral group II mGluR agonists suggests that reduction of excitatory transmission at the STN-SNr synapse may be involved in mediating this effect. However, in light of the finding that group II mGluR activation increases nigral dopamine release, it is possible that an increase in extracellular dopamine levels in the substantia nigra could also contribute to the antiparkinsonian effects, possibly via activation of D1 receptors in the substantia nigra (Mayorga et al. 1999).

While the effect of intranigral group II mGluR agonists indicates that activation of group II mGluRs in the substantia nigra at least partially mediates the reversal of PD-like motor impairments, several other sites of action may also contribute to antiparkinsonian effects. For example, increased corticostriatal transmission has been implicated in the pathogenesis of PD symptoms, so reduced excitatory transmission at corticostriatal synapses by group II mGluRs may contribute (Bonsi et al. 2007a). Further, the upregulation of group II mGluRs at corticostriatal synapses following dopamine depletion may increase the therapeutic potential for these receptors in the parkinsonian brain, whereas the reversal of this upregulation by chronic L-DOPA treatment suggests that combining group II mGluR agonists with L-DOPA may not be a viable therapeutic strategy (Picconi et al. 2002). Because increased striatal acetylcholine release relative to reduced dopamine levels is thought to contribute to the motor symptoms of PD, the ability of group II mGluR agonists to reduce acetylcholine release represents another possible mechanism by which activation of group II mGluRs could have antiparkinsonian effects (Pisani et al. 2003; Bonsi et al. 2007a). Finally, the finding that mGluR2 activation reduces excitatory transmission in the STN highlights an additional putative site of action for the reversal of motor impairments by group II mGluR agonists, because inhibiting excitatory drive onto STN neurons could reduce the hyperactivity of this nucleus (Shen and Johnson 2003). At this time, the contributions of striatal or subthalamic mGluRs to the antiparkinsonian effects of group II mGluR agonists have not been directly evaluated.

Although the antiparkinsonian actions of group II mGluR agonists have been demonstrated in certain behavioral models of PD, systemic administration of LY379268 fails to reverse motor deficits caused by chronic reserpine treatment or unilateral 6-hydroxydopamine lesion (Murray et al. 2002), raising concerns that this therapy may not be useful in a chronic state of dopamine depletion. The reduced ability of group II mGluR activation to inhibit excitatory transmission at the STN-SNr synapse in reserpinized brain slices suggests that pharmacotherapies for PD that target group II mGluR-mediated inhibition of excitatory transmission in the SNr may not be useful due to the loss of D1 receptor-mediated facilitation of group II mGluR function (Wittmann et al. 2002). In addition, because the potency of the group II mGluR-selective agonist LY354740 for depressing evoked acetylcholine release in the striatum is reduced following 6-hydroxydopamine lesion of the SNc (Marti et al. 2003), the therapeutic potential of mGluR2 activation in the context of cholinergic interneurons might be reduced. These findings highlight the importance of assessing the dependence of mGluR function on dopaminergic neurotransmission when evaluating novel therapeutic strategies for the treatment of PD.

1.9   GROUP II mGluR–MEDIATED PROTECTION AGAINST NIGROSTRIATAL DEGENERATION

Excitotoxicity due to excessive glutamatergic transmission at the STN-SNc synapse may contribute to the progressive degeneration of SNc neurons in PD and thus pharmacological manipulations that reduce STN-SNc transmission may confer neuroprotective benefits. Because activation of group II mGluRs at this synapse reduces excitatory transmission, the potential neuroprotective effects of group II mGluR agonists have been evaluated in multiple animal models of toxin-induced nigrostriatal degeneration. Several studies have demonstrated that systemic and intranigral administration of group II mGluR agonists (such as LY379268 and DCG-IV) reduce 6-hydroxydopamine-induced nigrostriatal degeneration in rats (Murray et al. 2002; Vernon et al. 2005) and MPTP-induced nigrostriatal degeneration in mice (Matarredona et al. 2001; Venero et al. 2002; Battaglia et al. 2003), suggesting that activation of group II mGluRs may provide protection against an excitotoxic component of SNc degeneration. However, other possible mechanisms for group II mGluR-mediated neuroprotection involving glial production of growth factors have also been identified; for example, recent *in vitro* studies demonstrated that group II mGluRs increase the production of the neuroprotective factor brain-derived neurotrophic factor (BDNF) in rat microglia and transforming growth factor-β (TGFβ) in mouse astrocytes (Bruno et al. 1997, 1998; D'Onofrio et al. 2001; Matarredona et al. 2001; Venero et al. 2002). Increased production of these growth factors may represent an alternative mechanism by which group II mGluR activation could confer neuroprotective benefits. Interestingly, *in vitro* and *in vivo* studies employing mice lacking mGluR2 and mGluR3 suggest that the neuroprotective effects of systemic LY379268 administration in MPTP-mice are specifically mediated by mGluR3, and that simultaneous activation of mGluR2 may counteract the neuroprotective effects of mGluR3 activation (Corti et al. 2007); thus, drugs targeting mGluR3 may be particularly beneficial for protection of SNc neurons. Further studies will be necessary to evaluate the relative contributions of effects on growth factor production by glial cells and effects on STN-SNc neurotransmission to the protection of dopamine neurons observed in these animal models.

1.10   MODULATION OF BASAL GANGLIA NEUROTRANSMISSION BY GROUP III mGluRs

Group III mGluR expression has been detected in multiple basal ganglia nuclei including the striatum, globus pallidus, and substantia nigra (Testa et al. 1994; Bradley et al. 1999a,b; Corti et al. 2002; Messenger et al. 2002). Similar to group II mGluRs, activation of group III mGluRs reduces corticostriatal transmission in the MSNs of the striatum by a presynaptic mechanism (Pisani et al. 1997a), although this effect was not observed in a prior study (Lovinger and McCool 1995). However, unlike the increase in potency of group II mGluR agonists in corticostriatal slices from 6-hydroxydopamine-lesioned rats, the potency of group III mGluR agonists does not change in response to nigrostriatal lesion (Picconi et al. 2002), suggesting that there are different mechanisms responsible for regulating the expression of various mGluR subtypes. Moreover, chronic L-DOPA treatment of rats with nigrostriatal lesions does not alter the effect of group III mGluR agonists on corticostriatal transmission (Picconi et al. 2002). These findings indicate that unlike group II mGluRs, striatal group III mGluR function is not regulated by dopaminergic transmission or the changes in corticostriatal glutamatergic transmission that result from nigrostriatal damage.

Immunohistochemical studies have detected high levels of mGluR4 expression in the rodent globus pallidus but not in the striatum (Bradley et al. 1999a,b), whereas mGluR4 mRNA has been detected in the striatum. These findings raise the possibility that mGluR4 activation could modulate striatopallidal neurotransmission by acting on presynaptic receptors. In agreement with this prediction, the group III mGluR agonist L-(+)-2-amino-4-phosphonobutyric acid (L-AP4) reduces the amplitude of evoked inhibitory postsynaptic currents (IPSCs) in rat brain slices by a presynaptic mechanism (Matsui and Kita 2003; Valenti et al. 2003). This effect is not mimicked by the mGluR8-preferring agonist (S)-3,4-dicarboxyphenylglycine (DCPG) and is absent in brain slices obtained from mice lacking mGluR4 (Valenti et al. 2003). In addition, the mGluR4-selective positive allosteric modulator N-phenyl-7-(hydroxyimino)cyclopropa[b]chromen-1a-carboxamide (PHCCC) potentiates the ability of a submaximal concentration of a group III agonist to reduce inhibitory transmission (Marino et al. 2003). Taken together, these results strongly suggest that mGluR4 is the receptor subtype that mediates the depression of inhibitory transmission in the GP. Because the ability of L-AP4 to reduce inhibitory transmission is not significantly altered by overnight reserpine treatment, it is unlikely that dopaminergic tone is required for mGluR4 function at the striatopallidal synapse. In addition to the effects of mGluR4 activation on inhibitory transmission in the rodent GP, L-AP4 also reduces evoked excitatory transmission in the GP by a presynaptic mechanism (Matsui and Kita 2003), indicating that activation of these receptors can simultaneously influence both excitatory and inhibitory transmission in the GP.

Anatomical studies have detected mGluR4 and mGluR7 immunoreactivity in the STN, suggesting that group III mGluR activation may also modulate excitatory or inhibitory synaptic transmission in this nucleus (Bradley et al. 1999a,b). Indeed, L-AP4 inhibits evoked excitatory transmission in the STN by a presynaptic mechanism (Awad-Granko and Conn 2001). Conversely, L-AP4 does not affect inhibitory transmission. Due to the lack of subtype-selective pharmacological tools at the time this study was performed, it is not known if one or both of these receptors contribute to the depression of excitatory transmission. In addition, the effect of disrupting dopaminergic transmission on group III mGluR function in the STN has not yet been determined.

Group III mGluRs are also expressed in the SNr (Bradley et al. 1999a,b; Messenger et al. 2002) and therefore have the potential to modulate basal ganglia output. Electrophysiological studies have demonstrated that activation of group III mGluRs reduces excitatory transmission at the STN-SNr synapse by a presynaptic mechanism (Wittmann et al. 2001b). In addition, recent microdialysis studies show that group III mGluR agonists L-AP4 and L-serine-O-phosphate (L-SOP) reduce KCl-evoked GABA release in the rat globus pallidus (Macinnes and Duty 2008). Group III mGluR agonists also reduce the activity of the direct pathway by depressing inhibitory transmission in the SNr (Wittmann et al. 2001b). Because the SNr receives major GABAergic input from the striatum, it is likely that group III mGluRs reduce inhibitory transmission in the SNr at least in part by reducing GABA release from striatonigral projections. The relatively high concentrations of L-AP4 that are necessary to produce a robust reduction in inhibitory transmission suggest that mGluR7 is likely to mediate or contribute to this effect. Indeed, anatomical studies indicate that mGluR7 is expressed presynaptically on both striatonigral and striatopallidal terminals (Kosinski et al. 1999), supporting the hypothesis that mGluR7 mediates the suppression of inhibitory transmission. Immunohistochemical evidence also suggests that mGluR7 may be the receptor subtype responsible for reducing excitatory transmission in the SNr (Kosinski et al. 1999).

1.10.1 DOPAMINERGIC MODULATION OF GROUP III mGluRs

Interestingly, electrophysiological recordings from brain slices of reserpinized animals suggest that dopamine differentially modulates the effects of mGluRs on excitatory and inhibitory transmission in the SNr. While dopamine depletion or dopamine receptor blockade impairs the ability of group II mGluRs to reduce excitatory transmission at the STN-SNr synapse, the effects of group III mGluRs on excitatory transmission in the SNr do not seem to be regulated by dopamine. In reserpinized brain slices, the ability of L-AP4 to reduce excitatory transmission is unaltered; conversely, the ability of L-AP4 to reduce inhibitory transmission is significantly diminished. Because selective inhibition of either D1 or D2 receptors is sufficient to impair the response to a group III mGluR agonist, the effect of group III mGluRs on inhibitory transmission is likely to depend on tonic activation of both D1 and D2 receptors. These findings highlight the complexity of dopaminergic regulation of mGluR function in the SNr; the ability of group II mGluRs to inhibit excitatory transmission is dopamine dependent, whereas the seemingly similar effect of group III mGluRs is not, suggesting differential mechanisms of regulation of these two subgroups of mGluRs in the same neuronal population. In addition, dopamine modulates the function of group III mGluRs at inhibitory but not excitatory synapses, suggesting that the mechanisms by which the group III mGluRs modulate neurotransmission at these two synapses are differentially regulated as well. Further studies will be required to determine the mechanisms by which dopaminergic tone acutely regulates the ability of presynaptic mGluRs to modulate neurotransmission in the SNr.

1.10.2 GROUP III mGluR MODULATION OF DOPAMINE

Group III mGluRs may regulate dopamine release in the basal ganglia by modulating excitatory transmission in SNc neurons. In rat brain slices, activation of group III mGluRs with agonists that do not distinguish between individual subtypes reduces excitatory transmission by a presynaptic mechanism (Wigmore and Lacey 1998; Valenti et al. 2005). Whilst the mGluR8-preferring agonist DCPG does not mimic this effect, the mGluR4-selective positive allosteric modulator PHCCC potentiates the reduction of excitatory transmission (Valenti et al. 2005), suggesting that this effect is at least partially mediated by mGluR4 but not by mGluR8. A notable species difference is that in slices obtained from mice, both mGluR4 and mGluR8 modulate excitatory transmission in the SNc (Valenti et al. 2005). Because the most prominent source of glutamatergic afferents in the SNc is STN neurons, and mGluR4 and mGluR8 mRNA has been detected in STN neurons (Testa et al. 1994; Messenger et al. 2002), it is likely that the group III mGluRs reduce excitatory transmission by reducing glutamate release from STN terminals. Interestingly, recent studies indicate that activation of presynaptic group III mGluRs also reduces inhibitory transmission in the SNc (Giustizieri et al. 2005). However, the effect of group III mGluR activation on dopamine release in the striatum and other basal ganglia nuclei has not been evaluated.

1.11   GROUP III mGluR AGONISTS IN ANIMAL MODELS OF PD

The ability of group III mGluRs to reduce transmission through the indirect pathway by reducing inhibitory striatopallidal transmission and reducing excitatory transmission in the STN and SNr suggests that activating one or more of the group III mGluRs may relieve the motor symptoms of PD. The finding that intracerebroventricular or systemic administration of group III mGluR agonists or mGluR4-selective positive allosteric modulators reverses motor deficits in both acute and chronic rodent models of PD provides strong support for activation of group III mGluRs as a therapeutic strategy (Marino et al. 2003; Valenti et al. 2003; MacInnes et al. 2004; Battaglia et al. 2006; Lopez et al. 2008; Niswender et al. 2008). Impressively, L-AP4 reverses the forelimb use asymmetry caused by unilateral 6-hydroxydopamine lesion to the same extent as L-DOPA, providing compelling evidence that targeting group III mGluRs may be highly efficacious in treating PD symptoms (Valenti et al. 2003). Because increased GABAergic transmission at the striatopallidal synapse is thought to contribute to the pathogenesis of PD-related motor deficits, the ability of mGluR4 activation to reduce striatopallidal transmission in both normal and dopamine-depleted brain slices makes mGluR4 an intriguing target for novel PD therapeutics (Marino et al. 2003; Valenti et al. 2003). Excitingly, numerous studies have demonstrated that intrapallidal infusion of group III mGluR agonists reverses akinetic deficits in several models of PD (MacInnes et al. 2004; Konieczny et al. 2007; Lopez et al. 2007; Sibille et al. 2007). The results of these studies provide strong evidence that reducing striatopallidal transmission by activating mGluR4 may be a promising approach for alleviating the motor symptoms of PD.

In contrast to the clear antiparkinsonian effects of group III mGluR activation at striatopallidal synapses, the ability of group III mGluRs to reverse motor deficits in PD models by reducing excitatory transmission in the STN or SNr is not as well established. Intranigral infusion of group III mGluR agonists reverses reserpine-induced akinesia and haloperidol-induced catalepsy in rats, indicating that group III mGluR activation in the substantia nigra may confer some antiparkinsonian benefits (MacInnes et al. 2004; Konieczny et al. 2007). However, a more recent study in rats demonstrated that intranigral infusion of group III mGluR agonists worsened akinetic deficits in a reaction time task caused by 6-hydroxydopamine, and intranigral infusion of the group III mGluR agonist ACPT-I did not robustly reduce haloperidol-induced catalepsy (Lopez et al. 2007). These findings suggest that directly targeting increased STN activity may not yield the expected alleviation of motor deficits, despite the fact that group III mGluR function at STN-SNr synapses is not reduced in the absence of dopamine (Wittmann et al. 2002).

1.12   GROUP III mGluR–MEDIATED PROTECTION AGAINST NIGROSTRIATAL DEGENERATION

Like the other mGluR subgroups, group III mGluRs have been implicated as targets for neuroprotection in toxin-based animal models of nigrostriatal degeneration. The ability of group III mGluR activation to reduce excitatory transmission at the STN-SNc synapse while simultaneously reducing disinhibition of the STN by depressing inhibitory striatopallidal transmission could reduce the increased excitatory transmission in the SNc that may lead to excitotoxic cell death. Recent *in vitro* and *in vivo* studies have shown that mGluR4 activation protects against NMDA-mediated neurotoxicity (Gasparini et al. 1999; Bruno et al. 2000; Flor et al. 2002), suggesting that mGluR4 may be an important mediator of protection against excitotoxic cell death in various neuronal populations. Interestingly, acute or subchronic intranigral administration of L-AP4 reduces nigrostriatal degeneration caused by 6-hydroxydopamine lesion in rats (Vernon et al. 2005, 2007), and combined treatment with L-AP4 and the mGluR5 antagonist MPEP confers more robust protection than either drug alone (Vernon et al. 2008). Further, systemic administration of the mGluR4-selective positive allosteric modulator PHCCC protects mice against MPTP-induced nigrostriatal degeneration (Battaglia et al. 2006). Taken together, these studies suggest that activation of group III mGluRs, particularly mGluR4, may be a promising therapeutic strategy for reducing SNc degeneration in the parkinsonian brain.

1.13   CONCLUDING REMARKS

The combined use of physiological, biochemical, and behavioral experimental approaches has allowed great advances in our understanding of the roles of mGluRs in the basal ganglia motor circuit, and the complexity of interactions between mGluRs and other neurotransmitter systems, particularly dopamine. While these studies have led to the identification of promising targets for the treatment of PD, it is likely that we are just beginning to understand the roles of mGluRs and their interactions with dopamine signaling. While the use of pharmacological agents and artificial manipulations of dopaminergic neurotransmission have provided insights into the roles of these receptors and their modulation by dopamine, our knowledge of the effects of mGluR activation by endogenous glutamate remains limited. In addition, currently available animal models of PD have yet to adequately recapitulate the human disease, placing another boundary on our ability to fully understand the changes in mGluR function caused by SNc degeneration. As our understanding of the cellular events leading to nigrostriatal degeneration advances, the ability of animal models to predict neuroprotective effects of mGluR ligands may improve. Furthermore, while many distinct roles of mGluR1 and mGluR5 have been determined, the previous lack of subtype-selective pharmacological tools that distinguish between group II and group III mGluR subtypes has restricted our ability to fully understand the roles of individual receptor subtypes. Therefore, much work remains to be done in order to fully elucidate the complex interactions of mGluRs and dopamine in the basal ganglia motor circuit.

REFERENCES

Aguirre, J. A., Kehr, J., Yoshitake, T. et al. 2005. Protection but maintained dysfunction of nigral dopaminergic nerve cell bodies and striatal dopaminergic terminals in MPTP-lesioned mice after acute treatment with the mGluR5 antagonist MPEP. *Brain Res* 1033:216–220.

Andersson, D. R., Nissbrandt, H., and Bergquist, F. 2006. Partial depletion of dopamine in substantia nigra impairs motor performance without altering striatal dopamine neurotransmission. *Eur J Neurosci* 24:617–624.

Armentero, M. T., Fancellu, R., Nappi, G. et al. 2006. Prolonged blockade of NMDA or mGluR5 glutamate receptors reduces nigrostriatal degeneration while inducing selective metabolic changes in the basal ganglia circuitry in a rodent model of Parkinson's disease. *Neurobiol Dis* 22:1–9.

Awad, H., Hubert, G. W., Smith, Y. et al. 2000. Activation of metabotropic glutamate receptor 5 has direct excitatory effects and potentiates NMDA receptor currents in neurons of the subthalamic nucleus. *J Neurosci* 20:7871–7879.

Awad-Granko, H. and Conn, P. J. 2001. Activation of groups I or III metabotropic glutamate receptors inhibits excitatory transmission in the rat subthalamic nucleus. *Neuropharmacology* 41:32–41.

Baker, D. A., Xi, Z. X., Shen, H. et al. 2002. The origin and neuronal function of in vivo nonsynaptic glutamate. *J Neurosci* 22:9134–9141.

Battaglia, G., Busceti, C. L., Molinaro, G. et al. 2004. Endogenous activation of mGlu5 metabotropic glutamate receptors contributes to the development of nigro-striatal damage induced by 1-methyl-4-phenyl-1,2,3,6-tetrahydropyridine in mice. *J Neurosci* 24:828–835.

Battaglia, G., Busceti, C. L., Molinaro, G. et al. 2006. Pharmacological activation of mGlu4 metabotropic glutamate receptors reduces nigrostriatal degeneration in mice treated with 1-methyl-4-phenyl-1,2,3,6-tetrahydropyridine. *J Neurosci* 26:7222–7229.

Battaglia, G., Busceti, C. L., Pontarelli, F. et al. 2003. Protective role of group-II metabotropic glutamate receptors against nigro-striatal degeneration induced by 1-methyl-4-phenyl-1,2,3,6-tetrahydropyridine in mice. *Neuropharmacology* 45:155–166.

Battaglia, G., Fornai, F., Busceti, C. L. et al. 2002. Selective blockade of mGlu5 metabotropic glutamate receptors is protective against methamphetamine neurotoxicity. *J Neurosci* 22:2135–2141.

Battaglia, G., Monn, J. A., and Schoepp, D. D. 1997. In vivo inhibition of veratridine-evoked release of striatal excitatory amino acids by the group II metabotropic glutamate receptor agonist LY354740 in rats. *Neurosci Lett* 229:161–164.

Bell, M. I., Richardson, P. J., and Lee, K. 2002. Functional and molecular characterization of metabotropic glutamate receptors expressed in rat striatal cholinergic interneurones. *J Neurochem* 81:142–149.

Bergman, H., Wichmann, T., Karmon, B. et al. 1994. The primate subthalamic nucleus. II. Neuronal activity in the MPTP model of parkinsonism. *J Neurophysiol* 72:507–520.

Beurrier, C., Congar, P., Bioulac, B. et al. 1999. Subthalamic nucleus neurons switch from single-spike activity to burst-firing mode. *J Neurosci* 19:599–609.

Bonsi, P., Cuomo, D., De Persis, C. et al. 2005. Modulatory action of metabotropic glutamate receptor (mGluR) 5 on mGluR1 function in striatal cholinergic interneurons. *Neuropharmacology* 49 (Suppl 1):104–113.

Bonsi, P., Cuomo, D., Picconi, B. et al. 2007a. Striatal metabotropic glutamate receptors as a target for pharmacotherapy in Parkinson's disease. *Amino Acids* 32:189–195.

Bonsi, P., Platania, P., Martella, G. et al. 2008. Distinct roles of group I mGlu receptors in striatal function. *Neuropharmacology* 55:392–395.

Bonsi, P., Sciamanna, G., Mitrano, D. A. et al. 2007b. Functional and ultrastructural analysis of group I mGluR in striatal fast-spiking interneurons. *Eur J Neurosci* 25:1319–1331.

Bradley, S. R., Marino, M. J., Wittmann, M. et al. 2000. Activation of group II metabotropic glutamate receptors inhibits synaptic excitation of the substantia Nigra pars reticulata. *J Neurosci* 20:3085–3094.

Bradley, S. R., Standaert, D. G., Levey, A. I. et al. 1999a. Distribution of group III mGluRs in rat basal ganglia with subtype-specific antibodies. *Ann NY Acad Sci* 868:531–534.

Bradley, S. R., Standaert, D. G., Rhodes, K. J. et al. 1999b. Immunohistochemical localization of subtype 4a metabotropic glutamate receptors in the rat and mouse basal ganglia. *J Comp Neurol* 407:33–46.

Breysse, N., Amalric, M., and Salin, P. 2003. Metabotropic glutamate 5 receptor blockade alleviates akinesia by normalizing activity of selective basal-ganglia structures in parkinsonian rats. *J Neurosci* 23:8302–8309.

Breysse, N., Baunez, C., Spooren, W. et al. 2002. Chronic but not acute treatment with a metabotropic glutamate 5 receptor antagonist reverses the akinetic deficits in a rat model of parkinsonism. *J Neurosci* 22:5669–5678.

Bruno, V., Battaglia, G., Casabona, G. et al. 1998. Neuroprotection by glial metabotropic glutamate receptors is mediated by transforming growth factor-beta. *J Neurosci* 18:9594–9600.

Bruno, V., Battaglia, G., Ksiazek, I. et al. 2000. Selective activation of mGlu4 metabotropic glutamate receptors is protective against excitotoxic neuronal death. *J Neurosci* 20:6413–6420.

Bruno, V., Sureda, F. X., Storto, M. et al. 1997. The neuroprotective activity of group-II metabotropic glutamate receptors requires new protein synthesis and involves a glial-neuronal signaling. *J Neurosci* 17:1891–1897.

Calabresi, P., Centonze, D., Gubellini, P. et al. 1998. Endogenous ACh enhances striatal NMDA-responses via M1-like muscarinic receptors and PKC activation. *Eur J Neurosci* 10:2887–2895.

Calabresi, P., Centonze, D., Pisani, A. et al. 1999. Metabotropic glutamate receptors and cell-type-specific vulnerability in the striatum: Implication for ischemia and Huntington's disease. *Exp Neurol* 158:97–108.

Calabresi, P., Gubellini, P., Centonze, D. et al. 2000. Dopamine and cAMP-regulated phosphoprotein 32kDa controls both striatal long-term depression and long-term potentiation, opposing forms of synaptic plasticity. *J Neurosci* 20:8443–8451.

Calabresi, P., Maj, R., Mercuri, N. B. et al. 1992a. Coactivation of D1 and D2 dopamine receptors is required for long-term synaptic depression in the striatum. *Neurosci Lett* 142:95–99.

Calabresi, P., Maj, R., Pisani, A. et al. 1992b. Long-term synaptic depression in the striatum: Physiological and pharmacological characterization. *J Neurosci* 12:4224–4233.

Calabresi, P., Mercuri, N. B., and Di Filippo, M. 2009. Synaptic plasticity, dopamine and Parkinson's disease: One step ahead. *Brain* 132:285–287.

Calabresi, P., Picconi, B., Tozzi, A. et al. 2007. Dopamine-mediated regulation of corticostriatal synaptic plasticity. *Trends Neurosci* 30:211–219.

Calabresi, P., Pisani, A., Mercuri, N. B. et al. 1992c. Long-term potentiation in the striatum is unmasked by removing the voltage-dependent magnesium block of NMDA receptor channels. *Eur J Neurosci* 4:929–935.

Calabresi, P., Pisani, A., Mercuri, N. B. et al. 1994. Post-receptor mechanisms underlying striatal long-term depression. *J Neurosci* 14:4871–4881.

Campusano, J. M., Abarca, J., Forray, M. I. et al. 2002. Modulation of dendritic release of dopamine by metabotropic glutamate receptors in rat substantia nigra. *Biochem Pharmacol* 63:1343–1352.

Cartmell, J., Salhoff, C. R., Perry, K. W. et al. 2000. Dopamine and 5-HT turnover are increased by the mGlu2/3 receptor agonist LY379268 in rat medial prefrontal cortex, nucleus accumbens and striatum. *Brain Res* 887:378–384.

Cepeda, C., Hurst, R. S., Altemus, K. L. et al. 2001. Facilitated glutamatergic transmission in the striatum of D2 dopamine receptor-deficient mice. *J Neurophysiol* 85:659–670.

Chen, J. J. and Swope, D. M. 2007. Pharmacotherapy for Parkinson's disease. *Pharmacotherapy* 27:161S–173S.

Choe, E. S., Chung, K. T., Mao, L. et al. 2002. Amphetamine increases phosphorylation of extracellular signal-regulated kinase and transcription factors in the rat striatum via group I metabotropic glutamate receptors. *Neuropsychopharmacology* 27:565–575.

Choe, E. S. and McGinty, J. F. 2001. Cyclic AMP and mitogen-activated protein kinases are required for glutamate-dependent cyclic AMP response element binding protein and Elk-1 phosphorylation in the dorsal striatum in vivo. *J Neurochem* 76:401–412.

Choe, E. S. and Wang, J. Q. 2001a. Group I metabotropic glutamate receptors control phosphorylation of CREB, Elk-1 and ERK via a CaMKII-dependent pathway in rat striatum. *Neurosci Lett* 313:129–132.

Choe, E. S. and Wang, J. Q. 2001b. Group I metabotropic glutamate receptor activation increases phosphorylation of cAMP response element-binding protein, Elk-1, and extracellular signal-regulated kinases in rat dorsal striatum. *Brain Res Mol Brain Res* 94:75–84.

Choi, S. and Lovinger, D. M. 1997. Decreased frequency but not amplitude of quantal synaptic responses associated with expression of corticostriatal long-term depression. *J Neurosci* 17:8613–8620.

Coccurello, R., Breysse, N., and Amalric, M. 2004. Simultaneous blockade of adenosine A2A and metabotropic glutamate mGlu5 receptors increase their efficacy in reversing Parkinsonian deficits in rats. *Neuropsychopharmacology* 29:1451–1461.

Conn, P. J., Battaglia, G., Marino, M. J. et al. 2005. Metabotropic glutamate receptors in the basal ganglia motor circuit. *Nat Rev Neurosci* 6:787–798.

Conn, P. J. and Pin, J. P. 1997. Pharmacology and functions of metabotropic glutamate receptors. *Annu Rev Pharmacol Toxicol* 37:205–237.

Corti, C., Aldegheri, L., Somogyi, P. et al. 2002. Distribution and synaptic localisation of the metabotropic glutamate receptor 4 (mGluR4) in the rodent CNS. *Neuroscience* 110:403–420.

Corti, C., Battaglia, G., Molinaro, G. et al. 2007. The use of knock-out mice unravels distinct roles for mGlu2 and mGlu3 metabotropic glutamate receptors in mechanisms of neurodegeneration/neuroprotection. *J Neurosci* 27:8297–8308.

Cozzi, A., Attucci, S., Peruginelli, F. et al. 1997. Type 2 metabotropic glutamate (mGlu) receptors tonically inhibit transmitter release in rat caudate nucleus: In vivo studies with (2S,1'S,2'S,3'R)-2-(2'-carboxy-3'-phenylcyclopropyl)glycine, a new potent and selective antagonist. *Eur J Neurosci* 9:1350–1355.

D'Onofrio, M., Cuomo, L., Battaglia, G. et al. 2001. Neuroprotection mediated by glial group-II metabotropic glutamate receptors requires the activation of the MAP kinase and the phosphatidylinositol-3-kinase pathways. *J Neurochem* 78:435–445.

Dawson, L., Chadha, A., Megalou, M. et al. 2000. The group II metabotropic glutamate receptor agonist, DCG-IV, alleviates akinesia following intranigral or intraventricular administration in the reserpine-treated rat. *Br J Pharmacol* 129:541–546.

De Leonibus, E., Manago, F., Giordani, F. et al. 2009. Metabotropic glutamate receptors 5 blockade reverses spatial memory deficits in a mouse model of Parkinson's disease. *Neuropsychopharmacology* 34:729–738.

Dekundy, A., Pietraszek, M., Schaefer, D. et al. 2006. Effects of group I metabotropic glutamate receptors blockade in experimental models of Parkinson's disease. *Brain Res Bull* 69:318–326.

Ferre, S., Karcz-Kubicha, M., Hope, B. T. et al. 2002. Synergistic interaction between adenosine A2A and glutamate mGlu5 receptors: Implications for striatal neuronal function. *Proc Natl Acad Sci USA* 99:11940–11945.

Fiorillo, C. D. and Williams, J. T. 1998. Glutamate mediates an inhibitory postsynaptic potential in dopamine neurons. *Nature* 394:78–82.

Flor, P. J., Battaglia, G., Nicoletti, F. et al. 2002. Neuroprotective activity of metabotropic glutamate receptor ligands. *Adv Exp Med Biol* 513:197–223.

Gasparini, F., Bruno, V., Battaglia, G. et al. 1999. (R,S)-4-phosphonophenylglycine, a potent and selective group III metabotropic glutamate receptor agonist, is anticonvulsive and neuroprotective in vivo. *J Pharmacol Exp Ther* 289:1678–1687.

Gerdeman, G. L., Ronesi, J., and Lovinger, D. M. 2002. Postsynaptic endocannabinoid release is critical to long-term depression in the striatum. *Nat Neurosci* 5:446–451.

Giustizieri, M., Bernardi, G., Mercuri, N. B. et al. 2005. Distinct mechanisms of presynaptic inhibition at GABAergic synapses of the rat substantia nigra pars compacta. *J Neurophysiol* 94:1992–2003.

Gubellini, P., Pisani, A., Centonze, D. et al. 2004. Metabotropic glutamate receptors and striatal synaptic plasticity: Implications for neurological diseases. *Prog Neurobiol* 74:271–300.

Gubellini, P., Saulle, E., Centonze, D. et al. 2001. Selective involvement of mGlu1 receptors in corticostriatal LTD. *Neuropharmacology* 40:839–846.

Gubellini, P., Saulle, E., Centonze, D. et al. 2003. Corticostriatal LTP requires combined mGluR1 and mGluR5 activation. *Neuropharmacology* 44:8–16.

Hanson, J. E. and Smith, Y. 1999. Group I metabotropic glutamate receptors at GABAergic synapses in monkeys. *J Neurosci* 19:6488–6496.

Hu, G., Duffy, P., Swanson, C. et al. 1999. The regulation of dopamine transmission by metabotropic glutamate receptors. *J Pharmacol Exp Ther* 289:412–416.

Ismayilova, N., Verkhratsky, A., and Dascombe, M. J. 2006. Changes in mGlu5 receptor expression in the basal ganglia of reserpinised rats. *Eur J Pharmacol* 545:134–141.

Kachroo, A., Orlando, L. R., Grandy, D. K. et al. 2005. Interactions between metabotropic glutamate 5 and adenosine A2A receptors in normal and parkinsonian mice. *J Neurosci* 25:10414–10419.

Kahn, L., Alonso, G., Robbe, D. et al. 2001. Group 2 metabotropic glutamate receptors induced long term depression in mouse striatal slices. *Neurosci Lett* 316:178–182.

Kaneda, K., Kita, T., and Kita, H. 2007. Repetitive activation of glutamatergic inputs evokes a long-lasting excitation in rat globus pallidus neurons in vitro. *J Neurophysiol* 97:121–133.

Kaneda, K., Tachibana, Y., Imanishi, M. et al. 2005. Down-regulation of metabotropic glutamate receptor 1alpha in globus pallidus and substantia nigra of parkinsonian monkeys. *Eur J Neurosci* 22:3241–3254.

Kearney, J. A., Becker, J. B., Frey, K. A. et al. 1998. The role of nigrostriatal dopamine in metabotropic glutamate agonist-induced rotation. *Neuroscience* 87:881–891.

Kearney, J. A., Frey, K. A., and Albin, R. L. 1997. Metabotropic glutamate agonist-induced rotation: A pharmacological, FOS immunohistochemical, and [14C]-2-deoxyglucose autoradiographic study. *J Neurosci* 17:4415–4425.

Konieczny, J., Ossowska, K., Wolfarth, S. et al. 1998. LY354740, a group II metabotropic glutamate receptor agonist with potential antiparkinsonian properties in rats. *Naunyn Schmiedebergs Arch Pharmacol* 358:500–502.

Konieczny, J., Wardas, J., Kuter, K. et al. 2007. The influence of group III metabotropic glutamate receptor stimulation by (1S,3R,4S)-1-aminocyclo-pentane-1,3,4-tricarboxylic acid
on the parkinsonian-like akinesia and striatal proenkephalin and prodynorphin mRNA
expression in rats. *Neuroscience* 145:611–620.

Kosinski, C. M., Risso Bradley, S., Conn, P. J. et al. 1999. Localization of metabotropic glutamate receptor 7 mRNA and mGluR7a protein in the rat basal ganglia. *J Comp Neurol*
415:266–284.

Kuwajima, M., Dehoff, M. H., Furuichi, T. et al. 2007. Localization and expression of group I
metabotropic glutamate receptors in the mouse striatum, globus pallidus, and subthalamic nucleus: Regulatory effects of MPTP treatment and constitutive Homer deletion.
J Neurosci 27:6249–6260.

Levandis, G., Bazzini, E., Armentero, M. T. et al. 2008. Systemic administration of an mGluR5
antagonist, but not unilateral subthalamic lesion, counteracts l-DOPA-induced dyskinesias in a rodent model of Parkinson's disease. *Neurobiol Dis* 29:161–168.

Liu, X., Ford-Dunn, H. L., Hayward, G. N. et al. 2002b. The oscillatory activity in the
Parkinsonian subthalamic nucleus investigated using the macro-electrodes for deep
brain stimulation. *Clin Neurophysiol* 113:1667–1672.

Liu, F., Ma, X. H., Ule, J. et al. 2001. Regulation of cyclin-dependent kinase 5 and
casein kinase 1 by metabotropic glutamate receptors. *Proc Natl Acad Sci USA*
98:11062–11068.

Liu, F., Virshup, D. M., Nairn, A. C. et al. 2002a. Mechanism of regulation of casein kinase I
activity by group I metabotropic glutamate receptors. *J Biol Chem* 277:45393–45399.

Lopez, S., Turle-Lorenzo, N., Acher, F. et al. 2007. Targeting group III metabotropic glutamate
receptors produces complex behavioral effects in rodent models of Parkinson's disease.
J Neurosci 27:6701–6711.

Lopez, S., Turle-Lorenzo, N., Johnston, T. H. et al. 2008. Functional interaction between
adenosine A2A and group III metabotropic glutamate receptors to reduce parkinsonian
symptoms in rats. *Neuropharmacology* 55:483–490.

Lovinger, D. M. and McCool, B. A. 1995. Metabotropic glutamate receptor-mediated presynaptic depression at corticostriatal synapses involves mGLuR2 or 3. *J Neurophysiol*
73:1076–1083.

MacInnes, N. and Duty, S. 2008. Group III metabotropic glutamate receptors act as heteroreceptors modulating evoked GABA release in the globus pallidus in vivo. *Eur J
Pharmacol* 580:95–99.

MacInnes, N., Messenger, M. J., and Duty, S. 2004. Activation of group III metabotropic
glutamate receptors in selected regions of the basal ganglia alleviates akinesia in the
reserpine-treated rat. *Br J Pharmacol* 141:15–22.

Mao, L., Conquet, F., and Wang, J. Q. 2001. Augmented motor activity and reduced striatal
preprodynorphin mRNA induction in response to acute amphetamine administration in
metabotropic glutamate receptor 1 knockout mice. *Neuroscience* 106:303–312.

Mao, L., Conquet, F., and Wang, J. Q. 2002. Impaired preprodynorphin, but not preproenkephalin, mRNA induction in the striatum of mGluR1 mutant mice in response
to acute administration of the full dopamine D(1) agonist SKF-82958. *Synapse*
44:86–93.

Mao, L. and Wang, J. Q. 2001. Selective activation of group I metabotropic glutamate receptors upregulates preprodynorphin, substance P, and preproenkephalin mRNA expression
in rat dorsal striatum. *Synapse* 39:82–94.

Mao, L., Yang, L., Tang, Q. et al. 2005. The scaffold protein Homer1b/c links metabotropic
glutamate receptor 5 to extracellular signal-regulated protein kinase cascades in neurons. *J Neurosci* 25:2741–2752.

Mao, L. M., Zhang, G. C., Liu, X. Y. et al. 2008. Group I metabotropic glutamate receptor-
mediated gene expression in striatal neurons. *Neurochem Res* 33:1920–1924.

Marino, M. J., Awad-Granko, H., Ciombor, K. J. et al. 2002. Haloperidol-induced alteration in the physiological actions of group I mGlus in the subthalamic nucleus and the substantia nigra pars reticulata. *Neuropharmacology* 43:147–159.

Marino, M. J., Williams, D. L., Jr., O'Brien, J. A. et al. 2003. Allosteric modulation of group III metabotropic glutamate receptor 4: A potential approach to Parkinson's disease treatment. *Proc Natl Acad Sci USA* 100:13668–13673.

Marino, M. J., Wittmann, M., Bradley, S. R. et al. 2001. Activation of group I metabotropic glutamate receptors produces a direct excitation and disinhibition of GABAergic projection neurons in the substantia nigra pars reticulata. *J Neurosci* 21:7001–7012.

Martella, G., Platania, P., Vita, D. et al. 2009. Enhanced sensitivity to group II mGlu receptor activation at corticostriatal synapses in mice lacking the familial parkinsonism-linked genes PINK1 or Parkin. *Exp Neurol* 215:388–396.

Marti, M., Paganini, F., Stocchi, S. et al. 2001. Presynaptic group I and II metabotropic glutamate receptors oppositely modulate striatal acetylcholine release. *Eur J Neurosci* 14:1181–1184.

Marti, M., Paganini, F., Stocchi, S. et al. 2003. Plasticity of glutamatergic control of striatal acetylcholine release in experimental parkinsonism: Opposite changes at group-II metabotropic and NMDA receptors. *J Neurochem* 84:792–802.

Matarredona, E. R., Santiago, M., Venero, J. L. et al. 2001. Group II metabotropic glutamate receptor activation protects striatal dopaminergic nerve terminals against MPP+-induced neurotoxicity along with brain-derived neurotrophic factor induction. *J Neurochem* 76:351–360.

Matsui, T. and Kita, H. 2003. Activation of group III metabotropic glutamate receptors presynaptically reduces both GABAergic and glutamatergic transmission in the rat globus pallidus. *Neuroscience* 122:727–737.

Mayorga, A. J., Trevitt, J. T., Conlan, A. et al. 1999. Striatal and nigral D1 mechanisms involved in the antiparkinsonian effects of SKF 82958 (APB): Studies of tremulous jaw movements in rats. *Psychopharmacology (Berl)* 143:72–81.

Mela, F., Marti, M., Dekundy, A. et al. 2007. Antagonism of metabotropic glutamate receptor type 5 attenuates l-DOPA-induced dyskinesia and its molecular and neurochemical correlates in a rat model of Parkinson's disease. *J Neurochem* 101:483–497.

Mercuri, N. B., Stratta, F., Calabresi, P. et al. 1992. Electrophysiological evidence for the presence of ionotropic and metabotropic excitatory amino acid receptors on dopaminergic neurons of the rat mesencephalon: An in vitro study. *Funct Neurol* 7:231–234.

Mercuri, N. B., Stratta, F., Calabresi, P. et al. 1993. Activation of metabotropic glutamate receptors induces an inward current in rat dopamine mesencephalic neurons. *Neuroscience* 56:399–407.

Messenger, M. J., Dawson, L. G., and Duty, S. 2002. Changes in metabotropic glutamate receptor 1-8 gene expression in the rodent basal ganglia motor loop following lesion of the nigrostriatal tract. *Neuropharmacology* 43:261–271.

Murray, T. K., Messenger, M. J., Ward, M. A. et al. 2002. Evaluation of the mGluR2/3 agonist LY379268 in rodent models of Parkinson's disease. *Pharmacol Biochem Behav* 73:455–466.

Ni, Z., Bouali-Benazzouz, R., Gao, D. et al. 2001a. Intrasubthalamic injection of 6-hydroxy-dopamine induces changes in the firing rate and pattern of subthalamic nucleus neurons in the rat. *Synapse* 40:145–153.

Ni, Z. G., Bouali-Benazzouz, R., Gao, D. M. et al. 2001b. Time-course of changes in firing rates and firing patterns of subthalamic nucleus neuronal activity after 6-OHDA-induced dopamine depletion in rats. *Brain Res* 899:142–147.

Nishi, A., Liu, F., Matsuyama, S. et al. 2003. Metabotropic mGlu5 receptors regulate adenosine A2A receptor signaling. *Proc Natl Acad Sci USA* 100:1322–1327.

Nishi, A., Watanabe, Y., Higashi, H. et al. 2005. Glutamate regulation of DARPP-32 phosphorylation in neostriatal neurons involves activation of multiple signaling cascades. *Proc Natl Acad Sci USA* 102:1199–1204.

Niswender, C. M., Johnson, K. A., Weaver, C. D. et al. 2008. Discovery, characterization, and antiparkinsonian effect of novel positive allosteric modulators of metabotropic glutamate receptor 4. *Mol Pharmacol* 74:1345–1358.

Ossowska, K., Konieczny, J., Wolfarth, S. et al. 2001. Blockade of the metabotropic glutamate receptor subtype 5 (mGluR5) produces antiparkinsonian-like effects in rats. *Neuropharmacology* 41:413–420.

Ossowska, K., Konieczny, J., Wolfarth, S. et al. 2005. MTEP, a new selective antagonist of the metabotropic glutamate receptor subtype 5 (mGluR5), produces antiparkinsonian-like effects in rats. *Neuropharmacology* 49:447–455.

Parelkar, N. K. and Wang, J. Q. 2003. Preproenkephalin mRNA expression in rat dorsal striatum induced by selective activation of metabotropic glutamate receptor subtype-5. *Synapse* 47:255–261.

Partridge, J. G., Tang, K. C., and Lovinger, D. M. 2000. Regional and postnatal heterogeneity of activity-dependent long-term changes in synaptic efficacy in the dorsal striatum. *J Neurophysiol* 84:1422–1429.

Pellegrino, D., Cicchetti, F., Wang, X. et al. 2007. Modulation of dopaminergic and glutamatergic brain function: PET studies on parkinsonian rats. *J Nucl Med* 48:1147–1153.

Phillips, J. M., Lam, H. A., Ackerson, L. C. et al. 2006. Blockade of mGluR glutamate receptors in the subthalamic nucleus ameliorates motor asymmetry in an animal model of Parkinson's disease. *Eur J Neurosci* 23:151–160.

Picconi, B., Pisani, A., Centonze, D. et al. 2002. Striatal metabotropic glutamate receptor function following experimental parkinsonism and chronic levodopa treatment. *Brain* 125:2635–2645.

Pin, J. P. and Acher, F. 2002. The metabotropic glutamate receptors: Structure, activation mechanism and pharmacology. *Curr Drug Targets CNS Neurol Disord* 1:297–317.

Pisani, A., Bonsi, P., Catania, M. V. et al. 2002. Metabotropic glutamate 2 receptors modulate synaptic inputs and calcium signals in striatal cholinergic interneurons. *J Neurosci* 22:6176–6185.

Pisani, A., Bonsi, P., Centonze, D. et al. 2001a. Functional coexpression of excitatory mGluR1 and mGluR5 on striatal cholinergic interneurons. *Neuropharmacology* 40:460–463.

Pisani, A., Bonsi, P., Centonze, D. et al. 2003. Targeting striatal cholinergic interneurons in Parkinson's disease: Focus on metabotropic glutamate receptors. *Neuropharmacology* 45:45–56.

Pisani, A., Calabresi, P., Centonze, D. et al. 1997a. Activation of group III metabotropic glutamate receptors depresses glutamatergic transmission at corticostriatal synapse. *Neuropharmacology* 36:845–851.

Pisani, A., Calabresi, P., Centonze, D. et al. 1997b. Enhancement of NMDA responses by group I metabotropic glutamate receptor activation in striatal neurones. *Br J Pharmacol* 120:1007–1014.

Pisani, A., Gubellini, P., Bonsi, P. et al. 2001b. Metabotropic glutamate receptor 5 mediates the potentiation of N-methyl-D-aspartate responses in medium spiny striatal neurons. *Neuroscience* 106:579–587.

Poisik, O. V., Mannaioni, G., Traynelis, S. et al. 2003. Distinct functional roles of the metabotropic glutamate receptors 1 and 5 in the rat globus pallidus. *J Neurosci* 23:122–130.

Poisik, O. V., Raju, D. V., Verreault, M. et al. 2005. Metabotropic glutamate receptor 2 modulates excitatory synaptic transmission in the rat globus pallidus. *Neuropharmacology* 49 (Suppl 1):57–69.

Poisik, O. V., Smith, Y., and Conn, P. J. 2007. D1- and D2-like dopamine receptors regulate signaling properties of group I metabotropic glutamate receptors in the rat globus pallidus. *Eur J Neurosci* 26:852–862.

Rouse, S. T., Marino, M. J., Bradley, S. R. et al. 2000. Distribution and roles of metabotropic glutamate receptors in the basal ganglia motor circuit: Implications for treatment of Parkinson's disease and related disorders. *Pharmacol Ther* 88:427–435.

Samadi, P., Gregoire, L., Morissette, M. et al. 2007. mGluR5 metabotropic glutamate receptors and dyskinesias in MPTP monkeys. *Neurobiol Aging* 29(7):1040–1051.

Samadi, P., Gregoire, L., Morissette, M. et al. 2008. Basal ganglia group II metabotropic glutamate receptors specific binding in non-human primate model of L-Dopa-induced dyskinesias. *Neuropharmacology* 54:258–268.

Sanchez-Pernaute, R., Wang, J. Q., Kuruppu, D. et al. 2008. Enhanced binding of metabotropic glutamate receptor type 5 (mGluR5) PET tracers in the brain of parkinsonian primates. *Neuroimage* 42:248–251.

Sergeeva, O. A., Doreulee, N., Chepkova, A. N. et al. 2007. Long-term depression of corticostriatal synaptic transmission by DHPG depends on endocannabinoid release and nitric oxide synthesis. *Eur J Neurosci* 26:1889–1894.

Shen, K. Z. and Johnson, S. W. 2003. Group II metabotropic glutamate receptor modulation of excitatory transmission in rat subthalamic nucleus. *J Physiol* 553:489–496.

Sibille, P., Lopez, S., Brabet, I. et al. 2007. Synthesis and biological evaluation of 1-amino-2-phosphonomethylcyclopropanecarboxylic acids, new group III metabotropic glutamate receptor agonists. *J Med Chem* 50:3585–3595.

Smith, Y. and Kieval, J. Z. 2000. Anatomy of the dopamine system in the basal ganglia. *Trends Neurosci* 23:S28–S33.

Spooren, W. P., Gasparini, F., Bergmann, R. et al. 2000. Effects of the prototypical mGlu(5) receptor antagonist 2-methyl-6-(phenylethynyl)-pyridine on rotarod, locomotor activity and rotational responses in unilateral 6-OHDA-lesioned rats. *Eur J Pharmacol* 406:403–410.

Stefani, A., Spadoni, F., and Bernardi, G. 1998. Group I mGluRs modulate calcium currents in rat GP: Functional implications. *Synapse* 30:424–432.

Sung, K. W., Choi, S., and Lovinger, D. M. 2001. Activation of group I mGluRs is necessary for induction of long-term depression at striatal synapses. *J Neurophysiol* 86:2405–2412.

Svenningsson, P., Nishi, A., Fisone, G. et al. 2004. DARPP-32: An integrator of neurotransmission. *Annu Rev Pharmacol Toxicol* 44:269–296.

Takeshita, Y., Harata, N., and Akaike, N. 1996. Suppression of K+ conductance by metabotropic glutamate receptor in acutely dissociated large cholinergic neurons of rat caudate putamen. *J Neurophysiol* 76:1545–1558.

Tallaksen-Greene, S. J., Kaatz, K. W., Romano, C. et al. 1998. Localization of mGluR1a-like immunoreactivity and mGluR5-like immunoreactivity in identified populations of striatal neurons. *Brain Res* 780:210–217.

Testa, C. M., Friberg, I. K., Weiss, S. W. et al. 1998. Immunohistochemical localization of metabotropic glutamate receptors mGluR1a and mGluR2/3 in the rat basal ganglia. *J Comp Neurol* 390:5–19.

Testa, C. M., Standaert, D. G., Young, A. B. et al. 1994. Metabotropic glutamate receptor mRNA expression in the basal ganglia of the rat. *J Neurosci* 14:3005–2518.

Turle-Lorenzo, N., Breysse, N., Baunez, C. et al. 2005. Functional interaction between mGlu 5 and NMDA receptors in a rat model of Parkinson's disease. *Psychopharmacology (Berl)* 179:117–127.

Valenti, O., Conn, P. J., and Marino, M. J. 2002. Distinct physiological roles of the Gq-coupled metabotropic glutamate receptors co-expressed in the same neuronal populations. *J Cell Physiol* 191:125–137.

Valenti, O., Mannaioni, G., Seabrook, G. R. et al. 2005. Group III metabotropic glutamate-receptor-mediated modulation of excitatory transmission in rodent substantia nigra pars compacta dopamine neurons. *J Pharmacol Exp Ther* 313:1296–1304.

Valenti, O., Marino, M. J., Wittmann, M. et al. 2003. Group III metabotropic glutamate receptor-mediated modulation of the striatopallidal synapse. *J Neurosci* 23:7218–7226.

Venero, J. L., Santiago, M., Tomas-Camardiel, M. et al. 2002. DCG-IV but not other group-II metabotropic receptor agonists induces microglial BDNF mRNA expression in the rat striatum. Correlation with neuronal injury. *Neuroscience* 113:857–869.

Vernon, A. C., Croucher, M. J., and Dexter, D. T. 2008. Additive neuroprotection by metabotropic glutamate receptor subtype-selective ligands in a rat Parkinson's model. *Neuroreport* 19:475–478.

Vernon, A. C., Palmer, S., Datla, K. P. et al. 2005. Neuroprotective effects of metabotropic glutamate receptor ligands in a 6-hydroxydopamine rodent model of Parkinson's disease. *Eur J Neurosci* 22:1799–1806.

Vernon, A. C., Zbarsky, V., Datla, K. P. et al. 2007. Subtype selective antagonism of substantia nigra pars compacta Group I metabotropic glutamate receptors protects the nigrostriatal system against 6-hydroxydopamine toxicity in vivo. *J Neurochem* 103:1075–1091.

Vila, M., Perier, C., Feger, J. et al. 2000. Evolution of changes in neuronal activity in the subthalamic nucleus of rats with unilateral lesion of the substantia nigra assessed by metabolic and electrophysiological measurements. *Eur J Neurosci* 12:337–344.

Voulalas, P. J., Holtzclaw, L., Wolstenholme, J. et al. 2005. Metabotropic glutamate receptors and dopamine receptors cooperate to enhance extracellular signal-regulated kinase phosphorylation in striatal neurons. *J Neurosci* 25:3763–3773.

Walter, B. L. and Vitek, J. L. 2004. Surgical treatment for Parkinson's disease. *Lancet Neurol* 3:719–728.

Wang, L., Kitai, S. T., and Xiang, Z. 2005. Modulation of excitatory synaptic transmission by endogenous glutamate acting on presynaptic group II mGluRs in rat substantia nigra compacta. *J Neurosci Res* 82:778–787.

Wang, H. and Pickel, V. M. 2002. Dopamine D2 receptors are present in prefrontal cortical afferents and their targets in patches of the rat caudate-putamen nucleus. *J Comp Neurol* 442:392–404.

Wichmann, T. and DeLong, M. R. 1996. Functional and pathophysiological models of the basal ganglia. *Curr Opin Neurobiol* 6:751–758.

Wichmann, T. and DeLong, M. R. 2006. Deep brain stimulation for neurologic and neuropsychiatric disorders. *Neuron* 52:197–204.

Wigmore, M. A. and Lacey, M. G. 1998. Metabotropic glutamate receptors depress glutamate-mediated synaptic input to rat midbrain dopamine neurones in vitro. *Br J Pharmacol* 123:667–674.

Wittmann, M., Hubert, G. W., Smith, Y. et al. 2001a. Activation of metabotropic glutamate receptor 1 inhibits glutamatergic transmission in the substantia nigra pars reticulata. *Neuroscience* 105:881–889.

Wittmann, M., Marino, M. J., Bradley, S. R. et al. 2001b. Activation of group III mGluRs inhibits GABAergic and glutamatergic transmission in the substantia nigra pars reticulata. *J Neurophysiol* 85:1960–1968.

Wittmann, M., Marino, M. J., and Conn, P. J. 2002. Dopamine modulates the function of group II and group III metabotropic glutamate receptors in the substantia nigra pars reticulata. *J Pharmacol Exp Ther* 302:433–441.

Yang, L., Mao, L., Tang, Q. et al. 2004. A novel Ca2+-independent signaling pathway to extracellular signal-regulated protein kinase by coactivation of NMDA receptors and metabotropic glutamate receptor 5 in neurons. *J Neurosci* 24:10846–10857.

Zhang, H. and Sulzer, D. 2003. Glutamate spillover in the striatum depresses dopaminergic transmission by activating group I metabotropic glutamate receptors. *J Neurosci* 23:10585–10592.

2 Ionotropic Glutamate Receptors in the Basal Ganglia

*Susan Jones, Shona L.C. Brothwell,
Isabel Huang-Doran, and James Hallett*

CONTENTS

2.1 INTRODUCTION

Ionotropic glutamate receptors (iGluRs) are ubiquitous throughout the mammalian CNS, where they mediate most excitatory synaptic transmission and participate in the plasticity of synaptic connections. Within the basal ganglia, alongside metabotropic glutamate receptors (Chapter 1), they respond to glutamate released from excitatory afferents to the striatum and to the midbrain nuclei, two pivotal points at which glutamatergic control influences the output of the basal ganglia.

In this chapter, we provide an overview of iGluR expression and function in the basal ganglia, focusing on the striatum and the midbrain dopaminergic nuclei. In the first part of the chapter, we review the general properties of iGluRs, primarily of the α-amino-3-hydroxy-5-methyl-4-isoxazolepropionic acid (AMPA) and *N*-methyl-D-aspartate (NMDA) subtypes. We then review the expression of these receptors in the basal ganglia circuitry and consider their functional and clinical significance.

2.2 IONOTROPIC GLUTAMATE RECEPTOR SUBTYPES

Three types of iGluR can be distinguished on the basis of subunit composition and the binding of selective agonists for which they are named (Dingledine et al., 1999): AMPA, kainate, and NMDA receptors. Most AMPA and kainate (non-NMDA) receptors are permeable to Na^+ and K^+ but not to Ca^{2+} and mediate the vast majority of fast excitatory synaptic transmission in the mammalian central nervous system. NMDA glutamate receptors are permeable to Ca^{2+} as well as to Na^+ and K^+ and play important roles in neural circuit development and ongoing neural circuit plasticity; they are one proposed molecular substrate for learning, adaptation and memory. NMDA receptors have two critically important properties: their Ca^{2+} permeability and a voltage-dependent block by Mg^{2+} ions that prevents the opening of the ion channel at hyperpolarized potentials even when agonist is bound.

2.2.1 IONOTROPIC GLUTAMATE RECEPTOR SUBUNITS

iGluRs form from tetrameric arrangements of subunit proteins and different subunit combinations give rise to receptors with different physiological and pharmacological properties. The genes encoding the iGluR subunits that co-assemble to form AMPA, NMDA, and kainate receptors are grouped on the basis of sequence identities (Dingledine et al., 1999). AMPA receptor subunits (GluR1–GluR4) and kainate receptor subunits (GluR5–GluR7, KA1 and KA2) are closely related but are distinct from the subunits which assemble to form NMDA receptors (NR1, NR2A-2D, NR3A, and NR3B).

The different iGluR subunits have similar hydrophobicity plots and are thought to have the same membrane topology: an extracellular N-terminal section, three transmembrane domains (M1, M3 and M4) with M2 forming a loop that both enters and exits the membrane on the cytoplasmic side, and a cytoplasmic C-terminal region (Hollmann and Heinemann, 1994; Dingledine et al., 1999). The ligand-binding site of iGluRs, which confers agonist specificity, is formed from two globular domains in each subunit (S1 and S2); ~150 amino acids each prior to the M1 transmembrane domain and the extracellular loop between M3 and M4 (Dingledine et al., 1999; Furukawa et al., 2005; Paoletti and Neyton, 2007).

AMPA receptors are made from combinations of GluR1–4 (previously named GluRA–D). The majority of AMPA receptors in the brain are monovalent cation channels with linear current–voltage ($I–V$) relationships that have reversal potentials close to 0 mV, as expected for receptors with approximately equal permeabilities to Na^+ and K^+ ions. AMPA receptors are generally around 100 times less permeable to Ca^{2+} than NMDA receptors (Mayer and Westbrook, 1987). In the 1990s, a population of AMPA receptors was found to have a higher permeability to Ca^{2+} (Iino et al., 1990; Ozawa and Iino, 1993). Work using heterologous expression of different AMPA receptor subunit combinations revealed that homomeric or heteromeric AMPA receptors containing GluR2 subunits are calcium impermeable and have a linear $I–V$ relationship. In contrast, those containing only GluR1, GluR3, or GluR4 subunits are calcium permeable and have nonlinear (inwardly rectifying)

I–V relationships due to voltage-dependent block by positively charged polyamines such as spermine (Bowie and Mayer, 1995; Kamboj et al., 1995; Cull-Candy et al., 2006). Jonas et al. (1994) demonstrated that AMPA receptor calcium permeability is inversely related to the amount of GluR2-specific mRNA. The effect of GluR2 on calcium permeability is mediated by a single amino acid change in the M2 region. At a key site, known as the Q/R site, the presence of neutral glutamine (Q) confers calcium permeability, while arginine (R), as found in the GluR2 subunit, causes calcium impermeability owing to its positive charge and long side chain (Burnashev et al., 1992).

NMDA receptors exist as tetrameric complexes comprising two NR1 and two NR2 and/or NR3 subunits (Cull-Candy and Leszkiewicz, 2004; Paoletti and Neyton, 2007). Ca^{2+} flux through NMDA receptors is approximately four times greater than that through Ca^{2+}-permeable AMPA receptors (Watanabe et al., 2002) and conveys to NMDA receptors their ability to regulate diverse processes including synaptic plasticity, gene expression, and cell survival (Hardingham and Bading, 2003; Soriano et al., 2006; Soriano and Hardingham, 2007; Papadia et al., 2008). Under pathological conditions, dysregulated calcium entry can activate signaling pathways leading to cell death (Choi et al., 1988; Soriano and Hardingham, 2007; Section 2.5).

The obligate glycine-binding NR1 subunit is expressed abundantly throughout the CNS. There are eight splice variants of the NR1 subunit resulting from alternative splicing of three exons (Hollmann and Heinemann, 1994). However, NMDA receptor properties are primarily determined by four NR2 subunits (A–D), encoded by four separate genes. The pore-forming re-entrant M2 domain of the NR2A and NR2B subunits have identical amino acid sequences, whereas there are four differences in this region between NR2A/NR2B and NR2C, leading to different ion permeation properties and Mg^{2+} sensitivity: at physiological concentrations of Mg^{2+} over a range of physiological membrane potentials, recombinant NR1/NR2A and NR1/NR2B diheteromeric receptors are blocked more strongly by Mg^{2+} than NR1/NR2C and NR1/NR2D receptors (Monyer et al., 1992, 1994; Kuner and Schoepfer, 1996). NR2 subunits also confer different single channel conductance, kinetics, and pharmacological properties (Cull-Candy and Leszkiewicz, 2004; Paoletti and Neyton, 2007).

More recently, two unusual NMDA receptor subunits have been described: NR3A (Ciabarra et al., 1995; Sucher et al., 1995; reviewed by Henson et al., 2010) and NR3B (Nishi et al., 2001; Chatterton et al., 2002; Matsuda et al., 2002). Recombinant NMDA receptors containing NR1/NR3 subunits form glycine-gated cation channels with reduced sensitivity to magnesium and reduced calcium permeability compared with NR2-containing NMDA receptors (Nishi et al., 2001; Sasaki et al., 2002; reviewed by Cavara and Hollmann, 2008; Tong et al., 2008; Low and Wee, 2010). Co-expression of NR1/NR2/NR3 results in receptors that are gated by glutamate and glycine but show reduced conductance compared with NR1/NR2 receptors (Tong et al., 2008), although NR2 and NR3 subunits may be unlikely to combine in the same receptor complex, preferring to form distinct populations of NR1/NR2 and NR1/NR3 (Ulbrich and Isacoff, 2008).

NR3A protein expression in the rat CNS—forebrain, cerebellum, brainstem, and spinal cord—is highest early in postnatal development and then declines into adulthood (Wong et al., 2002). NR3B protein expression in the adult rat CNS is notable in forebrain, cerebellum, and spinal cord (Wee et al., 2008). NR3B expression appears to peak later in postnatal development (around day 21) and persist in adults (Fukaya et al., 2005).

2.3 EXPRESSION OF iGluRs IN THE BASAL GANGLIA

Pioneering molecular genetic studies in the 1980s led to the isolation of genes encoding iGluR subunit proteins that form the receptors we know as AMPA, Kainate, and NMDA glutamate receptor subtypes (Hollmann and Heinemann, 1994). With this giant step came the ability to visualize the expression of iGluR subunit RNA or protein throughout the nervous system (for an early "gold standard" example, see Petralia et al., 1994), to confirm the presence of functional iGluR subtypes by the application of selective pharmacological ligands and to eliminate expression using increasingly sophisticated genetic approaches. Immunohistochemistry and in situ hybridization studies have shown that neurons throughout the basal ganglia, including the striatum and the midbrain nuclei, express iGluRs, although it should be noted that the reliability of expression depends on the selectivity of the probes used and the inclusion of robust control experiments (for a discussion of this, see review by Galvan et al., 2006).

2.3.1 iGLUTAMATE RECEPTOR EXPRESSION IN NEOSTRIATUM

In the principal output neurons of the striatum, the medium spiny neurons (MSNs), glutamatergic synaptic input is critical in driving the membrane potential to the "up state" from which action potentials can be generated more readily (for a recent review of the physiology of striatal neurons see Kreitzer, 2009). Thus, iGluR expression by striatal MSNs has been of great interest in basal ganglia research. In addition to mediating the effects of glutamatergic synaptic drive to the striatum (Section 2.4) and interactions with dopamine (Chapter 3), they are pivotal in basal ganglia plasticity (Chapters 4, 5, and 7) and learning (Chapter 9). Moreover, they are significant in basal ganglia disorders (Chapter 10). It is convenient to divide the MSN population into "direct" MSNs (projecting to substantia nigra pars reticulata (SNr) in rodents and internal globus pallidus plus SNr in primates) versus "indirect" MSNs (projecting to globus pallidus [external segment in primates]). Evidence suggests that glutamatergic drive differentially affects these two populations (reviewed by Kreitzer, 2009). Here we review the evidence for the expression of different iGluR subtypes, focusing primarily on MSNs and noting cell-specific patterns of expression.

An early quantitative autoradiographic study by Albin et al. (1992), using radiolabeled AMPA, kainate, and glutamate in sections through the adult rat basal ganglia, revealed labeling consistent with AMPA, kainate, and NMDA receptor expression in dorsal and ventral striatum, with detectable labeling of AMPA and NMDA receptors in globus pallidus, subthalamic nucleus (STN), substantia nigra, and ventral tegmental area (VTA). By degenerating striatonigral projection neurons (i.e., "direct" MSNs) and showing a corresponding decrease in autoradiographic labeling of AMPA, kainate, and NMDA receptors, the first indication of receptor localization to specific striatal projection neurons was made (Tallaksen-Greene et al., 1992).

The earliest immunohistochemical study used antibodies to AMPA receptor subunits (GluR1, GluR2/3, and GluR4) in an attempt to identify cell type-specific AMPA receptor subunit expression (Tallaksen-Greene and Albin, 1994). In this study, striatopallidal and striatonigral MSNs showed GluR2/3 labeling, while GluR1 appeared to be restricted to striatal interneurons. More recently, Deng et al. (2007) used retrograde and complex multilabeling methods with quantitative immunohistochemistry to assess striatal cell-dependent expression; they found that rat striatopallidal and striatonigral neurons all express GluR2, while GluR1 was present in many projection neurons (50%–75%, depending on target region). Intense GluR1 labeling of striatal GABAergic interneurons with sparse GluR2 labeling in this population supported the presence of GluR2-free, Ca^{2+}-permeable AMPA receptors in selected striatal interneurons. Using single-cell RT-PCR, MSNs were shown to express mRNA for GluR1 (Chen et al., 1998; Stefani et al., 1998) and these authors went on to show GluR1 immunolabeling of MSN dendrites in electron micrographs, suggesting that GluR1 receptors are preferentially localized to dendrites rather than somata in striatal projection neurons. Functional evidence for calcium-permeable AMPA receptors in MSNs has come from calcium imaging studies (Stefani et al., 1998; Carter and Sabatini, 2004; see Section 2.4.1).

In a comprehensive immunolabeling study of rat brain using NR2A/2B antibody, staining was moderately dense in basal ganglia structures including the striatum and substantia nigra (Petralia et al., 1994). Importantly, NMDA receptor NR2 subunit expression shows characteristic developmental changes in expression throughout the rat brain (Monyer et al., 1994). Thus, NR2B and NR2D mRNA expression occurs from embryonic day 14 while NR2A and NR2C mRNA appears around birth; NR2A expression increases in cortical areas during postnatal development, with NR2C expression increasing in the cerebellum during this period. In forebrain regions, NR2D expression reaches a peak by the first postnatal week; interestingly, NR2D mRNA persists in the midbrain beyond postnatal day 12.

The question of NMDA receptor expression in distinct populations of striatal neurons was addressed using double-label in situ hybridization in adult rats (Landwehrmeyer et al., 1995; Standaert et al., 1996, 1999; Kuppenbender et al., 1999). Projection neurons, as well as GABAergic and cholinergic interneurons, expressed NR1 subunit mRNA, with differential expression patterns of NR1 splice variants across the different cell populations. Intense signals for NR2B subunit mRNA were observed in projection neurons, less so in GABAergic interneurons, while the converse was true with NR2D subunit mRNA, which was barely detectable in projection neurons but observed in interneurons. NR2A subunit mRNA was uniformly relatively low across the striatal cell subpopulations, and NR2C subunit mRNA was undetected.

Early NMDA receptor subunit antibodies showed poor discrimination between different NR2 subunits; thus, an antibody to NR2A/NR2B was initially used to show that NMDA receptors composed of either NR2A or NR2B could be detected in the majority of MSN striatonigral projection neurons (Chen and Reiner, 1996).

NR3B subunit protein was detected in dorsal and ventral striatum (Wee et al., 2008). Data from co-immunoprecipitation suggest that NMDA receptors in the striatum can exist as NR1/NR2A or NR1/NR2B diheteromers or as NR1/NR2A/NR2B triheteromers (Dunah and Standaert, 2003). In human striatal tissue (Kuppenbender et al., 2000), strong hybridization signals for NR1 and NR2B suggest high levels in all striatal cell populations, similar to the rat striatum, while NR2A hybridization signal was strong in one population of interneurons, moderate in substance P-expressing MSNs and weak in enkephalin-expressing MSNs. NR2C and NR2D signals were low in all striatal cells except cholinergic interneurons.

An interesting question is the subcellular distribution of iGluRs, for example at synaptic versus extrasynaptic sites. This will not only influence the role a receptor plays in synaptic transmission but also in plasticity and, potentially, in cell survival (Galvan et al., 2006; Hardingham, 2006). An in-depth overview of the subcellular expression of AMPA, kainate, and NMDA receptors in the basal ganglia based on electron microscopy studies is provided by Galvan et al. (2006). Essentially, AMPA and NMDA receptor subunits are primarily localized to synaptic and (to a lesser degree) extrasynaptic sites but are not observed at presynaptic terminals, whereas kainate receptors are found at synaptic, extrasynaptic, and presynaptic sites.

2.3.2 iGLUTAMATE RECEPTOR EXPRESSION IN THE MIDBRAIN

As noted earlier, Albin et al. (1992) showed AMPA-, kainate-, and NMDA-binding sites in the substantia nigra using autoradiography. This was followed up with immunohistochemical localization of AMPA and NMDA receptor subunits in the adult rat substantia nigra pars compacta (SNc) (Albers et al., 1999). All tyrosine hydroxylase-positive (i.e., dopaminergic) cells expressed a particular splice variant of the NMDA receptor NR1 subunit as well as AMPA receptor GluR2/3 subunit. Around 50% of dopaminergic cells expressed GluR1. These data were consistent with functional studies reporting physiological responses of dopaminergic neurons to applications of AMPA and NMDA (Christoffersen and Meltzer, 1995). A more recent study of AMPA receptor subunit expression suggests that dopaminergic neurons in both SNc and VTA express GluR1, GluR2/3, and GluR4 (Chen et al., 2001). Similarly, in the squirrel monkey, the same AMPA receptor subunits are detected in dopaminergic neurons throughout the SNc and VTA, with some interesting regional differences in the level of expression of particular subunits; for example, higher levels of labeling of GluR2 and NR1 were detected in the ventral SNc than in VTA (Paquet et al., 1997).

Notable in the study by Monyer et al. (1994) was the finding that NR2D subunit expression persisted in the midbrain beyond the first postnatal week. Consistent with this, in situ hybridization detected high levels of NR2D and NR2C mRNA but lower levels of NR2A and NR2B in the substantia nigra (Standaert et al., 1994). Low levels of NR2A and NR2B protein were also reported by Albers et al. (1999). In the squirrel monkey, NR1 but not NR2A/B subunits were detected with immunolabeling (Paquet et al., 1997). Taken together, these data raised the intriguing possibility that midbrain dopaminergic neurons express NMDA receptors composed only of NR1/NR2C/NR2D subunits; that is, low conductance channels with distinct biophysical and pharmacological properties. More recent functional evidence suggests this is not entirely the case (see Section 2.4), and co-immunoprecipitation studies of NR2 sub-unit protein in the midbrain, in which NR2D subunits always co-precipitated with either NR2A or NR2B, suggest that midbrain NMDA receptors exist in triheteromeric configurations (Dunah et al., 1998). Interestingly, human dopaminergic neurons express high levels of NR2D subunit mRNA (Counihan et al., 1998).

2.4 SYNAPTIC ACTIVATION OF iGluRs IN BASAL GANGLIA NUCLEI

Glutamatergic input to the neostriatum arises from virtually the entire cerebral cortex and from the thalamus (Glees, 1944; Webster, 1961; Carman et al., 1963; Buchwald et al., 1973; Kitai et al., 1976; Kawaguchi et al., 1989). In addition, there is cortical glutamatergic drive to midbrain dopaminergic neurons (Section 2.4.2). The midbrain substantia nigra dopaminergic and GABAergic neurons receive additional glutamatergic input from the STN and, along with ventral tegmental neurons, from pedunculopontine glutamatergic fibers (Section 2.4.2).

2.4.1 SYNAPTIC iGluRs IN NEOSTRIATUM

Kitai et al. (1976) made intracellular recordings from individual neurons in the caudate nucleus *in vivo* and evoked monosynaptic (constant latency) excitatory postsynaptic potentials (EPSPs) by electrical stimulation of the cerebral cortex or thalamus. Post hoc processing of injected horseradish peroxidase (HRP) confirmed the recorded neurons as MSNs. Thus, MSNs receive direct excitation from cortex and thalamus. In a recent study, using mice expressing green fluorescent protein in either D1 receptor–expressing (direct) or D2 receptor–expressing (indirect) MSNs (Doig et al., 2010), cortical and thalamic terminals formed a similar proportion of synapses on both types of MSN, suggesting that the two main excitatory inputs play equally important roles.

Using *in vitro* rat brain slices containing neostriatum and preserving some corticostriatal input, Kawaguchi et al. (1989) examined passive and active membrane properties of HRP-labeled MSNs and confirmed cortically evoked EPSPs; however, these were insufficient to induce action potential firing unless depolarizing current was injected into the striatal neuron. Wilson and Kawaguchi (1996) proposed a two-state membrane potential in MSNs: the hyperpolarized or "down" state, from which EPSPs do not evoke spikes, and the depolarized or "up" state from which firing is more readily evoked. Wilson and Kawaguchi showed that in MSNs in anesthetized rats, the membrane potential spontaneously fluctuated between the down state ($\sim-75\,mV$), at which inwardly rectifying potassium current was proposed to clamp the membrane potential, and the up state ($\sim-54\,mV$), requiring sufficient excitatory synaptic input. The two-state membrane model is convenient for considering the role of intrinsic and synaptic conductances in governing MSN activity; a more detailed consideration of membrane transitions in MSNs, particularly in nonanesthetized animals, is provided in Chapter 7. In co-cultured organotypic slices from cortex and striatum, up and down state transitions are preserved and are blocked by the AMPA receptor antagonist CNQX (Plenz and Kitai, 1998). Thus, EPSPs play a critical role in the generation of the up state of MSNs, from which striatal output in the form of spikes is generated, making iGluRs fundamentally important for striatal function.

Closer inspection of the glutamate receptor subtypes involved in up state transitions was made in parasagittal striatal slices in which some cortical innervation was preserved (Vergara et al., 2003). Suprathreshold stimulation of cortical input evoked plateau potentials—sustained depolarization lasting several hundred milliseconds—followed by membrane potential oscillations lasting several seconds. The initiation (but not the maintenance) of cortical-evoked plateau potentials and oscillations depended on non-NMDA receptor–mediated synaptic transmission and was dendritic in origin. Interestingly, addition of the agonist NMDA during cortical stimulation enhanced the induction of plateau potentials and oscillations from otherwise subthreshold cortical stimulation; furthermore, spontaneous oscillations were blocked by the NMDA receptor antagonist D-AP5 (Vergara et al., 2003). The authors concluded that dendritic NMDA receptors act in conjunction with dendritic L-type Ca^{2+} current to promote membrane transitions to an up state *in vitro*. Early *in vivo* studies in anesthetized preparations had also indicated a role for NMDA receptors in up state transitions (e.g., Herrling et al., 1983, who used halothane anesthesia). More direct evidence for synaptically driven up state transitions mediated by NMDA receptors was provided by Pomata et al. (2008), using a microdialysis probe to deliver D-AP5 directly to the striatum of rats under urethane anesthesia while recording from individual MSNs or from multiple extracellular units. Spontaneous transitions to the up state were observed in control and during D-AP5 perfusion, but the amplitude of spontaneous up state transitions was reduced by D-AP5 and spontaneous firing of multiunits was inhibited by D-AP5, indicating that endogenous NMDA receptors contribute to activity during the up state *in vivo*, although not necessarily exclusively to the transition to the up state.

An important event in iGluR-mediated synaptic transmission is a rise in postsynaptic calcium via calcium-permeable AMPA and/or NMDA receptors. The contribution of AMPA and NMDA receptors to elevations in intracellular Ca^{2+} in the dendrites of MSNs was investigated by Carter and Sabatini (2004) in an elegant study using two-photon Ca^{2+} imaging combined with focal laser uncaging of glutamate over the dendrites. Because this approach abrogated the need for synaptic stimulation, the contribution of voltage-gated Ca^{2+} channels could be probed in the same study by the use of Ca^{2+} channel blockers. Currents with similar amplitudes to miniature excitatory postsynaptic currents (EPSCs) were evoked by uncaging glutamate in a spatially restricted manner; these currents were similar in amplitude at hyperpolarized (−80 mV) and depolarized (−50 mV) potentials, but at −50 mV the rate of decay of the current was slower and there was a corresponding increase in dendritic Ca^{2+}. Interestingly, calcium-permeable AMPA receptors were the predominant source of the initial, rapid glutamate-evoked Ca^{2+} increase in dendrites and spines in the down state, with NMDA receptors contributing a slower component. At depolarized states, NMDA receptors were the primary source of rapid and slow glutamate-evoked Ca^{2+} increase. Backpropagating action potentials generated Ca^{2+} signals in the dendrites that were mediated by low-threshold (T-type) Ca^{2+} channels at hyperpolarized potentials and L-type Ca^{2+} channels at depolarized potentials; pairing backpropagating action potentials with glutamate uncaging caused a nonlinear enhancement of Ca^{2+} entry via NMDA receptors in the dendrites.

iGluRs play a vital role in the synaptic plasticity of basal ganglia circuits, and glutamatergic synaptic plasticity in the striatum is considered in detail in Chapters 4 and 5. Furthermore, there is convincing evidence that striatal iGluRs, and in particular NMDA receptors, are important for learning functions of the basal ganglia (Chapter 9).

2.4.2 Synaptic iGluRs in Substantia Nigra

SNc dopaminergic neurons project primarily to the dorsal striatum, forming the nigrostriatal dopamine pathway, which together with other midbrain dopaminergic projections account for 70%–80% of dopamine in the brain (Grillner and Mercuri, 2002). In the direct/indirect model of basal ganglia processing, nigrostriatal dopamine differentially modulates direct versus indirect MSNs to tilt the balance in favor of disinhibition of thalamic and collicular circuits; without dopamine release, in this model corticostriatal synaptic transmission would be relatively ineffective on basal ganglia output. Given the pivotal role that dopamine release plays in basal ganglia output, one would expect very tightly regulated control of the midbrain dopaminergic circuits, particularly from the cortex. In this section, we review excitatory control of SNc dopamine neurons, and in the subsequent section we consider the control of the VTA.

The spontaneous activity of both SNc and VTA dopaminergic neurons *in vivo* exists as a spectrum of firing patterns, ranging from tonic regular and irregular firing of action potentials to transient bursts (lasting up to a few hundred milliseconds) of high-frequency firing (>10 Hz). Firing patterns reflect a combination of intrinsic conductances, notably Ca^{2+} currents and Ca^{2+}-dependent potassium currents (Grace and Onn, 1989; Seutin et al., 1993; Kitai et al., 1999; Shepard and Stump, 1999; Wolfart et al., 2001; Johnson and Wu, 2004), and the balance of excitatory and inhibitory afferent input (reviewed by Kitai et al., 1999). Focusing here on excitatory synaptic input to SNc dopamine neurons, these appear to arise primarily from the prefrontal cortex (PFC), STN, and pedunculopontine nucleus (PPN), as shown in both anatomical and physiological studies, although specific studies of the excitatory control of SNc neurons are few in number compared with studies of VTA neurons (Section 2.4.3). Indeed, the extent of cortical control of SNc dopamine neurons is somewhat unresolved, which is remarkable when considering the importance of the question.

Anatomical connectivity is well established in rats (Bunney and Aghajanian, 1976; Sesack et al., 1989), cats (Afifi et al., 1974), and primates (Ongür et al., 1998). However, Frankle et al. (2006) reported a "sparse" anatomical connection between the PFC and the midbrain SN/VTA dopamine neurons in primates. Ablation of frontal cortex results in a reduction of glutamate in the substantia nigra (Kornhuber et al., 1984), corroborating the idea of cortical control over glutamate release in the midbrain. One possibility is that cortical control is very effectively mediated through a spare connection; alternatively, cortical control may be exerted indirectly, either via nondopaminergic neurons in the midbrain (e.g., see Carr and Sesack, 2000; Section 2.4.3) or via other nuclei including the STN, brainstem nuclei, and inhibitory feedback loops from the striatum and globus pallidus. In anesthetized rats, simulation of PFC evoked burst firing of dopamine neurons in SNc and VTA (Gariano and Groves, 1988; Tong et al., 1996). This is fundamentally important to basal ganglia function; Gonon (1988) showed that dopamine release in response to burst firing is supralinear compared with regular spikes of a similar average frequency, and numerous studies have underscored the importance of the switch from tonic to phasic burst firing during basal ganglia-dependent behaviors (e.g., see Schultz and Romo, 1990; Schultz, 2000; Zweifel et al., 2009). Cortical induction of burst firing would demonstrate firm cortical control over dopamine release in the striatum. However, removal of GABAergic innervation to SNc dopamine neurons also facilitates burst firing (Tepper et al., 1995; Celada et al., 1999; Kitai et al., 1999), and direct cortical excitation may not be a requirement for tonic to phasic shifts.

An excitatory input from the STN to SNc is well established based on both anatomical (Van Der Kooy and Hattori, 1980; Chang et al., 1984; Kita and Kitai, 1987) and physiological studies (Hammond et al., 1978 and references below; reviewed by Kitai et al., 1999). Physiological studies suggest that subthalamic input plays important functional roles in SNc dopamine neurons, including regulation of firing patterns (Smith and Grace, 1992; Chergui et al., 1994; Lee et al., 2004), dopamine release (Rosales et al., 1994), and long-lasting potentiation of excitatory synaptic transmission (Overton et al., 1999) in SNc dopamine neurons.

Excitatory input arising in the brainstem PPN also innervates the mammalian SNc (Di Loreto et al., 1992; Charara et al., 1996; Forster and Blaha, 2003). As with stimulation of PFC, PPN stimulation also evokes burst firing in SNc dopamine neurons (Lokwan et al., 1999), and the authors proposed that the long-latency PFC-evoked bursts in their previous study (Tong et al., 1996) might be polysynaptic events mediated via the PPN; this would indicate a significant brainstem contribution to mammalian basal ganglia control.

Intranigral stimulation *in vitro* elicits EPSPs or EPSCs in SNc dopamine neurons and has allowed the identification of iGluR subtypes. Mereu et al. (1991) described EPSCs with both fast and slow components; the fast synaptic response was inhibited by non-NMDA receptor antagonists, while application of glycine, membrane depolarization, or the removal of Mg^{2+} from the extracellular solution enhanced, and NMDA receptor antagonists inhibited, the slow component. In a long-standing collaboration with Dr. Alasdair Gibb, we have investigated some of the properties of NMDA receptors in SN neurons. In both dopamine and nondopamine neurons, NMDA receptors show some unusual properties compared with those expressed in cortical neurons, including an apparent lack of NR2A subunits in late postnatal development, and the presence of NR2D subunits at both synaptic and extrasynaptic locations (Figure 2.1; Jones and Gibb, 2005; Brothwell et al., 2008; Suarez et al., 2010). Thus, NMDA receptors composed of NR2B and/or NR2D subunits but lacking NR2A subunits appear to be capable of inducing synaptic plasticity and burst firing patterns in substantia nigra dopamine neurons. In this regard, there appear to be some interesting distinctions with NMDA receptor subtypes in VTA dopamine neurons (see Section 2.4.3).

NMDA and non-NMDA receptor agonists increase the firing rate of SNc dopamine neurons and can evoke burst firing *in vitro* (Johnson et al., 1992; Wu et al., 1994; Mereu et al., 1997; Blythe et al., 2009) and *in vivo* (Overton and Clark, 1992; Zhang et al., 1994; Christoffersen and Meltzer, 1995). The observation in some studies (e.g., Zhang et al., 1994; Christoffersen and Meltzer, 1995) that application of glutamate,

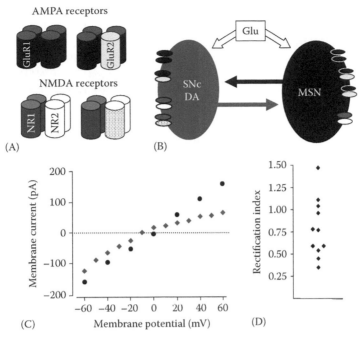

(A)   (B)   (C)   Membrane potential (mV)   (D)

FIGURE 2.1   AMPA and NMDA receptors in SNc and striatum. (A) Schematic of AMPA and NMDA receptor subunit composition. AMPA receptors are homomeric or heteromeric tetramers composed of GluR1–GluR4 subunits; the absence of GluR2 subunits (pale grey) confers calcium permeability. NMDA receptors are composed of NR1 (dark gray) and two identical or two different NR2 subunits (white/spotted white). (B) In midbrain dopamine neurons, synaptic and extrasynaptic NMDA receptors are composed of two types of NR2 subunits: NR2B and/or NR2D subunits; functional NR2A subunits appear to be absent (Jones and Gibb, 2005; Brothwell et al., 2008; Suarez et al., 2010). Synaptic AMPA subunits apparently form both calcium-permeable and calcium-impermeable (i.e., GluR2-containing) receptors. In striatal MSNs, calcium-permeable AMPA receptors have been detected in dendrites (Carter and Sabatini, 2004), but GluR2-containing AMPA receptor expression is more pronounced (Deng et al., 2007). (C) Calcium-permeable and calcium-impermeable AMPA receptors in VTA dopamine neurons can be detected in whole-cell patch-clamp recordings of AMPA–EPSCs measured at different membrane potentials; the resulting current–voltage relationship is linear in some cells, indicating a predominance of calcium-impermeable receptors (example cell, closed circles), while in other cells nonlinear (inwardly rectifying) relationships suggest that calcium-permeable AMPA receptors predominate (example cell, closed diamonds). (D) In 11 VTA dopamine neurons tested, seven had a rectification index of less than 1.0, indicative of a nonlinear, inwardly rectifying current–voltage relationship and suggestive of calcium-permeable AMPA receptors. (Data in panels C and D: Isabel Huang-Doran and Sue Jones, unpublished.)

NMDA, AMPA, or kainate can evoke burst firing of SNc dopaminergic neurons suggests that no single receptor subtype is entirely responsible for generating bursts; indeed, a powerful increase in firing rate (Zhang et al., 1994), alongside removal of inhibition and/or disruption of intrinsic conductances, may be more significant than a single population of receptors coupled to a specific cellular mechanism (Kitai et al.,

1999 for review). Evidence for a dominant role of NMDA receptors in burst firing is nonetheless very persuasive, with more studies reporting effects of NMDA receptor but not non-NMDA receptor ligands, rather than the converse (Suaud-Chagny et al., 1992; Chergui et al., 1993; Tong et al., 1996; Mereu et al., 1997). In studies where both AMPA and NMDA receptor ligands influence burst firing, NMDA receptors are sometimes more effective (e.g., Christoffersen and Meltzer, 1995), but this is not always the case (e.g., Zhang et al., 1994; Blythe et al., 2007).

Spontaneous burst firing of dopamine neurons is less frequently observed in midbrain slices, originally thought due to the loss of synaptic inputs (Johnson and North, 1992; Johnson et al., 1992; Mereu et al., 1997). However, synaptic stimulation *in vitro* can generate burst firing with action potential properties similar to those observed *in vivo*; application of either AMPA receptor or NMDA receptor antagonists reduces the capacity for burst firing, but the co-application of both antagonists is required in order to completely abolish the effects of synaptic stimulation (Blythe et al., 2007). Overall, it is likely that the orchestrated activity of inhibitory and excitatory afferents onto SNc dopaminergic neurons modulates their activity *in vivo* and that no single pathway is entirely responsible (Kitai et al., 1999).

2.4.3 VENTRAL TEGMENTAL AREA

Some important differences exist between dopamine neurons in VTA compared with those in SNc. First, the cell population appears to be more heterogeneous in VTA (Margolis et al., 2006), with dopamine neurons making up a smaller proportion (~60%) than in SNc (>90%). Second, VTA neurons both send and receive information from more disparate brain regions, including projections to and from prefrontal and cingulate cortices, ventral striatum, amygdala, and pontine nuclei, with additional projections out to the hypothalamus, cerebellum, and brainstem structures (reviewed by Bonci and Jones, 2007 and references therein). However, it is convenient to broadly classify VTA neurons as mesoaccumbens or mesocortical.

Substantial evidence exists for glutamatergic inputs from PFC to VTA dopamine neurons (Christie et al., 1985; Grenhoff et al., 1988; Svensson and Tung, 1989; Sesack and Pickel, 1992; Murase et al., 1993; Smith et al., 1996; Jones and Gutlerner, 2002 for a review). Interestingly, a careful triple-labeling approach revealed that PFC afferents to VTA neurons terminate on mesocortical dopamine neurons but not mesoaccumbens dopamine neurons; conversely, PFC neurons synapse on mesoaccumbens GABAergic neurons but not mesocortical GABAergic neurons (Carr and Sesack, 2000). Pontine nuclei, including the laterodorsal tegmentum and the PPN, also project to VTA dopamine and GABA neurons (Forster and Blaha, 2000; Omelchenko and Sesack, 2005; Lodge and Grace, 2006). Indeed, as for SNc neurons, it is suggested that subcortical sources of glutamatergic afferents to VTA dopamine neurons may be more significant than cortical inputs (Omelchenko and Sesack, 2007).

Tonic output of PFC evokes dopamine release in nucleus accumbens (NAc) via glutamate receptors located in VTA (Karreman and Moghaddam, 1996), possibly also involving the PPN (Lokwan et al., 1999; Carr and Sesack, 2000). Application of iGluR agonists (glutamate, NMDA, and quisqualate) to the VTA in anesthetized rats also leads to increased dopamine levels in NAc (Sauad-Chagny et al., 1992).

Fast and slow excitatory synaptic responses can be evoked in VTA dopamine neurons *in vitro* (Mereu et al., 1991; Johnson and North, 1992; Jones and Kauer, 1999), as in SNc dopamine neurons. However, differences in iGluR expression between VTA and SNc dopamine neurons may exist. Whereas SNc neurons express no detectable functional NR2A-containing NMDA receptors, in VTA dopamine neurons NMDA receptors are composed of NR2A as well as NR2B subunits (Borgland et al., 2006). AMPA receptors appear to be composed of GluR2-containing calcium-impermeable as well as non-GluR2 calcium-permeable subtypes (I. Huang-Doran and S. Jones, unpublished data, Figure 2.1; Bellone and Lüscher, 2005).

Thus, glutamatergic input from both cortical and subcortical sources influence the rate and pattern of firing of midbrain dopamine neurons, impacting on dopamine release in the dorsal and ventral striatum. In the striatum, released dopamine shows important interactions with corticostriatal and corticothalamic glutamate transmission, notably in the plasticity of these synapses (Chapters 4 and 5) and in learning and action (Chapter 9).

2.5 NMDA RECEPTORS, EXCITOTOXICITY, AND PARKINSON'S DISEASE

Over 1% of the 65+ population are diagnosed with Parkinson's disease (PD), a progressive neurological disorder characterized by motor deficits including bradykinesia, tremor, and rigidity. The striking pathology of PD is a substantial degeneration of SNc dopaminergic neurons: striatal dopamine levels are reduced to <40%, giving rise to motor signs of the disease (Hallett and Standaert, 2004). Although treatment options exist to ameliorate the symptoms of PD, none of these prevent the progressive degeneration of SNc dopaminergic neurons. NMDA receptors on SNc dopaminergic neurons may be involved in the neurodegenerative processes underlying PD. Furthermore, changes in NMDA receptor expression have been reported in post-mortem tissue from PD patients and in animal models of PD, although these changes are complex and often contradictory (Hallett and Standaert, 2004) and are not considered further here.

The discovery that application of D-APV prior to the induction of forebrain ischemia has a neuroprotective effect (Simon et al., 1984) led to studies revealing a specific role of the NMDA receptor in glutamate excitotoxicity (Choi et al., 1988). Tight regulation of intracellular Ca^{2+} is fundamental to neuronal survival and function (Choi et al., 1988; Lipton and Kater, 1989; Hardingham and Bading, 2003) and so the permeability of NMDA receptors to Ca^{2+} neurons is the potential trigger for apoptotic and necrotic cell death (Blandini et al., 2000).

NMDA receptors at synaptic versus extrasynaptic sites may be linked to different neuronal functions, such as pro-survival signaling at synapses versus pro-excitotoxic signaling at extrasynaptic sites (Hardingham, 2006; Soriano and Hardingham, 2007) because intracellular proteins and signaling cascades linked to the NMDA receptor include specific anti-apoptotic versus pro-excitotoxic signaling pathways (Hardingham et al., 2002; Hardingham, 2006; Ivanov et al., 2006; Soriano and Hardingham, 2007). Following excessive glutamate release during pathophysiological conditions, the increased activation of NMDA receptors linked to pro-excitotoxic pathways, potentially coupled with other factors such as damaged energy metabolism or overwhelmed homeostatic mechanisms that usually control Ca^{2+} levels, may tip the balance in favor of neuronal death. Moreover, overactivity of STN glutamatergic afferents to SNc dopamine neurons is an early hallmark of PD (Blandini et al., 2000).

Clearly, NMDA receptors have important physiological roles in the mammalian brain and do not necessarily engender harmful excitotoxic effects. However, it is feasible that environmental toxins might induce or exacerbate NMDA receptor–mediated cell death (Blandini et al., 2000). For example, the mitochondrial toxin rotenone enhances NMDA receptor responses and NMDA plus rotenone have synergistic effects on cell death (Marey-Semper et al., 1995; Wu and Johnson, 2007, 2009). In addition, as dopamine neurons degenerate, reduced dopamine release results in increased excitatory drive from the STN to SNc (Blandini et al., 2000), which may cause additional stress in dopamine neurons. One can imagine a spiral of excessive NMDA receptor activity driven initially by exposure to environmental toxins and perpetuated by compromised cell metabolism and over-activity of glutamatergic pathways.

Not surprisingly then, NMDA receptors on SNc dopaminergic neurons are possible targets for neuroprotective drug treatment in PD patients. NMDA receptor antagonists reduce some of the motor signs in animal models of PD and may have neuroprotective potential (Hallett and Standaert, 2004) and, importantly, the noncompetitive use-dependent NMDA receptor channel-blocking agent memantine is used by some European PD patients (Chen and Lipton, 2006). However, certain problematic issues must be resolved; clinical efficacy is limited by unwanted side effects and complicated by increasing evidence that different subtypes of NMDA receptors may be linked to pro-survival roles, such that functionally inhibiting them may not be a good strategy.

2.6   GLUTAMATE RECEPTORS AND DRUGS OF ABUSE

Substantial evidence supports the involvement of midbrain and striatal iGluRs in the persistent behavioral effects of addictive drugs (for reviews see Kauer, 2004; Jones and Bonci, 2005). A flurry of recent reviews cover the detailed changes in iGluR function that have been documented (Kauer and Malenka, 2007; Russo et al., 2010; Luscher and Malenka, 2011; Sulzer, 2011).

Work in the early noughties consistently and convincingly supported a case for forms of excitatory synaptic plasticity in response to addictive drugs, both in midbrain dopamine neurons (e.g., Jones et al., 2000; Mansvelder and McGehee, 2000; Ungless et al., 2001; Saal et al., 2003; Dong et al., 2004; Faleiro et al., 2004; Liu and Poo, 2005) and in NAcc (e.g., Thomas et al., 2001; Brebner et al., 2005; Bamford et al., 2008). Interestingly, changes in iGluR function can be evoked not only in response to drug administration, but also in response to cues that predict the presence of rewards (Stuber et al., 2008).

More recently, the details of iGluR subunit-specific requirements and changes for drug-induced synaptic plasticity have been the focus of attention (Engblom et al., 2008; Zweifel et al., 2008). The following findings are very robust in midbrain dopamine neurons: drug-induced plasticity involves a change in the ratio of AMPA receptor–mediated synaptic transmission relative to NMDA receptor–mediated synaptic transmission; changes in AMPA receptor transmission involves a switch in subunit composition, most likely an upregulation of calcium-permeable AMPA receptors; NMDA receptor transmission may be downregulated. These changes depend on NMDA receptor activation, as conditional and cell-specific knockout of the NR1 subunit abrogates the physiological and behavioral effects of cocaine (Engblom et al., 2008; Zweifel et al., 2008). Inherent plasticity at excitatory synapses is altered by these adaptations (Ungless et al., 2001). The precise mechanisms of induction, persistence, and transfer through the basal ganglia nuclei are yet to be resolved; recent evidence suggests that disruption of NMDA receptor-dependent plasticity in VTA dopamine neurons prevents drug-induced synaptic changes in NAcc (Mameli et al., 2009).

2.7 CONCLUSIONS

AMPA and NMDA iGluRs are expressed in basal ganglia nuclei, notably in the striatum and in the midbrain dopaminergic nuclei, where they mediate excitatory control of these regions. In the striatum, cell-specific patterns of expression of both AMPA and NMDA receptor subunit proteins are well documented. Functionally, calcium-permeable AMPA receptors have been demonstrated in MSN dendrites, while less is known about functional NMDA receptor subunit composition. Synaptic NMDA receptors are likely to make a significant contribution to the active "up state" of MSNs. In midbrain dopaminergic nuclei, expression studies suggest that all AMPA receptor subunits are present, while NMDA receptor NR1 and NR2D subunits show high levels of expression. Functionally, glutamatergic synapses may express either calcium-permeable or calcium-impermeable AMPA receptors; synaptic NMDA receptors form from NR2B and NR2D subunits but not NR2A subunits in SNc. Both AMPA and NMDA receptors are likely to contribute to burst firing of dopamine neurons, for example in response to salient stimuli. In basal ganglia disorders, iGluRs in both the striatum and the midbrain are potential therapeutic targets. Drugs of abuse cause persistent changes in iGluR expression and function in the ventral striatum and in the VTA, while NMDA receptor–mediated excitotoxicity is one putative mechanism of dopamine cell death in PD. Subsequent chapters elaborate on the functional roles of iGluRs in the basal ganglia, at the level of synaptic plasticity and modulation and in the context of circuit activity, learning, and behavior.

ACKNOWLEDGMENTS

Our work has been supported by the BBSRC and by Parkinson's UK.

REFERENCES

Afifi AK, Bahuth NB, Kaelber WW, Mikhael E, Nassar S (1974). The cortico-nigral fibre tract. An experimental Fink-Heimer study in cats. *J Anat* 118(Pt 3):469–476.

Albers DS, Weiss SW, Ladarola MJ, Standaert DG (1999). Immunohistochemical localization of N-methyl-D-aspartate and alpha-amino-3-hydroxy-5-methyl-4-isoxazolepropionate receptor subunits in the substantia nigra pars compacta of the rat. *Neuroscience* 89:209–220.

Albin RL, Makowiec RL, Hollingsworth ZR, Dure LS 4th, Penney JB, Young AB (1992). Excitatory amino acid binding sites in the basal ganglia of the rat: A quantitative autoradiographic study. *Neuroscience* 46:35–48.

Bamford NS, Zhang H, Joyce JA, Scarlis CA, Hanan W, Wu NP, André VM, Cohen R, Cepeda C, Levine MS, Harleton E, Sulzer D (2008). Repeated exposure to methamphetamine causes long-lasting presynaptic corticostriatal depression that is renormalized with drug readministration. *Neuron* 58(1):89–103.

Bellone C, Lüscher C (2005). mGluRs induce a long-term depression in the ventral tegmental area that involves a switch of the subunit composition of AMPA receptors. *Eur J Neurosci* 21(5):1280–1288.

Blandini F, Nappi G, Tassorelli C, Martignoni E (2000). Functional changes of the basal ganglia circuitry in Parkinson's disease. *Prog Neurobiol* 62:63–88.

Blythe SN, Atherton JF, Bevan MD (2007). Synaptic activation of dendritic AMPA and NMDA receptors generates transient high-frequency firing in substantia nigra dopamine neurons in vitro. *J Neurophysiol* 97:2837–2850.

Blythe SN, Wokosin D, Atherton JF, Bevan MD (2009). Cellular mechanisms underlying burst firing in substantia nigra dopamine neurons. *J Neurosci* 29:15531–15541.

Bonci A, Jones, S (2007). The mesocortical dopamine system. In: *The Human Frontal Lobes.* Eds: B. Miller and J.L. Cummings, Guilford Publications, New York.

Borgland SL, Taha SA, Sarti F, Fields HL, Bonci A (2006). Orexin A in the VTA is critical for the induction of synaptic plasticity and behavioral sensitization to cocaine. *Neuron* 49:589–601.

Bowie D, Mayer ML (1995). Inward rectification of both AMPA and kainate subtype glutamate receptors generated by polyamine-mediated ion channel block. *Neuron* 15:453–462.

Brebner K, Wong TP, Liu L, Liu Y, Campsall P, Gray S, Phelps L, Phillips AG, Wang YT (2005). Nucleus accumbens long-term depression and the expression of behavioral sensitization. *Science* 310(5752):1340–1343.

Brothwell SL, Barber JL, Monaghan DT, Jane DE, Gibb AJ, Jones S (2008). NR2B- and NR2D-containing synaptic NMDA receptors in developing rat substantia nigra pars compacta dopaminergic neurones. *J Physiol* 586:739–750.

Buchwald NA, Price DD, Vernon L, Hull CD (1973). Caudate intracellular response to thalamic and cortical inputs. *Exp Neurol* 38(2):311–323.

Bunney BS, Aghajanian GK (1976). The precise localization of nigral afferents in the rat as determined by a retrograde tracing technique. *Brain Res* 117(3):423–435.

Burnashev N, Monyer H, Seeburg PH, Sakmann B (1992). Divalent ion permeability of AMPA receptor channels is dominated by the edited form of a single subunit. *Neuron* 8:189–198.

Carman JB, Cowan WM, Powell TP (1963). The organization of cortico-striate connexions in the rabbit. *Brain* 86:525–562.

Carr DB, Sesack SR (2000). Projections from the rat prefrontal cortex to the ventral tegmental area: Target specificity in the synaptic associations with mesoaccumbens and mesocortical neurons. *J Neurosci* 20:3864–3873.

Carter AG, Sabatini BL (2004). State-dependent calcium signaling in dendritic spines of striatal medium spiny neurons. *Neuron* 44:483–493.

Cavara NA, Hollmann M (2008). Shuffling the deck anew: How NR3 tweaks NMDA receptor function. *Mol Neurobiol* 38(1):16–26.

Celada P, Paladini CA, Tepper JM (1999). GABAergic control of rat substantia nigra dopaminergic neurons: Role of globus pallidus and substantia nigra pars reticulata. *Neuroscience* 89:813–825.

Chang HT, Kita H, Kitai ST (1984). The ultrastructural morphology of the subthalamic-nigral axon terminals intracellularly labeled with horseradish peroxidase. *Brain Res* 299(1):182–185.

Charara A, Smith Y, Parent A (1996). Glutamatergic inputs from the pedunculopontine nucleus to midbrain dopaminergic neurons in primates: Phaseolus vulgaris-leucoagglutinin anterograde labeling combined with postembedding glutamate and GABA immunohistochemistry. *J Comp Neurol* 364:254–266.

Chatterton JE, Awobuluyi M, Premkumar LS, Takahashi H, Talantova M, Shin Y, Cui J, Tu S, Sevarino KA, Nakanishi N, Tong G, Lipton SA, Zhang D (2002). Excitatory glycine receptors containing the NR3 family of NMDA receptor subunits. *Nature* 415:793–798.

Chen HS, Lipton SA (2006). The chemical biology of clinically tolerated NMDA receptor antagonists. *J Neurochem* 97(6):1611–1626.

Chen Q, Reiner A (1996). Cellular distribution of the NMDA receptor NR2A/2B subunits in the rat striatum. *Brain Res* 743:346–352.

Chen Q, Veenman L, Knopp K, Yan Z, Medina L, Song WJ, Surmeier DJ, Reiner A (1998). Evidence for the preferential localization of glutamate receptor-1 subunits of AMPA receptors to the dendritic spines of medium spiny neurons in rat striatum. *Neuroscience* 83:749–761.

Chen LW, Wei LC, Lang B, Ju G, Chan YS (2001). Differential expression of AMPA receptor subunits in dopamine neurons of the rat brain: A double immunocytochemical study. *Neuroscience* 106(1):149–160.

Chergui K, Akaoka H, Charléty PJ, Saunier CF, Buda M, Chouvet G (1994). Subthalamic nucleus modulates burst firing of nigral dopamine neurones via NMDA receptors. *Neuroreport* 5(10):1185–1188.

Chergui K, Charléty PJ, Akaoka H, Saunier CF, Brunet JL, Buda M, Svensson TH, Chouvet G (1993). Tonic activation of NMDA receptors causes spontaneous burst discharge of rat midbrain dopamine neurons in vivo. *Eur J Neurosci* 5:137–144.

Choi DW, Koh JY, Peters S (1988). Pharmacology of glutamate neurotoxicity in cortical cell culture: Attenuation by NMDA antagonists. *J Neurosci* 8:185–196.

Christie MJ, Bridge S, James LB, Beart PM (1985). Excitotoxin lesions suggest an aspartatergic projection from rat medial prefrontal cortex to ventral tegmental area. *Brain Res* 333(1):169–172.

Christoffersen CL, Meltzer LT (1995). Evidence for N-methyl-D-aspartate and AMPA subtypes of the glutamate receptor on substantia nigra dopamine neurons: Possible preferential role for N-methyl-D-aspartate receptors. *Neuroscience* 67:373–381.

Ciabarra AM, Sullivan JM, Gahn LG, Pecht G, Heinemann S, Sevarino KA (1995). Cloning and characterization of chi-1: A developmentally regulated member of a novel class of the ionotropic glutamate receptor family. *Neuroscience* 15(10):6498–6508.

Counihan TJ, Landwehrmeyer GB, Standaert DG, Kosinski CM, Scherzer CR, Daggett LP, Veliçelebi G, Young AB, Penney JB Jr (1998). Expression of N-methyl-D-aspartate receptor subunit mRNA in the human brain: Mesencephalic dopaminergic neurons. *J Comp Neurol* 390:91–101.

Cull-Candy S, Kelly L, Farrant M (2006). Regulation of Ca^{2+}-permeable AMPA receptors: Synaptic plasticity and beyond. *Curr Opin Neurobiol* 16:288–297.

Cull-Candy SG, Leszkiewicz DN (2004). Role of distinct NMDA receptor subtypes at central synapses. *Sci STKE* 2004(2ss), re16.

Deng YP, Xie JP, Wang HB, Lei WL, Chen Q, Reiner A (2007). Differential localization of the GluR1 and GluR2 subunits of the AMPA-type glutamate receptor among striatal neuron types in rats. *J Chem Neuroanat* 33(4):167–192.

Di Loreto S, Florio T, Scarnati E (1992). Evidence that non-NMDA receptors are involved in the excitatory pathway from the pedunculopontine region to nigrostriatal dopaminergic neurons. *Exp Brain Res* 89:79–86.

Dingledine R, Borges K, Bowie D, Traynelis SF (1999). The glutamate receptor ion channels. *Pharmacol Rev* 51:7–61.

Doig NM, Moss J, Bolam JP (2010). Cortical and thalamic innervation of direct and indirect pathway medium-sized spiny neurons in mouse striatum. *J Neurosci* 30(44):14610–14618.

Dong Y, Saal D, Thomas M, Faust R, Bonci A, Robinson T, Malenka RC (2004). Cocaine induced potentiation of synaptic strength in dopamine neurons: Behavioral correlates in GluRA(-/-) mice. *Proc Natl Acad Sci USA* 101:14282–14287.

Dunah AW, Luo J, Wang YH, Yasuda RP, Wolfe BB (1998). Subunit composition of N-methyl-D-aspartate receptors in the central nervous system that contain the NR2D subunit. *Mol Pharmacol* 53:429–437.

Dunah AW, Standaert DG (2003). Subcellular segregation of distinct heteromeric NMDA glutamate receptors in the striatum. *J Neurochem* 85:935–943.

Engblom D, Bilbao A, Sanchis-Segura C, Dahan L, Perreau-Lenz S, Balland B, Parkitna JR, Luján R, Halbout B, Mameli M, Parlato R, Sprengel R, Lüscher C, Schütz G, Spanagel R (2008). Glutamate receptors on dopamine neurons control the persistence of cocaine seeking. *Neuron* 59(3):497–508.

Faleiro LJ, Jones S, Kauer JA (2004). Rapid synaptic plasticity of glutamatergic synapses on dopamine neurons in the ventral tegmental area in response to acute amphetamine injection. *Neuropsychopharmacology* 29:2115–2125.

Forster GL, Blaha CD (2000). Laterodorsal tegmental stimulation elicits dopamine efflux in the rat nucleus accumbens by activation of acetylcholine and glutamate receptors in the ventral tegmental area. *Eur J Neurosci* 12:3596–3604.

Forster GL, Blaha CD (2003) Pedunculopontine tegmental stimulation evokes striatal dopamine efflux by activation of acetylcholine and glutamate receptors in the midbrain and pons of the rat. *Eur J Neurosci* 17:751–762.

Frankle WG, Laruelle M, Haber SN (2006). Prefrontal cortical projections to the midbrain in primates: Evidence for a sparse connection. *Neuropsychopharmacology* 31(8):1627–1636.

Fukaya M, Hayashi Y, Watanabe M (2005). NR2 to NR3B subunit switchover of NMDA receptors in early postnatal motoneurons. *Eur J Neurosci* 21:1432–1436.

Furukawa H, Singh SK, Mancusso R, Gouaux E (2005). Subunit arrangement and function in NMDA receptors. *Nature* 438:185–192.

Galvan A, Kuwajima M, Smith Y (2006). Glutamate and GABA receptors and transporters in the basal ganglia: What does their subsynaptic localization reveal about their function? *Neuroscience* 143(2):351–375.

Gariano RF, Groves PM (1988). Burst firing induced in midbrain dopamine neurons by stimulation of the medial prefrontal and anterior cingulate cortices. *Brain Res* 462(1):194–198.

Glees P (1944). The anatomical basis of cortico-striate connexions. *J Anat* 78(Pt 1–2):47–51.

Gonon FG (1988). Nonlinear relationship between impulse flow and dopamine released by rat midbrain dopaminergic neurons as studied by in vivo electrochemistry. *Neuroscience* 24:19–28.

Grace AA, Onn SP (1989). Morphology and electrophysiological properties of immunocytochemically identified rat dopamine neurons recorded in vitro. *J Neurosci* 9(10):3463–3481.

Grenhoff J, Tung CS, Svensson TH (1988). The excitatory amino acid antagonist kynurenate induces pacemaker-like firing of dopamine neurons in rat ventral tegmental area in vivo. *Acta Physiol Scand* 134(4):567–568.

Grillner P, Mercuri NB (2002). Intrinsic membrane properties and synaptic inputs regulating the firing activity of the dopamine neurons. *Behav Brain Res* 130:149–169.

Hallett PJ, Standaert DG (2004). Rationale for and use of NMDA receptor antagonists in Parkinson's disease. *Pharmacol Ther* 102:155–174.

Hammond C, Deniau JM, Rizk A, Feger J (1978). Electrophysiological demonstration of an excitatory subthalamonigral pathway in the rat. *Brain Res* 151(2):235–244.

Hardingham GE (2006). Pro-survival signalling from the NMDA receptor. *Biochem Soc Trans* 34:936–938.

Hardingham GE, Bading H (2003). The Yin and Yang of NMDA receptor signalling. *Trends Neurosci* 26:81–89.

Hardingham GE, Fukunaga Y, Bading H (2002). Extrasynaptic NMDARs oppose synaptic NMDARs by triggering CREB shut-off and cell death pathways. *Nat Neurosci* 5:405–414.

Henson MA, Roberts AC, Pérez-Otaño I, Philpot BD (2010). Influence of the NR3A subunit on NMDA receptor functions. *Prog Neurobiol* 91(1):23–37.

Herrling PL, Morris R, Salt TE (1983). Effects of excitatory amino acids and their antagonists on membrane and action potentials of cat caudate neurones. *J Physiol* 339:207–222.

Hollmann M, Heinemann S (1994). Cloned glutamate receptors. *Annu Rev Neurosci* 17:31–108.

Iino M, Ozawa S, Tsuzuki K (1990). Permeation of calcium through excitatory amino acid receptor channels in cultured rat hippocampal neurones. *J Physiol* 424:151–165.

Ivanov A, Pellegrino C, Rama S, Dumalska I, Salyha Y, Ben-Ari Y, Medina I (2006). Opposing role of synaptic and extrasynaptic NMDA receptors in regulation of the extracellular signal-regulated kinases (ERK) activity in cultured rat hippocampal neurons. *J Physiol* 572:789–798.

Johnson SW, North RA (1992). Two types of neurone in the rat ventral tegmental area and their synaptic inputs. *J Physiol* 450:455–468.

Johnson SW, Seutin V, North RA (1992). Burst firing in dopamine neurons induced by N-methyl-D-aspartate: Role of electrogenic sodium pump. *Science* 258:665–667.

Johnson SW, Wu YN (2004). Multiple mechanisms underlie burst firing in rat midbrain dopamine neurons in vitro. *Brain Res* 1019:293–296.

Jonas P, Racca C, Sakmann B, Seeburg PH, Monyer H (1994). Differences in Ca^{2+} permeability of AMPA-type glutamate receptor channels in neocortical neurons caused by differential GluR-B subunit expression. *Neuron* 12:1281–1289.

Jones S, Bonci A (2005). Synaptic plasticity and drug addiction. *Curr Opin Pharmacol* 5:20–25.

Jones S, Gibb AJ (2005). Functional NR2B- and NR2D-containing NMDA receptor channels in rat substantia nigra dopaminergic neurones. *J Physiol* 569:209–221.

Jones S, Gutlerner JL (2002). Addictive drugs modify excitatory synaptic control of midbrain dopamine cells. *Neuroreport* 13(2):A29–A33.

Jones S, Kauer JA (1999). Amphetamine depresses excitatory synaptic transmission via serotonin receptors in the ventral tegmental area. *J Neurosci* 19(22):9780–9787.

Jones S, Kornblum JL, Kauer JA (2000). Amphetamine blocks long-term synaptic depression in the ventral tegmental area. *J Neurosci* 20(15):5575–5580.

Kamboj SK, Swanson GT, Cull-Candy SG (1995). Intracellular spermine confers rectification on rat calcium-permeable AMPA and kainate receptors. *J Physiol* 486(Pt 2):297–303.

Karreman M, Moghaddam B (1996). The prefrontal cortex regulates the basal release of dopamine in the limbic striatum: An effect mediated by ventral tegmental area. *J Neurochem* 66(2):589–598.

Kauer JA (2004). Learning mechanisms in addiction: Synaptic plasticity in the ventral tegmental area as a result of exposure to drugs of abuse. *Annu Rev Physiol* 66:447–475.

Kauer JA, Malenka RC (2007). Synaptic plasticity and addiction. *Nat Rev Neurosci* 8:844–858.

Kawaguchi Y, Wilson CJ, Emson PC (1989). Intracellular recording of identified neostriatal patch and matrix spiny cells in a slice preparation preserving cortical inputs. *J Neurophysiol* 62(5):1052–1068.

Kita H, Kitai ST (1987). Efferent projections of the subthalamic nucleus in the rat: Light and electron microscopic analysis with the PHA-L method. *J Comp Neurol* 260(3):435–452.

Kitai ST, Kocsis JD, Preston RJ, Sugimori M (1976). Monosynaptic inputs to caudate neurons identified by intracellular injection of horseradish peroxidase. *Brain Res* 109(3):601–606.

Kitai ST, Shepard PD, Callaway JC, Scroggs R (1999). Afferent modulation of dopamine neuron firing patterns. *Curr Opin Neurobiol* 9:690–697.

Kreitzer AC (2009). Physiology and pharmacology of striatal neurons. *Annu Rev Neurosci* 32:27–47.

Kornhuber J, Kim JS, Kornhuber ME, Kornhuber HH (1984). The cortico-nigral projection: Reduced glutamate content in the substantia nigra following frontal cortex ablation in the rat. *Brain Res* 322(1):124–126.

Kuner T, Schoepfer R (1996). Multiple structural elements determine subunit specificity of Mg^{2+} block in NMDA receptor channels. *J Neurosci* 16:3549–3558.

Küppenbender KD, Albers DS, Iadarola MJ, Landwehrmeyer GB, Standaert DG (1999). Localization of alternatively spliced NMDAR1 glutamate receptor isoforms in rat striatal neurons. *J Comp Neurol* 415(2):204–217.

Küppenbender KD, Standaert DG, Feuerstein TJ, Penney JB Jr, Young AB, Landwehrmeyer GB (2000). Expression of NMDA receptor subunit mRNAs in neurochemically identified projection and interneurons in the human striatum. *J Comp Neurol* 419(4):407–421.

Landwehrmeyer GB, Standaert DG, Testa CM, Penney JB Jr, Young AB (1995). NMDA receptor subunit mRNA expression by projection neurons and interneurons in rat striatum. *J Neurosci* 157:5297–5307.

Lee KH, Chang SY, Roberts DW, Kim U (2004). Neurotransmitter release from high-frequency stimulation of the subthalamic nucleus. *J Neurosurg* 101(3):511–517.

Lipton SA, Kater SB (1989). Neurotransmitter regulation of neuronal outgrowth, plasticity and survival. *Trends Neurosci* 12:265–270.

Liu QS, Pu L, Poo MM (2005). Repeated cocaine exposure in vivo facilitates LTP induction in midbrain dopamine neurons. *Nature* 437(7061):1027–1031.

Lodge DJ, Grace AA (2006). The laterodorsal tegmentum is essential for burst firing of ventral tegmental area dopamine neurons. *Proc Natl Acad Sci USA* 103(13):5167–5172.

Lokwan SJ, Overton PG, Berry MS, Clark D (1999). Stimulation of the pedunculopontine tegmental nucleus in the rat produces burst firing in A9 dopaminergic neurons. *Neuroscience* 92:245–254.

Low CM, Wee KS (2010). New insights into the not-so-new NR3 subunits of N-methyl-D-aspartate receptor: Localization, structure, and function. *Mol Pharmacol* 78(1):1–11.

Lüscher C, Malenka RC (2011). Drug-evoked synaptic plasticity in addiction: From molecular changes to circuit remodeling. *Neuron* 69(4):650–663.

Mameli M, Halbout B, Creton C, Engblom D, Parkitna JR, Spanagel R, Lüscher C (2009). Cocaine-evoked synaptic plasticity: Persistence in the VTA triggers adaptations in the NAc. *Nat Neurosci* 12(8):1036–1041.

Mansvelder HD, McGehee DS (2000). Long-term potentiation of excitatory inputs to brain reward areas by nicotine. *Neuron* 27(2):349–357.

Marey-Semper I, Gelman M, Levi-Strauss M (1995). A selective toxicity toward cultured mesencephalic dopaminergic neurons is induced by the synergistic effects of energetic metabolism impairment and NMDA receptor activation. *J Neurosci* 15:5912–5918.

Margolis EB, Lock H, Hjelmstad GO, Fields HL (2006). The ventral tegmental area revisited: Is there an electrophysiological marker for dopaminergic neurons? *J Physiol* 577:907–924.

Matsuda K, Kamiya Y, Matsuda S, Yuzaki M (2002). Cloning and characterization of a novel NMDA receptor subunit NR3B: A dominant subunit that reduces calcium permeability. *Brain Res Mol Brain Res* 100(1–2):43–52.

Mayer ML, Westbrook GL (1987). Permeation and block of N-methyl-D-aspartic acid receptor channels by divalent cations in mouse cultured central neurones. *J Physiol* 394:501–527.

Mereu G, Costa E, Armstrong DM, Vicini S (1991). Glutamate receptor subtypes mediate excitatory synaptic currents of dopamine neurons in midbrain slices. *J Neurosci* 11:1359–1366.

Mereu G, Lilliu V, Casula A, Vargiu PF, Diana M, Musa A, Gessa GL (1997). Spontaneous bursting activity of dopaminergic neurons in midbrain slices from immature rats: Role of N-methyl-D-aspartate receptors. *Neuroscience* 77:1029–1036.

Monyer H, Burnashev N, Laurie DJ, Sakmann B, Seeburg PH (1994). Developmental and regional expression in the rat brain and functional properties of four NMDA receptors. *Neuron* 12:529–540.

Monyer H, Sprengel R, Schoepfer R, Herb A, Higuchi M, Lomeli H, Burnashev N, Sakmann B, Seeburg PH (1992). Heteromeric NMDA receptors: Molecular and functional distinction of subtypes. *Science* 256:1217–1221.

Murase S, Grenhoff J, Chouvet G, Gonon FG, Svensson TH (1993). Prefrontal cortex regulates burst firing and transmitter release in rat mesolimbic dopamine neurons studied in vivo. *Neurosci Lett* 157(1):53–56.

Nishi M, Hinds H, Lu HP, Kawata M, Hayashi Y (2001). Motoneuron-specific expression of NR3B, a novel NMDA-type glutamate receptor subunit that works in a dominant-negative manner. *J Neurosci* 21(23):RC185.

Omelchenko N, Sesack SR (2005). Laterodorsal tegmental projections to identified cell populations in the rat ventral tegmental area. *J Comp Neurol* 483(2):217–235

Omelchenko N, Sesack SR (2007). Glutamate synaptic inputs to ventral tegmental area neurons in the rat derive primarily from subcortical sources. *Neuroscience* 146(3):1259–1274.

Ongür D, An X, Price JL (1998). Prefrontal cortical projections to the hypothalamus in macaque monkeys. *J Comp Neurol* 401(4):480–505.

Overton P, Clark D (1992). Iontophoretically administered drugs acting at the N-methyl-D-aspartate receptor modulate burst firing in A9 dopamine neurons in the rat. *Synapse* 10:131–140.

Overton PG, Richards CD, Berry MS, Clark D (1999). Long-term potentiation at excitatory amino acid synapses on midbrain dopamine neurons. *Neuroreport* 10:221–226.

Ozawa, S, Iino M (1993). Two distinct types of AMPA responses in cultured rat hippocampal neurons. *Neurosci lett* 155:187–190.

Paoletti P, Neyton J (2007). NMDA receptor subunits: Function and pharmacology. *Curr Opin Pharmacol* 7:39–47.

Papadia S, Soriano FX, Léveillé F, Martel MA, Dakin KA, Hansen HH, Kaindl A, Sifringer M, Fowler J, Stefovska V, McKenzie G, Craigon M, Corriveau R, Ghazal P, Horsburgh K, Yankner BA, Wyllie DJ, Ikonomidou C, Hardingham GE (2008). Synaptic NMDA receptor activity boosts intrinsic antioxidant defenses. *Nat Neurosci* 11(4):476–487.

Paquet M, Tremblay M, Soghomonian JJ, Smith Y (1997). AMPA and NMDA glutamate receptor subunits in midbrain dopaminergic neurons in the squirrel monkey: An immunohistochemical and in situ hybridization study. *J Neurosci* 17(4):1377–1396.

Petralia RS, Wang YX, Wenthold RJ (1994). The NMDA receptor subunits NR2A and NR2B show histological and ultrastructural localization patterns similar to those of NR1. *J Neurosci* 14(10):6102–6120.

Plenz D, Kitai ST (1998). Up and down states in striatal medium spiny neurons simultaneously recorded with spontaneous activity in fast-spiking interneurons studied in cortex-striatum-substantia nigra organotypic cultures. *J Neurosci* 18(1):266–283.

Pomata PE, Belluscio MA, Riquelme LA, Murer MG (2008). NMDA receptor gating of information flow through the striatum in vivo. *J Neurosci* 28(50):13384–13389.

Rosales MG, Flores G, Hernández S, Martínez-Fong D, Aceves J (1994). Activation of subthalamic neurons produces NMDA receptor-mediated dendritic dopamine release in substantia nigra pars reticulata: A microdialysis study in the rat. *Brain Res* 645(1–2):335–337.

Russo SJ, Dietz DM, Dumitriu D, Morrison JH, Malenka RC, Nestler EJ (2010). The addicted synapse: Mechanisms of synaptic and structural plasticity in nucleus accumbens. *Trends Neurosci* 33(6):267–276.

Saal D, Dong Y, Bonci A, Malenka RC (2003). Drugs of abuse and stress trigger a common synaptic adaptation in dopamine neurons. *Neuron* 37:577–582.

Sasaki YF, Rothe T, Premkumar LS, Das S, Cui J, Talantova MV, Wong HK, Gong X, Chan SF, Zhang D, Nakanishi N, Sucher NJ, Lipton SA (2002). Characterization and comparison of the NR3A subunit of the NMDA receptor in recombinant systems and primary cortical neurons. *J Neurophysiol* 87(4):2052–2063.

Schultz W (2000). Multiple reward signals in the brain. *Nat Rev Neurosci* 1(3):199–207.

Schultz W, Romo R (1990). Dopamine neurons of the monkey midbrain: Contingencies of responses to stimuli eliciting immediate behavioral reactions. *J Neurophysiol* 63(3):607–624.

Sesack SR, Deutch AY, Roth RH, Bunney BS (1989). Topographical organization of the efferent projections of the medial prefrontal cortex in the rat: An anterograde tract-tracing study with Phaseolus vulgaris leucoagglutinin. *J Comp Neurol* 290(2):213–242.

Sesack SR, Pickel VM (1992). Prefrontal cortical efferents in the rat synapse on unlabeled neuronal targets of catecholamine terminals in the nucleus accumbens septi and on dopamine neurons in the ventral tegmental area. *J Comp Neurol* 320(2):145–160.

Seutin V, Johnson SW, North RA (1993). Apamin increases NMDA-induced burst-firing of rat mesencephalic dopamine neurons. *Brain Res* 630(1–2):341–344.

Shepard PD, Stump D (1999). Nifedipine blocks apamin-induced bursting activity in nigral dopamine-containing neurons. *Brain Res* 817(1–2):104–109.

Simon RP, Swan JH, Griffiths T, Meldrum BS (1984). Blockade of N-methyl-D-aspartate receptors may protect against ischemic damage in the brain. *Science* 226:850–852.

Smith Y, Charara A, Parent A (1996). Synaptic innervation of midbrain dopaminergic neurons by glutamate-enriched terminals in the squirrel monkey. *J Comp Neurol* 364(2):231–253.

Smith ID, Grace AA (1992). Role of the subthalamic nucleus in the regulation of nigral dopamine neuron activity. *Synapse* 12(4):287–303.

Soriano FX, Hardingham GE (2007). Compartmentalized NMDA receptor signalling to survival and death. *J Physiol* 584:381–387.

Soriano FX, Papadia S, Hofmann F, Hardingham NR, Bading H, Hardingham GE (2006). Preconditioning doses of NMDA promote neuroprotection by enhancing neuronal excitability. *J Neurosci* 26(17):4509–4518.

Standaert DG, Friberg IK, Landwehrmeyer GB, Young AB, Penney JB Jr (1999). Expression of NMDA glutamate receptor subunit mRNAs in neurochemically identified projection and interneurons in the striatum of the rat. *Brain Res Mol Brain Res* 64(1):11–23.

Standaert DG, Landwehrmeyer GB, Kerner JA, Penney JB Jr, Young AB (1996). Expression of NMDAR2D glutamate receptor subunit mRNA in neurochemically identified interneurons in the rat neostriatum, neocortex and hippocampus. *Brain Res Mol Brain Res* 42(1):89–102.

Standaert DG, Testa CM, Young AB, Penney JB Jr (1994). Organization of N-methyl-D-aspartate glutamate receptor gene expression in the basal ganglia of the rat. *J Comp Neurol* 343:1–16.

Stefani A, Chen Q, Flores-Hernandez J, Jiao Y, Reiner A, Surmeier DJ (1998). Physiological and molecular properties of AMPA/Kainate receptors expressed by striatal medium spiny neurons. *Dev Neurosci* 20:242–252.

Stuber GD, Klanker M, de Ridder B, Bowers MS, Joosten RN, Feenstra MG, Bonci A (2008). Reward-predictive cues enhance excitatory synaptic strength onto midbrain dopamine neurons. *Science* 321(5896):1690–1692.

Suárez F, Zhao Q, Monaghan DT, Jane DE, Jones S, Gibb AJ (2010). Functional heterogeneity of NMDA receptors in rat substantia nigra pars compacta and reticulata neurones. *Eur J Neurosci* 32(3):359–367.

Suaud-Chagny MF, Chergui K, Chouvet G, Gonon F (1992). Relationship between dopamine release in the rat nucleus accumbens and the discharge activity of dopaminergic neurons during local in vivo application of amino acids in the ventral tegmental area. *Neuroscience* 49(1):63–72.

Sucher NJ, Akbarian S, Chi CL, Leclerc CL, Awobuluyi M, Deitcher DL, Wu MK, Yuan JP, Jones EG, Lipton SA (1995). Developmental and regional expression pattern of a novel NMDA receptor-like subunit (NMDAR-L) in the rodent brain. *J Neurosci* 15(10):6509–6520.

Sulzer D (2011). How addictive drugs disrupt presynaptic dopamine neurotransmission. *Neuron* 69(4):628–649.

Svensson TH, Tung CS (1989). Local cooling of pre-frontal cortex induces pacemaker-like firing of dopamine neurons in rat ventral tegmental area in vivo. *Acta Physiol Scand* 136(1):135–136.

Tallaksen-Greene SJ, Albin RL (1994). Localization of AMPA-selective excitatory amino acid receptor subunits in identified populations of striatal neurons. *Neuroscience* 61(3):509–519.

Tallaksen-Greene SJ, Wiley RG, Albin RL (1992). Localization of striatal excitatory amino acid binding site subtypes to striatonigral projection neurons. *Brain Res* 594(1):165–1670.

Tepper JM, Martin LP, Anderson DR (1995). GABAA receptor-mediated inhibition of rat substantia nigra dopaminergic neurons by pars reticulata projection neurons. *J Neurosci* 15(4):3092–3103.

Thomas MJ, Beurrier C, Bonci A, Malenka RC (2001). Long-term depression in the nucleus accumbens: A neural correlate of behavioral sensitization to cocaine. *Nat Neurosci* 4(12):1217–1223.

Tong ZY, Overton PG, Clark D (1996). Antagonism of NMDA receptors but not AMPA/kainate receptors blocks bursting in dopaminergic neurons induced by electrical stimulation of the prefrontal cortex. *J Neural Transm* 103:889–904.

Tong G, Takahashi H, Tu S, Shin Y, Talantova M, Zago W, Xia P, Nie Z, Goetz T, Zhang D, Lipton SA, Nakanishi N (2008). Modulation of NMDA receptor properties and synaptic transmission by the NR3A subunit in mouse hippocampal and cerebrocortical neurons. *J Neurophysiol* 99(1):122–132.

Ulbrich MH, Isacoff EY (2008). Rules of engagement for NMDA receptor subunits. *Proc Natl Acad Sci USA* 105(37):14163–14168.

Ungless MA, Whistler JL, Malenka RC, Bonci A (2001). Single cocaine exposure in vivo induces long-term potentiation in dopamine neurons. *Nature* 411:583–587.

Van Der Kooy D, Hattori T (1980). Single subthalamic nucleus neurons project to both the globus pallidus and substantia nigra in rat. *J Comp Neurol* 192(4):751–768.

Vergara R, Rick C, Hernández-López S, Laville JA, Guzman JN, Galarraga E, Surmeier DJ, Bargas J (2003). Spontaneous voltage oscillations in striatal projection neurons in a rat corticostriatal slice. *J Physiol* 553(Pt 1):169–182.

Watanabe J, Beck C, Kuner T, Premkumar LS, Wollmuth LP (2002). DRPEER: A motif in the extracellular vestibule conferring high Ca^{2+} flux rates in NMDA receptor channels. *J Neurosci* 22:10209–10216.

Webster KE (1961). Cortico-striate interrelations in the albino rat. *J Anat* 95:532–544.

Wee KS, Zhang Y, Khanna S, Low CM (2008). Immunolocalization of NMDA receptor subunit NR3B in selected structures in the rat forebrain, cerebellum, and lumbar spinal cord. *J Comp Neurol* 509(1):118–135.

Wilson CJ, Kawaguchi Y (1996). The origins of two-state spontaneous membrane potential fluctuations of neostriatal spiny neurons. *J Neurosci* 16(7):2397–2410.

Wolfart J, Neuhoff H, Franz O, Roeper J (2001). Differential expression of the small-conductance, calcium-activated potassium channel SK3 is critical for pacemaker control in dopaminergic midbrain neurons. *J Neurosci* 21(10):3443–3456.

Wong HK, Liu XB, Matos MF, Chan SF, Pérez-Otaño I, Boysen M, Cui J, Nakanishi N, Trimmer JS, Jones EG, Lipton SA, Sucher NJ (2002). Temporal and regional expression of NMDA receptor subunit NR3A in the mammalian brain. *J Comp Neurol* 450:303–317.

Wu YN, Johnson SW (2007). Rotenone potentiates NMDA currents in substantia nigra dopamine neurons. *Neurosci Lett* 421(2):96–100.

Wu YN, Johnson SW (2009). Rotenone reduces Mg^{2+}-dependent block of NMDA currents in substantia nigra dopamine neurons. *Neurotoxicology* 30(2):320–325.

Wu HQ, Schwarcz R, Shepard PD (1994). Excitatory amino acid-induced excitation of dopamine-containing neurons in the rat substantia nigra: Modulation by kynurenic acid. *Synapse* 16(3):219–230.

Zhang J, Chiodo LA, Freeman AS (1994). Influence of excitatory amino acid receptor subtypes on the electrophysiological activity of dopaminergic and nondopaminergic neurons in rat substantia nigra. *J Pharmacol Exp Ther* 269:313–321.

Zweifel LS, Argilli E, Bonci A, Palmiter RD (2008). Role of NMDA receptors in dopamine neurons for plasticity and addictive behaviors. *Neuron* 59(3):486–496.

Zweifel LS, Parker JG, Lobb CJ, Rainwater A, Wall VZ, Fadok JP, Darvas M, Kim MJ, Mizumori SJ, Paladini CA, Phillips PE, Palmiter RD (2009). Disruption of NMDAR-dependent burst firing by dopamine neurons provides selective assessment of phasic dopamine-dependent behavior. *Proc Natl Acad Sci USA* 106(18):7281–7288.

3 Dopamine Receptors and their Interactions with NMDA Receptors

Alasdair J. Gibb and Huaxia Tong

CONTENTS

3.1 INTRODUCTION

Dopamine exerts crucial effects on many aspects of brain functioning, from control of movement to higher cognitive processes, and hence on animal behavior. Research on the actions of dopamine in the basal ganglia has been a major area of activity for the past 50 years since the discovery of the loss of dopamine in Parkinson's patients (Hornykiewicz, 2001; Marsden, 2006; Bjorklund and Dunnett, 2007). The neuropathological hallmarks of Parkinson's disease (PD) include the loss of the melanin-pigmented dopaminergic neurones of the substantia nigra. Loss of nigral dopaminergic neurones results in a severe deficit of dopamine signaling in the striatum which can account for many of the clinical features of the disease. By the time parkinsonian symptoms become apparent, more than 75% of dopaminergic innervation to the striatum has been lost (Hornykiewicz, 2001). The loss of dopamine signaling alters the behavior of the basal ganglia circuit in complex ways that are now understood within the conceptual framework provided by direct and indirect pathways of the basal ganglia circuit (Albin et al., 1989; Alexander and Crutcher, 1990; Surmeier et al., 2010). These changes involve changes in both synaptic strengths and neurone firing properties and yet at the cellular level, the mechanisms involved are not yet fully understood, and further questions are raised by the development of dyskinesias in a significant number of PD patients, particularly following prolonged treatment with L-DOPA. Thus, a major aim of the neuroscience community must be to search for the cellular mechanisms underlying these changes and hence raise the possibility of improved symptomatic treatments in PD.

3.2   DOPAMINE RECEPTORS AND THEIR SIGNALING PATHWAYS

Dopamine is one of the three crucial monoamine neurotransmitters in the brain (along with norepinephrine and serotonin) and, like the other monoamines, is produced by relatively few neurones that exist in specific nuclei in the brain. Compared to the hundreds of millions of glutamatergic and GABAergic neurones in the brain, dopaminergic neurones are numbered in hundreds of thousands. However, the influence of dopaminergic neurones is much greater than their relative paucity in number might suggest, partly due to the impressive axon arborization of dopaminergic neurone projections in the striatum, and partly due to the exquisite anatomical arrangement of dopaminergic neurone terminals which in the striatum serves to place dopamine release (Bolam et al., 2000) at precisely the locations needed to regulate the excitatory input to the neurone.

3.2.1   DOPAMINE RECEPTORS

The effects of dopamine are mediated by the activation of G-protein-coupled receptors (GPCRs) of the D1 class (subtypes D1 and D5) and D2 class (subtypes D2, D3, and D4).

1. The *D1 class*, blocked by SCH-23390 ($Ki = 0.3\,nM$), is composed of D1—activated by SKF-82958 ($Ka = 4\,nM$)—and D5—which has 10-fold higher affinity for dopamine than D1 receptors ($200\,nM$).
2. The *D2 class*, blocked by spiperone ($Ki = 0.1\,nM$), selectively activated by quinpirole ($Ka = 4\,nM$), is composed of $D2_{short}$ and $D2_{long}$, D3, and D4 receptors.

3.3   DOPAMINE RECEPTOR EXPRESSION IN STRIATAL MEDIUM SPINY NEURONES

While there have been conflicting reports regarding coexpression of D1 and D2 receptors in striatal neurones, molecular genetics has confirmed that in mouse the main functional distinctions among medium spiny projection neurones (MSNs) are that D2 receptors are associated with indirect pathway neurones while D1 receptors are expressed by direct pathway neurones (Gerfen et al., 1990; Shuen et al., 2008; Gerfen and Surmeier, 2010). These neurones also express substance-P and muscarinic m4 receptors and provide the direct pathway GABAergic projection to the internal globus pallidus and substantia nigra. D5 receptors on the other hand are mainly found in the prefrontal cortex and hippocampus and substantia nigra.

D2 class receptors, the main target of antipsychotic D2 receptor antagonists and anti-parkinsonian dopamine receptor agonists like ropinirole, are highly expressed in indirect pathway MSNs. These neurones make short projections to the lateral globus pallidus and express enkephalin. $D2_{short}$ and $D2_{long}$ receptors are generated by alternate splicing of a 29-amino acid insert in the third intracellular loop (the region of the receptor responsible for G-protein coupling and β-arrestin binding). Both receptor forms activate Gi/o G-proteins and β-arrestin signaling.

3.3.1 Dopamine Receptors and DARPP-32 Signaling in the Striatum

Classically, dopamine receptors couple to G_s and $G_{i/o}$ class G-proteins (Figure 3.1). Upon receptor activation following agonist binding, the G-protein splits into α and βγ subunits. The α subunits may then bind to adenylyl cyclase and βγ subunits to ion channels or other effector proteins, altering the effector activity and giving rise to a change in the activity of the neurone. Typical examples of effector mechanism are as follows:

- D1 class activation of G_s—Stimulation of adenylyl cyclase raising cAMP and hence increasing PKA activity. PKA can phosphorylate many target proteins including L-type calcium channels and DARPP-32 (the dopamine and cyclic AMP-regulated phosphoprotein of M_r 32,000) which inhibits protein phosphatase PP1, resulting in increased levels of target protein phosphorylation (Greengard, 2001).
- D2 class activation of $G_{i/o}$—Decreasing cAMP, hence reducing PKA activity and DARPP-32 phosphorylation, activation of inward rectifier potassium channels (GIRK channels), and inhibition of voltage-dependent Ca^{2+} channels (N and P/Q type). DARPP-32 dephosphorylation is by the calcium- and calmodulin-dependent protein phosphatase, calcineurin, or PP2B. In some cell types, activation of D2 receptors can result in phospholipase-C activation and a rise in intracellular calcium.

3.3.2 Nonclassical Dopamine Receptor Signaling via β-Arrestins

In the classical model of GPCR signaling, arrestins terminate receptor signaling by binding to receptors phosphorylated by specific G-protein receptor kinases (GRKs) (Pearce and Lefkowitz, 2001). Arrestin binding prevents G-protein activation and initiates dynamin-dependent receptor internalization because the arrestin functions as an adaptor that links the receptor to AP2 and hence to clathrin leading to internalization via clathrin-coated pits. It is now clear though that arrestin signaling is in itself a key aspect of GPCR function (Violin and Lefkowitz, 2007) leading to

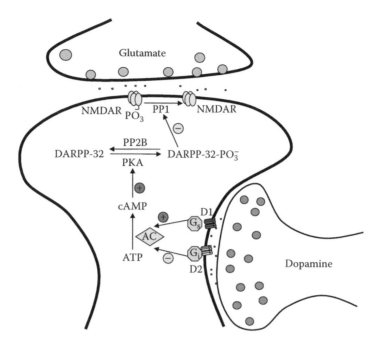

FIGURE 3.1 **(See color insert.)** Cartoon representation of dopamine receptor modulation of NMDA receptors. In this classical pathway, D1 receptors couple to G_s G-proteins, raising cAMP and resulting in protein kinase A (PKA) phosphorylation of DARPP-32 (the dopamine- and cyclic AMP-regulated phosphoprotein of Mr 32000) and inhibition of protein phosphatase-1 (PP1). The calcium- and calmodulin-dependent phosphatase, calcineurin (PP2B), dephosphorylates DARPP-32. NMDA receptor phosphorylation may increase the NMDA response, by directly increasing channel open probability or by stabilizing receptors at the cell surface. D2 receptor stimulation would be expected to give the opposite effect compared to D1 activation and the mechanism should be blocked by inhibitors of G-protein signaling or PKA. Because PKA can also directly phosphorylate the NMDA receptor on the GluN1 C-terminal, the balance of effect on NMDA responses is complicated to predict.

MAP kinase, Src kinase, phosphatidylinositol 3-kinase, and Akt (protein kinase B)—glycogen synthase kinase-3 (GSK-3) signaling following dopamine D2 receptor activation (Beaulieu and Gainetdinov, 2011; Mannoury la Cour et al., 2011).

β-arrestin signaling is also triggered by D1 receptor activation, coupling D1 receptors to ERK activation via the β-arrestin and MAP kinase complex (Chen et al., 2009; Urs et al., 2011). Phosphorylation of ERK by the MAP kinase MEK has been shown in the striatum to be reversed by the striatal-enriched tyrosine phosphatase (STEP). STEP activity depends on the activity of the phosphatase PP1 which is inhibited by DARPP-32. Thus D1 receptor activation can result in the inactivation of STEP and so enhanced ERK signaling. These results therefore link the classical DARPP-32 path with β-arrestin signaling.

Finally, in striatal neurones, D1 receptor activation has been shown to cause an Src family kinase (fyn)–dependent trafficking of N-methyl-D-aspartate (NMDA) receptors to the synapse (Dunah et al., 2004; Hallett et al., 2006), and although a direct role for β-arrestin was not determined in this case, it follows from other studies that this was the likely mechanism.

3.4 DOPAMINE RECEPTOR–NMDA RECEPTOR INTERACTIONS

NMDA receptors play a crucial role in excitatory synaptic transmission throughout the brain. Understanding how NMDA receptors behave and how they are regulated is important for a number of reasons. As a cellular model for learning and information storage in the central nervous system, NMDA receptor–dependent long-term potentiation (LTP) and long-term depression (LTD) have been described in several brain areas. But surprisingly, LTP and LTD in the striatum are not usually NMDA receptor dependent (Calabresi et al., 1992; Charpier and Denieu, 1997; Kreitzer and Malenka, 2008) but depend on dopamine receptor activation, although NMDA receptor–dependent LTP in young animals has been described (Partridge et al., 2000). In addition, NMDA receptor–mediated excitotoxicity contributes to the death of striatal neurons under some pathological conditions such as Huntington's disease where there is increased surface expression of GluN2B-containing receptors (Fan and Raymond, 2007).

3.4.1 NMDA RECEPTOR MODULATION

NMDA receptor activity can be regulated by protein kinases (PKA, PKC, and tyrosine kinases) (Blank et al., 1997; Lu et al., 1999; Xiong et al., 1999; Lei et al., 2002) and protein phosphatases (PP1 and calcineurin; PP2B) (Lieberman and Mody, 1994; Morishita et al., 2001; Krupp et al., 2002; Rycroft and Gibb, 2004). There is also good evidence showing that dopamine receptors can modulate striatal NMDA receptors (Levine et al., 1996; Blank et al., 1997; Cepeda et al., 1998; Lee et al., 2002; Dunah et al., 2004; Hallett et al., 2006; Tong and Gibb, 2008; Beaulieu and Gainetdinov, 2011).

3.4.2 D1 DOPAMINE RECEPTORS

Some studies have shown that D1 receptor activation enhanced NMDA receptor responses by the classical adenylate cyclase pathway (Blank et al., 1997; Cepeda et al., 1998; Snyder et al., 1998; Flores-Hernandez et al., 2002). In this classical pathway, D1 receptors couple to Gs G-proteins, raising cAMP and resulting in phosphorylation of DARPP-32 and inhibition of protein phosphatase-1 (Blank et al., 1997; Snyder et al., 1998; Greengard, 2001). Thus, as summarized in Figure 3.1, there are established mechanisms in striatal MSNs that could modulate NMDA receptors.

On the other hand, some studies have shown that dopamine can attenuate NMDA-mediated currents (Law-Tho et al., 1994; Lee et al., 2002; Lin et al., 2003). In particular, Lee et al. (2002) demonstrated the inhibition of NMDA responses by a direct protein–protein interaction between the dopamine D1 receptor and GluN2A subunit C-termini while direct interaction between the GluN1-1a C-terminus and D1 receptor C-terminus mediated a PI-3 kinase–dependent protective effect from NMDA-induced excitotoxicity in hippocampal cultures. A summary of the literature where D1 receptor activation was found to modulate NMDA receptor currents is shown in Table 3.1. Although the relationship is not

TABLE 3.1

A Summary of Literature Showing D1 Receptor Modulation of NMDA Receptors

Authors	Neuron Type	Animal Age	D1 Receptor Modulation of NMDA Receptors
Blank et al. (1997)	Oocytes injected with striatal and hippocampal mRNA	mRNA from adult rats	DARPP-32 mediated cAMP-dependent potentiation of NMDA responses
Cepeda et al. (1998)	Striatal neurons in slice	12–18 day old rats	L-type Ca^{2+} channel contribution to D1 receptor-induced potentiation of NMDA whole-cell current
Flores-Hernandez et al. (2002)	Dissociated medium spiny neurons from the striatum	21–40 day old rats	D1 potentiation of NMDA whole-cell current mediated by DARPP-32 but not L-type Ca^{2+} channels
Chen et al. (2004)	Dissociated prefrontal cortex neurons	3–5 week old rats	D1 potentiation of NMDA whole-cell current involved PKC and reduction of Ca^{2+}/calmodulin-dependent inactivation of NMDA receptors
Seamans et al. (2001)	Prefrontal cortex slice	14–20 day old rats	D1 activation selectively increased the NMDA component of EPSCs
Synder et al. (1998)	Nucleus accumbens slice	150–200 g rats	D1 activation increased NR1 phosphorylation. DARPP-32 and PKA involved
Levine et al. (1996)	Neostriatum slice D1-deficient mutant mice		D1 agonist increased responses mediated by activation of NMDA receptors
Lee et al. (2002)	Cultured hippocampal and striatal neurons		D1 inhibition of NMDA currents was dependent on direct protein–protein interaction between the C-terminus of the D1 receptor (D1-t3) and the NR2A subunit
Lin et al. (2003)	Dissociated medium spiny neurons from the striatum	14–24 day old rats	D1 agonist reduced NMDA induced currents
Tong and Gibb (2008)	Medium spiny neurons in acute brain slices	7 day old rats	D1 inhibition of NMDA receptor whole-cell current by G-protein-independent but tyrosine kinase- and dynamin-dependent pathway

clear-cut, D1 inhibitions have generally been described in relatively young animals (<3 weeks) and potentiation in older animals.

In addition to the classical pathway, Dunah et al. (2004) have shown that deletion of the gene for the protein tyrosine kinase, Fyn, inhibits dopamine D1 receptor-induced enhancement of the abundance of GluN1, GluN2A, and GluN2B subunits in the synaptosomal membrane fraction of adult mice striatal homogenates. Such a shift in receptor distribution would likely be reflected in increased NMDA receptor current during synaptic transmission.

3.4.3 D2 DOPAMINE RECEPTOR MODULATION OF NMDA RECEPTORS

There are also some reports of D2 family dopamine receptor inhibition of NMDA receptor transmission (Kotecha et al., 2002; Wang et al., 2003; Li et al., 2009). Kotecha et al. (2002) suggested that the activation of D2 class receptors in hippocampal CA1 neurons depressed NMDA receptor activity and excitatory NMDAR-mediated synaptic transmission via the transactivation of PDGFRs (PDGF: platelet-derived growth factor) followed by mobilization of intracellular Ca^{2+} and Ca^{2+}-dependent inactivation of NMDA receptors. In prefrontal cortex, Wang et al. (2003) showed that D4 receptor activation decreases the NMDA receptor–mediated current by means of the inhibition of PKA followed by the activation of PP1 and the ensuing inhibition of Ca^{2+}/calmodulin-dependent protein kinase II (CaMKII). Also in prefrontal cortex, Li et al. (2009) demonstrated the role of GSK-3β phosphorylation of β-catenin in the inhibition of NMDA currents following D2 receptor activation.

3.4.4 NMDA RECEPTOR TRAFFICKING MECHANISMS

NMDA receptors were thought to be stable at the excitatory postsynaptic membrane with a very slow turnover rate (Luscher et al., 1999). However, over the past 10 years, it has been shown that the synaptic expression of NMDA receptors can be changed rapidly by export and internalization (Roche et al., 2001; Snyder et al., 2001; Grosshans et al., 2002; Bellone and Nicoll, 2007), and so it is interesting to consider that dopamine receptor activation could result in changes in the number and type of NMDA receptor expressed at synapses in the striatum.

The NMDA receptor GluN2 subunits play an essential role in regulating receptor expression on the cell surface. Temporal variation of GluN2 (NR2) subunit gene expression may underlie the developmental changes observed in NMDA receptor trafficking and synaptic expression. Lavezzari et al. (2004) showed distinct intracellular pathways for GluN2A and GluN2B endocytosis, and in mature neurons internalization of GluN2B is more robust than of GluN2A. Roche et al. (2001) reported a progressive decrease of NMDA receptor endocytosis as neurons mature.

NMDA receptor trafficking and stability at the cell surface is determined by interaction with several proteins such as the PSD-95 family and AP-2 adaptor protein which influence the targeting and localization of the receptor to the synapse.

In addition, some proteins target phosphorylation sites on NMDA receptors and affect receptor trafficking as well as function (Wenthold et al., 2003).

Roche et al. (2001) showed that coexpression with PSD-95 blocked the GluN2B receptor internalization and when cultured hippocampal neurons were transfected with a truncated GluN2B (lacking the PDZ-binding domain), receptors exhibited robust internalization. Thus they concluded that disruption of the NMDA receptor–PSD-95 complex destabilized the NMDA receptor allowing receptor internalization. Chung et al. (2004) also revealed that the phosphorylation of a serine residue within the PDZ-binding domain of GluN2B disrupts the interaction with PSD-95 and SAP102 and decreases surface NR2B expression. Moreover, the PDZ-binding motif in the NR2A subunit has been reported to be crucial for the PSD-95 potentiation of NMDA current and the ability of PSD-95 to increase NMDA receptor surface expression (Lin et al., 2004).

Many studies have reported a clathrin-dependent mechanism for NMDA receptor endocytosis as NMDA receptor internalization can be blocked by coexpression of a dominant-negative dynamin (Roche et al., 2001). Lavezzari et al. (2004) showed that the medium chain of AP2 ($\mu 2$) interacts with GluN2B C-terminus directly, and mutation of tyrosine motif (YEKL) wipes out the interaction. Unlike GluN2B, GluN2A binds to a different adaptor medium chain ($\mu 1$) of AP1, and it traffics to distinct intracellular compartments (Lavezzari et al., 2004).

3.4.5 SRC FAMILY TYROSINE KINASE MODULATION OF NMDA RECEPTORS

The Src family of non-receptor tyrosine kinases are attractive candidates linking dopamine receptors to NMDA receptor activity because dopamine receptor β-arrestin signaling can alter both Src tyrosine kinase activity and the activity of the tyrosine phosphatase STEP (Beaulieu and Gainetdinov, 2011). The family consists of Src, Fyn, Lyn, Lck, and Yes and they have been shown to regulate NMDA receptor function and expression (Salter and Kalia, 2004). As described earlier, tyrosine motifs in the GluN2 subunit C-terminus regulate NMDA receptor internalization. Roche et al. (2001) demonstrated that the removal of the tyrosine motif (YEKL) of GluN2B subunits dramatically inhibited the receptor internalization. Moreover Dunah et al. (2004) showed that deletion of the gene for the protein tyrosine kinase Fyn remarkably reduces the basal content of tyrosine-phosphorylated GluN2A and GluN2B in the striatum and inhibits dopamine D1 receptor-induced enhancement of the abundance of GluN1, GluN2A, and GluN2B subunits in the synaptosomal membrane fraction.

GluN2B has been reported to be the most prominently tyrosine-phosphorylated protein in the postsynaptic density (Moon et al., 1994), and its interaction with PSD proteins may contribute to hold NMDA receptors at the synapse. PSD-95 associates with the last four amino acids (ESDV) of the GluN2 subunit (Kornau et al., 1995; Niethammer et al., 1996), and this binding site is very close to Y1472 (Nakazawa et al., 2001), the main site of tyrosine phosphorylation. Roche et al. (2001) suggested that tyrosine phosphorylation could inhibit the receptor's interaction with PSD-95 and while a common observation is that tyrosine kinase activation promotes GluN2B receptor retention at the synapse, there are some reports suggesting that tyrosine phosphorylation may disrupt the interaction with PSD-95 and SAP102, and therefore decreases surface GluN2B expression (Chung et al., 2004).

3.4.6 D1 DOPAMINE RECEPTOR MODULATION OF STRIATAL NMDA RECEPTORS

Figure 3.2 illustrates D1 receptor inhibition of the whole-cell NMDA current by the partial agonist, SKF-82958 (20 nM), in neonatal rat striatal slices. In these experiments, a 2 min bath application of NMDA (10 μM) and glycine (10 μM) is used to

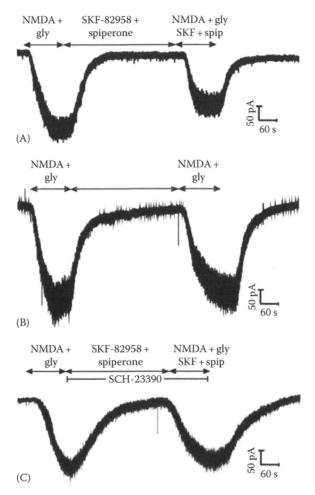

FIGURE 3.2 Whole-cell NMDA receptor currents are inhibited by D1 receptor activation in striatal neurons from 7 day old rats. (A) Response to NMDA (10 μM) and glycine (10 μM) applied for 2 min, followed by 5 min SKF-82958 (20 nM) and spiperone (2 nM), and finally NMDA (10 μM), glycine (10 μM), SKF-82958 (20 nM), and spiperone (2 nM) for 2 min. (B) Control response to NMDA and glycine recorded in the absence of dopamine receptor stimulation. (C) D1 inhibition of response to NMDA and glycine is blocked by the D1 antagonist, SCH-23390 (300 nM).

(*continued*)

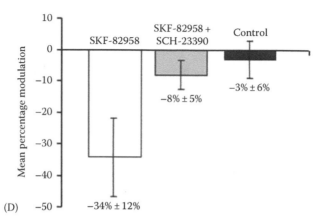

FIGURE 3.2 (continued)   (D) Mean percent change (±SE) in NMDA current in the presence of SKF-82598. (Reproduced from Tong, H. and Gibb, A.J., *J. Physiol.*, 586, 4693, 2008.)

measure the response of both synaptic and extrasynaptic NMDA receptors. This is followed by a 5 min application of SKF-82958 (20 nM) and spiperone (2 nM) in the absence of NMDA and glycine and then another 2 min in the presence of NMDA and glycine. Spiperone, a D2 receptor antagonist, was used to rule out the effect of D2 receptor activation on the NMDA responses. ATP (1 mM) and GTP (1 mM) were included in the pipette solution in order to support kinase activity and G-protein signaling.

The data in Figure 3.3 illustrate that there is a developmental change in the modulation of the NMDA response by the D1 agonist. While SKF-82958 produced a reduction in the NMDA current at P7, at P21 there is no significant change and at P28 SKF-82958 enhances the NMDA response.

In further experiments in neonatal striatum (Tong and Gibb, 2008), D1 inhibition of the NMDA current was shown to be not sensitive to inhibitors of PKA and G-proteins, while the tyrosine kinase inhibitors lavendustin A and PP2 both blocked D1 inhibition. Intracellular application of a dynamin inhibitory peptide showed that D1 inhibition was dependent on dynamin, suggesting that the mechanism of inhibition involved internalization of NMDA receptors. In addition, in the presence of dynamin inhibitory peptide, NMDA responses were significantly increased compared to all other control NMDA responses. Control currents recorded with our standard pipette solution averaged 251 ± 27 pA ($n = 32$), whereas in the presence of intracellular dynamin inhibitory peptide, control currents averaged 497 ± 62 pA ($n = 18$). D1 inhibition of NMDA responses was abolished by intracellular dynamin inhibitory peptide; the average percentage inhibition decreased from $34.3\% \pm 12.3\%$ to $-1.1\% \pm 8.2\%$ ($n = 9$ cells). These data illustrate that in striatal neurones there is a constant turnover of NMDA receptors between intracellular and cell surface compartments. Modulation of receptor trafficking processes is one possible mechanism for dopaminergic modulation of ion channel receptor currents.

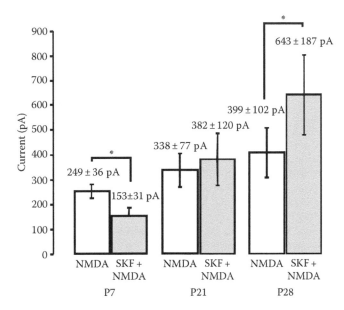

FIGURE 3.3 D1 receptor modulation of whole-cell NMDA receptor currents is developmentally regulated in striatal neurons from 7 day old (P7), 21 day old (P21), and 28 day old (P28) rats. While a significant inhibition of the NMDA current is produced in P7 neurones, no significant change is observed at P21 (*n* = 9) and a significant potentiation of the NMDA response is seen at P28 (*n* = 6). (Reproduced from Tong, H. and Gibb, A.J., *J. Physiol.*, 586, 4693, 2008.)

3.5   CONCLUSIONS

Studies investigating dopamine receptor modulation, both whole-cell and synaptic NMDA currents, suggest that there are multiple mechanisms at play. In developing striatal neurones, the balance of these mechanisms is for D1 receptor activation to cause a decrease in NMDA current, while in more mature cells D1 activation causes potentiation (Figure 3.3, Table 3.1). Whole-cell currents reflect both synaptic and extrasynaptic receptor activity. Moreover, distinct subunit compositions of NMDA receptors have been found in synapses and extrasynaptic areas in, for example, hippocampal neurones where GLUN2B receptors predominate at extrasynaptic sites, while GluN2A receptors predominate at mature synapses (e.g., Tovar and Westbrook, 1999; Groc et al., 2006). During synaptic development, the subunit composition of synaptic NMDARs changes from heterodimers containing predominantly GLUN2B subunits at early stages to heterodimers containing GluN1/GluN2B, GluN1/GluN2A, and GluN1/GluN2A/GluN2B subunits at mature stages. In young animals, the balance of the evidence suggests that β-arrestin signaling dominates to cause an Src kinase–dependent internalization of GluN2B-type NMDA receptors (in neonatal striatum there is no evidence for GluN2A-containing receptors), while in older animals D1 receptor activation may generate stabilization of GluN2A-type receptors at the cell surface and at synapses. It follows from these considerations that in mature animals there might be differential regulation of synaptic and extrasynaptic receptors by dopamine signaling. The significance of these changes for the development of dopamine-dependent behaviors in the young animal remains to be determined (Beutler et al., 2011; Surmeier et al., 2010).

REFERENCES

Albin RL, Young AB, Penney JB. (1989). The functional anatomy of basal ganglia disorders. *Trends Neurosci* **12**:366–375.

Alexander GE, Crutcher ME. (1990). Functional architecture of basal ganglia circuits: Neural substrates of parallel processing. *Trends Neurosci* **13**:266–271.

Beaulieu JM, Gainetdinov RR. (2011). The physiology, signaling, and pharmacology of dopamine receptors. *Pharmacol Rev* **63(1)**:182–217.

Bellone C, Nicoll RA. (2007). Rapid bidirectional switching of synaptic NMDA receptors. *Neuron* **55(5)**:779–785.

Beutler LR, Wanat MJ, Quintana A, Sanz E, Bamford NS, Zweifel LS, Palmiter RD. (2011). Balanced NMDA receptor activity in dopamine D1 receptor (D1R)- and D2R-expressing medium spiny neurons is required for amphetamine sensitization. *PNAS* **108**:4206–4211.

Bjorklund A, Dunnett SB. (2007). Fifty years of dopamine research. *Trends Neurosci* **30**:185–187.

Blank T, Nijholt I, Teichert U, Kugler H, Behrsing H, Fienberg A, Greengard P, Spiess J. (1997). The phosphoprotein DARPP-32 mediates cAMP-dependent potentiation of striatal N-methyl-D-aspartate responses. *Proc Natl Acad Sci* **94**:14859–14864.

Bolam JP, Hanley JJ, Booth PA, Bevan MD. (2000). Synaptic organisation of the basal ganglia. *J Anat* **196**:527–542.

Calabresi P, Pisani A, Mercuri NB, Bernardi G. (1992). Long-term potentiation in the striatum is unmasked by removing the voltage-dependent magnesium block of NMDA receptor channels. *Eur J Neurosci* **4**:929–935.

Cepeda C, Colwell CS, Itri JN, Chandler SH, Levine MS. (1998). Dopaminergic modulation of NMDA-induced whole cell currents in neostriatal neurons in slices: Contribution of calcium conductances. *J Neurophysiol* **79**:82–94.

Charpier S, Deniau JM. (1997). In vivo activity-dependent plasticity at cortico-striatal connections: Evidence for physiological long-term potentiation. *Proc Natl Acad Sci* **94**:7036–7040.

Chen G, Greengard P, Yan Z. (2004). Potentiation of NMDA receptor currents by dopamine D1 receptors in prefrontal cortex. *Proc Natl Acad Sci* **101**:2596–2600.

Chen J, Rusnak M, Lombroso PJ, Sidhu A. (2009). Dopamine promotes striatal neuronal apoptotic death via ERK signaling cascades. *Eur J Neurosci* **29**:287–306.

Chung HJ, Huang YH, Lau LF, Huganir RL. (2004). Regulation of the NMDA receptor complex and trafficking by activity-dependent phosphorylation of the GLUN2B subunit PDZ ligand. *J Neurosci* **24**:10248–10259.

Dunah AW, Sirianni AC, Fienberg AA, Bastia E, Schwarzschild MA, Standaert DG. (2004). Dopamine D1-dependent trafficking of striatal N-methyl-D-aspartate glutamate receptors requires Fyn protein tyrosine kinase but not DARPP-32. *Mol Pharmacol* **65**:121–129.

Fan MM, Raymond, LA. (2007). N-Methyl-d-aspartate (NMDA) receptor function and excitotoxicity in Huntington's disease. *Prog Neurobiol* **81**:272–293.

Flores-Hernandez J, Cepeda C, Hernandez-Echeagaray E, Calvert CR, Jokel ES, Fienberg AA, Greengard P, Levine MS. (2002). Dopamine enhancement of NMDA currents in dissociated medium-sized striatal neurons: Role of D1 receptors and DARPP-32. *J Neurophysiol* **88**:3010–3020.

Gerfen CR, Engber TM, Mahan LC, Susel Z, Chase TN, Monsma FJ Jr., Sibley DR. (1990). D1 and D2 dopamine receptor-regulated gene expression of striatonigral and striatopallidal neurons. *Science* **250**:1429–1432.

Gerfen CR, Surmeier DJ. (2010). Modulation of striatal projection systems by dopamine. *Annu Rev Neurosci* [Epub ahead of print].

Greengard P. (2001). The neurobiology of slow synaptic transmission. *Science* **294**:1024–1030.

Groc L, Heine M, Cousins SL, Stephenson FA, Lounis B, Cognet L, Choquet D. (2006). NMDA receptor surface mobility depends on NR2A-2B subunits. *Proc Natl Acad Sci* **103(49)**:18769–18774.

Grosshans DR, Clayton DA, Coultrap SJ, Browning MD. (2002). LTP leads to rapid surface expression of NMDA but not AMPA receptors in adult rat CA1. *Nat Neurosci* **5**:27–33.

Hallett PJ, Spoelgen R, Hyman BT, Standaert DG, Dunah AW. (2006). Dopamine D1 activation potentiates striatal NMDA receptors by tyrosine phosphorylation-dependent subunit trafficking. *J Neurosci* **26**:4690–4700.

Hornykiewicz, O. (2001). Parkinson's disease. eLS (http://www.els.net) (accessed on 2011).

Kornau HC, Schenker LT, Kennedy MB, Seeburg PH. (1995). Domain interaction between NMDA receptor subunits and the postsynaptic density protein PSD-95. *Science* **269**:1737–1740.

Kotecha SA, Oak JN, Jackson MF, Perez Y, Orser BA, Van Tol HH, MacDonald JF. (2002). A D2 class dopamine receptor transactivates a receptor tyrosine kinase to inhibit NMDA receptor transmission. *Neuron* **35**:1111–1122.

Kreitzer AC, Malenka RC. (2008). Striatal plasticity and basal ganglia circuit function. *Neuron* **60**:543–554.

Krupp JJ, Vissel B, Thomas CG, Heinemann SF, Westbrook GL. (2002). Calcineurin acts via the C-terminus of GLUN2A to modulate desensitization of NMDA receptors. *Neuropharmacology* **42**:593–602.

Lavezzari G, McCallum J, Dewey CM, Roche KW. (2004). Subunit-specific regulation of NMDA receptor endocytosis. *J Neurosci* **24(28)**:6383–6391.

Law-Tho D, Hirsch JC, Crepel F. (1994). Dopamine modulation of synaptic transmission in rat prefrontal cortex: An in vitro electrophysiological study. *Neurosci Res* **21**:151–160.

Lee FJ, Xue S, Pei L, Vukusic B, Chery N, Wang Y, Wang YT, Niznik HB, Yu XM, Liu F. (2002). Dual regulation of NMDA receptor functions by direct protein-protein interactions with the dopamine D1 receptor. *Cell* **111**:219–302.

Lei G, Xue S, Chery N, Liu Q, Xu J, Kwan CL, Fu YP, Lu YM, Liu M, Harder KW, Yu XM. (2002). Gain control of N-methyl-D-aspartate receptor activity by receptor-like protein tyrosine phosphatase alpha. *EMBO J* **21**:2977–2989.

Levine MS, Altemus KL, Cepeda C, Cromwell HC, Crawford C, Ariano MA, Drago J, Sibley DR, Westphal H. (1996). Modulatory actions of dopamine on NMDA receptor-mediated responses are reduced in D1A-deficient mutant mice. *J Neurosci* **16**:5870–5882.

Li YC, Xi D, Roman J, Huang YQ, Gao WJ. (2009). Activation of glycogen synthase kinase-3 beta is required for hyperdopamine and D2 receptor-mediated inhibition of synaptic NMDA receptor function in the rat prefrontal cortex. *J Neurosci* **29**:15551–15563.

Lieberman DN, Mody I. (1994). Regulation of NMDA channel function by endogenous Ca^{2+}-dependent phosphatase. *Nature* **369**:235–239.

Lin JY, Dubey R, Funk GD, Lipski J. (2003). Receptor subtype-specific modulation by dopamine of glutamatergic responses in striatal medium spiny neurons. *Brain Res* **959**:251–625.

Lin Y, Skeberdis VA, Francesconi A, Bennett MV, Zukin RS. (2004). Postsynaptic density protein-95 regulates NMDA channel gating and surface expression. *J Neurosci* **24**:10138–10148.

Lu WY, Xiong ZG, Lei S, Orser BA, Dudek E, Browning MD, MacDonald JF. (1999). G-protein-coupled receptors act via protein kinase C and Src to regulate NMDA receptors. *Nat Neurosci* **2**:331–238.

Luscher C, Xia H, Beattie EC, Carroll RC, von Zastrow M, Malenka RC, Nicoll RA. (1999). Role of AMPA receptor cycling in synaptic transmission and plasticity. *Neuron* **24**:649–658.

Mannoury la Cour C, Salles MJ, Pasteau V, Millan MJ. (2011). Signaling pathways leading to phosphorylation of Akt and GSK-3β by activation of cloned human and rat cerebral D_2 and D_3 receptors. *Mol Pharmacol* **79**:91–105.

Marsden CA. (2006). Dopamine: The rewarding years. *Br J Pharmacol* **147**:S136–S144.

Moon IS, Apperson ML, Kennedy MB. (1994). The major tyrosine-phosphorylated protein in the postsynaptic density fraction is N-methyl-D-aspartate receptor subunit 2B. *Proc Natl Acad Sci* **91**:3954–3958.

Morishita W, Connor JH, Xia H, Quinlan EM, Shenolikar S, Malenka RC. (2001). Regulation of synaptic strength by protein phosphatase 1. *Neuron* **32**:1133–1148.

Nakazawa T, Komai S, Tezuka T, Hisatsune C, Umemori H, Semba K, Mishina M, Manabe T, Yamamoto T. (2001). Characterization of Fyn-mediated tyrosine phosphorylation sites on GluR epsilon 2 (GLUN2B) subunit of the N-methyl-D-aspartate receptor. *J Biol Chem* **276**:693–699.

Niethammer M, Kim E, Sheng M. (1996). Interaction between the C terminus of NMDA receptor subunits and multiple members of the PSD-95 family of membrane-associated guanylate kinases. *J Neurosci* **16**:2157–2163.

Partridge JG, Tang KC, Lovinger DM. (2000). Regional and postnatal heterogeneity of activity-dependent long-term changes in synaptic efficacy in the dorsal striatum. *J Neurophysiol* **84**:1422–1429.

Pierce KL, Lefkowitz RJ. (2001). Classical and new roles of β-arrestins in the regulation of G-protein coupled receptors. *Nat Rev Neuro* **2**:727–733.

Roche KW, Standley S, McCallum J, Dune Ly C, Ehlers MD, Wenthold RJ. (2001). Molecular determinants of NMDA receptor internalization. *Nat Neurosci* **4**:794–802.

Rycroft BK, Gibb AJ. (2004). Inhibitory interactions of calcineurin (phosphatase 2B) and calmodulin on rat hippocampal NMDA receptors. *Neuropharmacology* **47**:505–514.

Salter MW, Kalia LV. (2004). Src kinases: A hub for NMDA receptor regulation. *Nat Rev Neurosci* **5(4)**:317–328.

Seamans JK, Durstewitz D, Christie BR, Stevens CF, Sejnowski TJ. (2001). Dopamine D1/D5 receptor modulation of excitatory synaptic inputs to layer V prefrontal cortex neurons. *Proc Natl Acad Sci* **98**:301–306.

Shuen JA, Chen M, Gloss B, Calakos N. (2008). Drd1a-tdTomato BAC transgenic mice for simultaneous visualization of medium spiny neurons in the direct and indirect pathways of the basal ganglia. *J Neurosci* **28**:2681–2685.

Snyder GL, Fienberg AA, Huganir RL, Greengard P. (1998). A dopamine/D1 receptor/protein kinase A/dopamine- and cAMP-regulated phosphoprotein (Mr 32 kDa)/protein phosphatase-1 pathway regulates dephosphorylation of the NMDA receptor. *J Neurosci* **18**:10297–10303.

Surmeier DJ, Shen W, Day M, Gertler T, Chan S, Tian X, Plotkin JL. (2010). The role of dopamine in modulating the structure and function of striatal circuits. *Prog Brain Res* **183**:149–167.

Tong H, Gibb AJ. (2008). Dopamine D1 receptor inhibition of NMDA receptor currents mediated by tyrosine kinase-dependent receptor trafficking in neonatal rat striatum. *J Physiol* **586**:4693–4707.

Tovar KR, Westbrook GL. (1999). The incorporation of NMDA receptors with a distinct subunit composition at nascent hippocampal synapses in vitro. *J Neurosci* **19**:4180–4188.

Urs NM, Daigle TL, Caron MG. (2011). A dopamine D1 receptor-dependent β-arrestin signaling complex potentially regulates morphine-induced psychomotor activation but not reward in mice. *Neuropsychopharmacology* **36(3)**:551–558.

Violin JD, Lefkowitz RJ. (2007). β-arrestin biased ligands at seven transmembrane receptors. *TIPS* **28**:416–422.

Wang X, Zhong P, Gu Z, Yan Z. (2003). Regulation of NMDA receptors by dopamine D4 signaling in prefrontal cortex. *J Neurosci* **23**:9852–9861.

Wenthold RJ, Prybylowski K, Standley S, Sans N, Petralia RS. (2003). Trafficking of NMDA receptors. *Annu Rev Pharmacol Toxicol* **43**:335–358.

Xiong ZG, Pelkey KA, Lu WY, Lu YM, Roder JC, MacDonald JF, Salter MW. (1999). Src potentiation of NMDA receptors in hippocampal and spinal neurons is not mediated by reducing zinc inhibition. *J Neurosci* **19**:RC37.

4 Synaptic Triad in the Neostriatum
Dopamine, Glutamate, and the MSN

Jason R. Klug, Ariel Y. Deutch,
Roger J. Colbran, and Danny G. Winder

CONTENTS

4.1 INTRODUCTION

The basal ganglia are a collection of subcortical nuclei that subserve a variety of behaviors, including motor planning, procedural learning, and motivation state. Although there is no universal agreement on which nuclei comprise the basal ganglia (early anatomists used basal ganglia to refer to all large subcortical masses, including the thalamus), the term today generally refers to the neostriatum (caudate nucleus and putamen) and its projection targets.

As knowledge has accrued concerning the anatomical organization of the striatum, including its characteristic cell types and efferent projections, the term neostriatum has reduced to the simpler term striatum. This striatum encompasses a dorsal portion (with its pallidal and nigral targets) as well as the ventral striatum, including the nucleus accumbens and its targets in the ventral pallidum. While anatomical studies have revealed an overarching theme for the dorsal and ventral striatum, subdivisions within each of these structures have become apparent, with corresponding functional specialization. For example, within the dorsal striatum, the dorsomedial aspect appears to be associated with goal-directed learning in rodents, while the dorsolateral striatum is associated with habit learning (Yin and Knowlton, 2006). The ventral striatum includes the multicompartment nucleus accumbens (shell, core, and septal pole) and the olfactory tubercle and plays a critical role in translating motivational states into goal-directed behavior (Robbins et al., 2008).

Given the diverse behaviours in which striatal function appears critical, it is not surprising that there are a number of disorders that are associated with basal ganglia pathology. Among these are Parkinson's and Huntington's diseases, dystonia, and Tourette's syndrome. Several psychiatric conditions, including obsessive-compulsive disorder, substance abuse, and schizophrenia, also appear in part to be reflecting changes in basal ganglia function.

4.2   BASAL GANGLIA CIRCUITRY

As noted earlier, a formal definition of the basal ganglia is lacking. Our current informal use of the term has evolved over the past century as advances in our understanding of the anatomy of the brain have grown. Today most consider the basal ganglia to be a collection of subcortical structures with interconnected neurons. Such a loose definition encompasses the striatum, containing both interneurons and medium spiny neurons (MSNs), the latter type of cell projecting to several downstream sites. These targets of the rodent dorsal striatum include the globus pallidus (GP), subthalamic nucleus, and substantia nigra (SN). From these targets, projections are directed to thalamic relay nuclei, which in turn innervate various neocortical regions. Finally, the cortex projects in a topographically defined manner back onto the striatum. Differences in the topographical organization of these projection systems are thought to relate to different functional attributes of the striatum.

The afferents that innervate the striatum are not typically included in the definition of basal ganglia. This is a somewhat surprising omission, particularly because all parts of the cortex innervate the striatum, thus rendering the striatum a funnel through which cortical information is relayed back to the cortex to hone subsequent motor and cognitive acts. Another major input to the dorsal striatum of the rodent originates in the thalamic central medial complex (including the central medial, centrolateral, and paracentral nuclei) and the parafascicular nucleus. These thalamostriatal neurons, like their corticostriatal counterparts, are glutamatergic (Smith et al., 2004). The third and final major input to the striatum originates in the ventral midbrain dopamine cell groups of the SN, retrorubral field, and ventral tegmental area. There are a restricted number of other "minor" afferents, including those from the GP, ventral pallidum and contiguous basal forebrain, as well as the serotonin neurons of the dorsal and median raphe in the upper brainstem. The ventral striatum, including the nucleus accumbens, which abuts both the septum and the striatum, receives inputs from sites adjacent to those that innervate the dorsal striatum, thus defining a parallel set of ventral basal ganglia structures. Although one might guess that the connections of the various parts of the basal ganglia are fully known, it is becoming clear that even after the advent of the modern era of neuroanatomical tract-tracing methods, there are gaps in our knowledge.

The major type of striatal neuron is the MSN. Approximately 95% of striatal neurons are MSNs, with a cell body of medium size (~9–17 μm in diameter) and radially extending dendrites that are densely studded with dendritic spines. These GABAergic MSNs are the projection neurons of the striatum (Tepper et al., 2007).

Different portions of the relatively long dendrites of MSNs, with their high density of dendritic spines, are targeted by various inputs to the neuron. Recurrent collaterals of one MSN often terminate on the soma or proximal-most dendrite of another MSN. Striatal interneurons similarly synapse onto the soma or most proximal dendritic segment of MSNs. In contrast, cortical afferents synapse onto spines located more distally on the dendrite, as do the terminals of dopamine neurons in the SN. A single cortical terminal generally synapses with a single dendritic spine. Thus, even modest changes in the numbers of MSN dendritic spines, particularly those on distal portions of the dendrite, can have major effects on the ability of afferent volleys to be propagated to the cell body and thence to projection targets of MSNs.

MSNs, despite their morphological similarities, are not a homogeneous population of cells. Indeed, subtle variations in the dendritic tree of MSNs have led one group to characterize five different types of MSNs on structural grounds, although the great majority of MSNs fall into one class (Chang et al., 1982). MSNs can also be differentiated on the basis of their projection targets, the receptors they express, and the co-transmitters in addition to GABA that they possess; these three factors co-segregate to yield two major populations of MSNs. In rodents, one type of MSN sends a dense projection to the GP (equivalent to external GP in primates), with the other type densely innervating the SN (equivalent to internal GP in primates). Retrograde tract tracing studies from these sites have consistently revealed two nonoverlapping populations of MSNs. The MSNs that innervate the SN form the so-called direct pathway, while those that project to the GP, neurons of which in turn send their projections to the SN, form the indirect pathway. These two populations of MSNs differ on the basis of the dopamine receptor they express, with the direct pathway cells expressing the D1 receptor and the striatopallidal indirect pathway cells expressing the D2 receptor. Expression of other dopamine receptors, including the D3 receptor of the D2 class of dopamine receptors follows the D1/D2 segregation to different MSNs under basal conditions but not in pathophysiological states. Finally, MSNs can be differentiated on the basis of their co-transmitter. Direct pathway D1-expressing MSNs uses tachykinin peptides such as substance P as a co-transmitter, while D2-expressing striatopallidal MSNs contain enkephalin.

The direct and indirect pathway MSNs have figured prominently in models of basal ganglia function, with these two striatofugal pathways contributing to opposing effects on thalamic output to the cortex. Activation of the direct pathway disinhibits the thalamus, increasing output to the cortex and promoting intended movement, while activation of the indirect pathway further inhibits thalamocortical neurons preventing unintended movements. These ideas have figured prominently in our approach to diseases of the basal ganglia. For example, it is hypothesized that there is an imbalance of these two pathways in Parkinson's disease, tipping the balance in favor of the indirect pathway and therefore inhibiting voluntary movement. Recently, *in vivo* activation of direct or indirect pathways via an optogenetic approach utilizing Cre-dependent viral expression of channelrhodopsin-2 in BAC transgenic D1- or D2-driven Cre lines lends evidence in support of this hypothesis. Activation of direct pathway neurons reduces freezing and promotes locomotor activity, while activation of indirect pathway neurons increases freezing behavior and decreases locomotor initiations (Kravitz et al., 2010).

While this model has been invaluable in guiding research into the pathophysiology of parkinsonism, over the years it has become clear that various aspects of the model are incorrect and that the model often lacks predictive validity (DeLong and Wichmann, 2009). Moreover, the structural basis for the differences in MSNs now appears to reflect a limitation of the retrograde-tract tracing methods originally used to label the two types of cells. Thus, more recent single cell labeling studies have revealed that essentially all MSNs project to both the GP and SN, but with markedly different axonal arbors (Wu et al., 2000). The direct pathway MSNs have a large axonal plexus in the SN, with a very small axon terminal array in the GP—so small that it is insufficient to accumulate significant amounts of the retrograde tracer. Conversely, the axons of indirect pathway MSNs collateralize extensively in the GP but have a very small axonal arbor in the SN. While strictly speaking these data belie the concept of the direct and indirect pathways, functional data indicate that striatal efferent projections do indeed segregate into two classes, with one type of MSN primarily modulating the SN and the other the GP (Squire, 2003). This has best been shown by the ability of D1 and D2 agonists and antagonists to differentially modulate preprotachykinin and preproenkephalin mRNAs in a regionally specific manner.

A very different type of segregation of MSNs from the direct and indirect pathway distinction can be seen when examining the intrastriatal spatial localization of MSNs. In contrast to MSNs that contribute to the direct and indirect pathways, which are intermingled throughout the striatum, distinct clusters of MSNs that express the mu opioid receptor can be seen as patches throughout the striatum, superimposed over a larger diffuse matrix of MSNs that do not express the mu opioid receptor and appear at different times in development (Johnston et al., 1990). These mu opioid receptor-expressing clusters of cells have been called striosomes or patches and occupy about 15% of the volume of the striatum, with the much larger diffuse compartment named the matrix. MSNs that contribute to the direct and indirect pathways are found in both compartments. Although these two compartments were at one time thought to receive inputs from spatially segregated cortical cells, more recent data have suggested that one cannot define these compartments based strictly on the origin of their cortical innervations. Nonetheless, important functional differences have emerged, with relative activity in MSNs across the two compartments suggested to be disturbed in several basal ganglia disorders, including those involving disturbances of habit, such as obsessive-compulsive disorder and Tourette's syndrome (Graybiel, 2008).

4.3   DENDRITIC SPINE: GATEWAY TO THE MEDIUM SPINY NEURON

Dendritic spines on striatal MSNs receive excitatory drive from the cortex and thalamus, the former relaying the consequences of higher order processing and the latter of ascending reticular drive to the MSN. The cortical influence over MSNs has been intensively studied over the past generation. The heads of MSN dendritic spines receive a single excitatory input from a cortical neuron, with the neck of the spine typically often being the site at which dopamine axons from the SN synapse, forming a synaptic triad. At the individual spine level, this close spatial arrangement allows dopamine to modulate incoming excitatory glutamatergic drive (Kemp and Powell, 1971; Freund et al., 1984; Smith and Kieval, 2000). While recent reviews have suggested that the frequency of dopaminergic synapses onto MSN spines is lower than originally proposed, even in those cases where dopamine axons do not synapse onto the spine one finds that dopamine terminals are located within $1.0\,\mu m$ of a spine, allowing volume (paracrine) transmission and emphasizing temporal as well as spatial regulation of MSNs (Arbuthnott and Wickens, 2007).

In addition to dopaminergic modulation of cortical drive onto MSNs at the level of the MSN spine, D2 heteroreceptors on glutamatergic corticostriatal terminals can also tonically inhibit glutamate release from these terminals (Bamford et al., 2004). The cortical influence over striatal cells is critical because MSNs have spontaneous fluctuations in their membrane potential, ranging from a relatively hyperpolarized "downstate" around $-80\,mV$ to a more depolarized "upstate" around $-50\,mV$. While the downstate is maintained by a rapidly activating inwardly rectifying potassium current, transition to the upstate appears to be determined largely by strong, correlated release of glutamate from corticostriatal glutamatergic terminals (Wilson and Kawaguchi, 1996; Plenz and Kitai, 1998). These differing MSN membrane potential states dictate if an incoming excitatory volley, simultaneous activation of many cortical afferents at various points on the MSN dendritic tree, will depolarize MSNs from a hyperpolarized potential to action potential firing (Wilson and Kawaguchi, 1996; Carter and Sabatini, 2004; Carter et al., 2007).

4.4   ROLE OF DOPAMINE IN BASAL GANGLIA FUNCTION

A unique feature of the striatum is the extraordinarily dense dopaminergic innervation that it receives. A recent remarkable study employing a viral vector to target green fluorescent protein to the membranes of neurons has found that single dopamine neurons in the SN give rise to remarkably long intrastriatal axons (up to $780,000\,\mu m$, i.e., 780 cm in length), and that the dense portion of the striatal axonal plexus derived from one nigral dopamine neuron can cover up to 5.7% of the total striatal volume (Matsuda et al., 2009). As such, even one dopamine neuron can influence a very large number of striatal MSNs, and it is therefore not surprising that striatal dopamine plays a critical role in modulating motor behavior and learning.

While the timing and magnitude of dopamine release is important for normal voluntary movement, the firing patterns of dopamine neurons do not correlate with voluntary movements. Instead dopamine neurons of the ventral tegmental area and SN play a central role in positive reinforcement learning (Mirenowicz and Schultz, 1994, 1996). Dopamine neurons show phasic or burst-like firing patterns following unexpected reward. However, after repeated conditioning sessions with an environmental cue (e.g., a light or tone) preceding the reward, phasic firing of dopamine neurons can be induced by the cue alone, as a predictor of a reward (Schultz, 1998), whereas actual delivery of reward without the cue now does not lead to an increase firing. In contrast, aversive events inhibit dopamine cell firing. These findings have led to the hypothesis that dopamine acts as a natural reward prediction error signal, acting to gauge between expected and actual reward (Schultz, 1997). Natural rewards, as well as many drugs of abuse (Wise, 2004), elevate dopamine concentrations in the nucleus accumbens (Ahn and Phillips, 2007) and striatum (Nakazato, 2005).

Canonically, D1 and D5 receptors normally enhance cAMP levels while D2, D3, and D4 receptors inhibit the production of cAMP. Phasic bursts of dopamine are hypothesized to activate low-affinity D1 receptors, while tonic release of dopamine would favor high-affinity D2 receptor binding (Lovinger et al., 2003). In addition to D1- and D2-receptor-expressing cells in the dorsal striatum, expression of D3, D4, and D5 receptors is present at varying extents. While the expression levels of these dopamine receptors are lower than D1 and D2 receptors, they most likely have important functional roles. While there is little expression of D4 and D5 receptor in MSNs (Bergson et al., 1995), the D3 receptor is expressed at significant levels in a subpopulation of D1 receptor–expressing cells. In fact, many drugs that are intended to be specific for D2 receptors actually bind to D3 receptors as well. Interestingly, following dopamine depletion and subsequent levodopa administration D3 receptors are upregulated in the dorsal striatum (Bordet et al., 1997). Additionally in dyskinetic monkeys, or monkeys exhibiting abnormal involuntary movements following repeated dopamine replacement therapies, D3 receptor binding is higher than in non-dyskinetic or control monkeys following dopamine depletion and levodopa administration. Also D3 receptor–binding levels in the striatum correlated with the severity of levodopa-induced dyskinesias (Guigoni et al., 2005). Brain-derived neurotrophic factor (BDNF) is important in D3 receptor expression and the maintenance of its expression. Moreover, the administration of a D3 receptor partial agonist strongly attenuates levodopa-induced dyskinesias (Guillin et al., 2003). These data exemplify the importance of examining other less characterized dopamine receptors in the striatum and their possible connection to basal ganglia–related disorders.

4.5   SYNAPTIC TRIAD: A KEY SITE FOR MODIFICATION OF CIRCUIT BEHAVIOR

Glutamatergic synapses on MSN dendritic spines are a major site of long-lasting neuroadaptations and likely a key site underlying the neural correlates of motor learning. Dopamine also plays an important role in regulating plasticity in the striatum. In addition to dopamine, the striatum receives highly topographic glutamatergic thalamostriatal projections from intralaminar nuclei, such as the centromedian and parafascicular nuclei, and non-intralaminar nuclei (Smith et al., 2004). Thalamic inputs, unlike cortical inputs, target mainly dendritic shafts in both rats and monkeys

(Smith and Bolam, 1990; Sadikot et al., 1992; Smith et al., 1994; Sidibe and Smith, 1996). In monkeys, centromedian projections preferentially synapse onto "direct" pathway neurons versus "indirect" pathway neurons (Sidibe and Smith, 1996). In addition to innervating distinct targets, there is evidence that synaptic transmission is not similar between these inputs. For example, the probability of release at cortical and thalamic glutamatergic synapses on MSNs appears to be different (Smeal et al., 2007; Ding et al., 2008). Additionally, a recent study suggests that thalamic stimulation leads to burst firing of cholinergic interneurons which is followed by a pause in the striatum similar to what is seen *in vivo* following presentation of salient stimuli (Ding et al., 2010). The initial burst release of acetylcholine leads to the suppression of MSN firing via M2R presynaptic inhibition of glutamate release followed by a slower M1R-mediated postsynaptic facilitation of responses in D2R MSNs. This pause may account for the transient inhibition in behavior following salient stimulation presentation. Further studies utilizing techniques to better separate these distinct inputs are necessary to characterize their unique properties.

While electron microscopy analysis of D2R immunohistochemical localization on glutamatergic synaptic terminals is controversial, revealing exceedingly rare or low levels of D2R (Fisher et al., 1994; Sesack et al., 1994; Hersch et al., 1995; Wang and Pickel, 2002), an elegant study by Sulzer's group showed functional evidence of dopamine acting presynaptically to reduce glutamate release (Bamford et al., 2004a,b). Utilizing a styryl dye (FM1–43) to load glutamatergic synaptic vesicles they were able to monitor release via destaining coupled with electrochemical recordings to directly measure the effects of dopamine at the level of individual presynaptic terminals. Interestingly, another study showed that the inhibition of glutamate release by D2R stimulation was frequency dependent, resulting in inhibition of glutamate release at high-frequency stimulation (HFS) (20 Hz), but not at lower frequencies (1 Hz) (Yin and Lovinger, 2006). This inhibition was dependent on CB1R, mGluRs, and rises in internal calcium levels. These data suggest a postsynaptic mechanism for D2R-mediated control of glutamatergic synaptic transmission, but they leave open the question of exactly where the D2Rs are localized that mediate this inhibition. A recent study combining optogenetics, two-photon microscopy, and glutamate uncaging suggests that D2R signaling reduces corticostriatal glutamate release and additionally inhibits calcium influx through primarily NMDAR and R-type VGCCs (Higley and Sabatini, 2010). The D2R-mediated reduction in synaptic calcium influx is PKA dependent while D2R actions on VGCC-mediated transients are PKA and A2AR independent. These results suggest that D2R activation engages at least two divergent signaling cascades.

4.5.1 Glutamate Synapse

Glutamate synapses in the striatum contain a presynaptic terminal that releases glutamate and a postsynaptic dendritic spine filled with receptors that bind glutamate. These receptors come in two varieties: ionotropic and metabotropic glutamate receptors (Chapters 1 and 2). Two major ionotropic glutamate receptors are the AMPA and NMDA receptor subtypes, which are defined by their specific pharmacological sensitivity to the respective agonists and are tethered to the postsynaptic density (PSD) by associated anchoring and scaffolding proteins. AMPA receptors (AMPARs), which mediate the majority of fast excitatory transmission in the brain, are tetramers that are made up of some combination of GluA1–GluA4 (GluR1–GluR4) AMPAR subunits. Most AMPARs respond to synaptic glutamate release by gating a mon-ovalent cation current that helps depolarize the postsynaptic terminal. Tetrameric NMDA receptors (NMDARs) are thought to contain two obligatory GluN1 (NR1) subunits and some combination of GluN2A-2D (NR2A-2D) subunit and GluN3A or 3B (NR3A or 3B). Binding of glutamate and glycine to GluN2 and GluN1 subunits, respectively, gates a cation channel that additionally (a) fluxes calcium and (b) is blocked at hyperpolarized membrane potentials by a magnesium ion. The magne-sium block of NMDARs is relieved by a strong depolarization greater than what is achieved via upstate transitions. Thus, NMDARs act as coincidence detectors, allowing calcium influx following coincident presynaptic release of neurotransmit-ters coupled with AMPA-mediated postsynaptic depolarization. The consequential rise in intracellular calcium is a critical trigger for long-term changes in the efficacy of transmission at these synapses.

4.5.2 NMDAR-DEPENDENT SYNAPTIC PLASTICITY: LESSONS FROM THE HIPPOCAMPUS

Glutamatergic synapses throughout the brain have a unique ability to undergo long-lasting changes in synaptic efficacy in response to very transient signals. The pio-neering work of Bliss and Lomo in the perforant-pathway dentate gyrus synapse using HFS to induce a long-lasting potentiation of excitatory transmission (Bliss and Lomo, 1973) sets the stage for much of the subsequent work published on neural plas-ticity, including in the striatum. Two classic forms of synaptic plasticity—long-term potentiation (LTP) and long-term depression (LTD)—which involve long-lasting changes in glutamatergic transmission, were characterized initially in the hippo-campus, and later described in the striatum, as discussed in Sections 4.6 and 4.7. Key principles of hippocampal synaptic plasticity are reviewed here, to highlight commonalities and differences in hippocampal and basal ganglia plasticity.

NMDA receptor-dependent LTP and LTD in the CA1 region of the hippocam-pus has been the most widely studied form of plasticity. These activity-dependent, long-lasting synaptic adaptations are hypothesized to play key roles in learning and memory. HFS or stimulation paired with postsynaptic depolarization leads to the induction of an NMDAR-dependent form of LTP in the CA1 region of the hippo-campus. Additionally, LTP has been shown to persist for at least hours *in vitro* and for months *in vivo* (Abraham et al., 2002). While late phases of LTP, like learn-ing and memory, are dependent on gene transcription and new protein synthesis (Madison et al., 1991), a transient (2–3 s) elevation of calcium via NMDARs seems to be sufficient for induction of this form of LTP (Lynch et al., 1983; Malenka et al., 1988; Malenka et al., 1992). One important mediator of NMDAR-dependent LTP in the hippocampus, no matter the induction mechanism, is the postsynaptic calcium-activated kinase, CaMKII (Lisman et al., 2002).

A major locus of potentiation is postsynaptic via an activity-dependent increase in function and/or number of AMPARs at the synapse (Malenka and Nicoll, 1999;

Malinow and Malenka, 2002; Song and Huganir, 2002; Bredt and Nicoll, 2003). Potentiation can proceed by the addition of new AMPARs at the synapse or by the recruitment of AMPARs to synapses that initially only contain NMDARs, or so-called "silent" synapses, which are thought to potentially be key sites for NMDAR-dependent LTP. The unsilencing of synapses in adulthood shares its NMDAR and CaMKII dependence, and data suggest that LTP is induced in part by the unsilencing of synapses (Liao et al., 1992; Manabe et al., 1992; Liao et al., 1995). Additionally, LTP at hippocampal synapses is also thought to involve enlargement of the post-synaptic dendritic spine (Lisman and Harris, 1993; Lisman and Zhabotinsky, 2001). Indeed, induction of LTP at single dendritic spines by local uncaging of glutamate results in actin-dependent structural enlargement (Fukazawa et al., 2003; Matsuzaki et al., 2004). Previous studies have shown that introduction of constitutively active CaMKII into hippocampal neurons is sufficient to induce spine growth (Jourdain et al., 2003).

In addition to LTP, NMDAR activation can also produce LTD at glutamate synapses in the hippocampus. Typically, this LTD is induced in the CA1 region of the hippocampus via prolonged low-frequency stimulation (LFS) (0.5–3 Hz) (Dudek and Bear, 1992; Mulkey and Malenka, 1992). This LTD is dependent on NMDARs, increases in postsynaptic calcium, and activation of serine-threonine protein phos-phatases (Mulkey and Malenka, 1992; Mulkey et al., 1993, 1994), which are thought to drive internalization of AMPARs (Carroll et al., 1999; Beattie et al., 2000). Dephosphorylation at GluR1 Ser845, a site that seems to be constitutively phosphor-ylated under basal conditions, accompanies LTD induction (Lee et al., 1998).

In addition to NMDAR-dependent LTD, an mGluR-dependent form of LTD (Bolshakov and Siegelbaum, 1994; Oliet et al., 1997) can be elicited in several brain regions by paired-pulse low-frequency stimulation (PP-LFS) or bath application of the group 1 selective agonist (R,S)-3,5-dihydroxyphenylglycine (DHPG) (Ito et al., 1982; Kano and Kato, 1987; Huber et al., 2000). Both forms of mGluR-LTD via PP-LFS or DHPG occlude each other suggesting similar expression mechanisms (Huber et al., 2001), but there are notable mechanistic differences from NMDAR-mediated LTD. mGluR-LTD was first characterized at parallel fiber–Purkinje cell synapses of the cerebellum and is primarily dependent on mGluR1 activation and the resulting activation of PKC. Within the cerebellum, PKC phosphorylates GluR2 at Ser880, leading to the clathrin-dependent removal of GluR2/3 containing AMPA receptors from the synapse (Wang and Linden, 2000; Chung et al., 2003).

Induction of hippocampal mGluR-LTD is independent of NMDARs and does not occlude the induction of NMDAR-dependent LTD, suggesting that it utilizes differ-ent mechanisms (Oliet et al., 1997). Hippocampal mGluR-LTD has been shown to be dependent on intracellular calcium concentration in certain experiments (Oliet et al., 1997) yet independent of intracellular calcium concentration in others (Fitzjohn et al., 1999). In the hippocampus, mGluR-LTD seems to depend on both mGluR5 and mGluR1 receptors: a combination of mGluR1 and mGluR5 antagonists is required to block its induction (Huber et al., 2001), but hippocampal mGluR-LTD is totally blocked in mGluR5-KO mice and only partially blocked in mGluR1KO mice (Volk et al., 2006). This suggests that activation of these two receptors produces synergis-tic responses to induce LTD. Group 1 mGluRs, like mGluR1 or mGluR5, couple to effectors via Gαq G proteins. The addition of GDPβS to the patch pipette or recordings in Gαq$^{-/-}$ mice inhibited the induction of mGluR-LTD. Activation of group 1 mGluRs leads to the activation of phospholipase C, increasing IP3 and DAG levels. Activation of these two signaling proteins leads to the liberation of calcium from internal stores, which can activate PKC. In experiments where a PKC inhibitor was added to the patch pipette, the induction of mGluR-LTD in both the hippocampus and cerebellum was blocked (Linden and Connor, 1991; Wang et al., 2007). Group I mGluR-LTD also involves both the mitogen-activated protein kinase/extracellular signal-regulated kinase (MAPK/ERK) pathway. Indeed ERK inhibitors block the induction of mGluR-LTD in the hippocampus (Huber et al., 2000, 2001; Gallagher et al., 2004; Volk et al., 2006; Ronesi and Huber, 2008) and cerebellum (Ahn et al., 1999; Kawasaki et al., 1999).

ERK signaling can regulate translation machinery, which suggests that mGluR-LTD may regulate translation of new proteins. In fact this seems to be the case. There is much evidence suggesting that mGluR-LTD requires new protein synthesis (Huber et al., 2001), yet some reports suggest it may be independent of new protein synthesis (Huber et al., 2000; Moult et al., 2008). The putative new protein synthesized and needed during mGluR-LTD is possibly the AMPA receptor GluA2 subunit, due to the fact that mGluR-LTD is blocked following pretreatment with siRNA and oligonucleotides blocking GluA2 translation (Mameli et al., 2007). Group I mGluR-LTD induction is also thought to be dependent on activation of postsynaptic protein tyrosine phosphatases, specifically striatal-enriched tyrosine phosphatase (STEP), which dephosphorylates the GluA2 subunit of AMPARs triggering lateral diffusion and subsequent endocytosis (Moult et al., 2002, 2006; Huang and Hsu, 2006; Gladding et al., 2009). However, conflicting results do not allow a conclusive statement about the primary locus for the induction of mGluR-LTD (presynaptic, postsynaptic, or both). Inconsistencies could be possibly explained by the use of different experimental bath temperatures and/or animal ages. Additional work will be needed to delineate complex mechanisms underlying the induction and expression of mGluR-LTD. NMDA receptor– and mGlu receptor–dependent forms of synaptic plasticity in the striatum, also involving complex cascades of second messengers, are discussed in Section 4.6.

4.5.3 CaMKII AND PLASTICITY

CaMKII is a calcium/calmodulin-activated kinase that is highly expressed throughout the brain and is enriched in the PSD of glutamatergic synapses (Lisman et al., 2002). The predominant neuronal CaMKIIα and CaMKIIβ isoforms are enriched in the forebrain and cerebellum, respectively (Yamagata et al., 2009). One or more CaMKII isoforms assemble to form a dodecameric holoenzyme. Many lines of evidence show that CaMKII responds to changes in intracellular calcium levels to promote biochemical signaling cascades that lead to potentiated synaptic transmission. While signaling via CaMKII is an important mechanism underlying synaptic potentiation in many regions of the brain, alternate signaling pathways like cAMP-PKA and MAPK signaling pathways may also be engaged depending on the induction parameters, but this extensive literature will not be reviewed here (Thomas and

Huganir, 2004). Again, insight into striatal mechanisms may be derived from the hippocampus, where CaMKII is activated during LTP induction and its activation is necessary and sufficient for LTP (Malinow et al., 1989; Tokumitsu et al., 1990; Otmakhov et al., 1997; Lisman et al., 2002; Colbran, 2004; Sanhueza et al., 2007). A recent study found that a knock-in mutation of CaMKIIα to inactivate the kinase disrupts LTP induction, the associated enlargement of spines, and spatial learning (Yamagata et al., 2009). Following robust calcium entry into the postsynaptic neuron, CaMKII is rapidly autophosphorylated at threonine 286, converting the enzyme from a calcium-dependent into a calcium-independent form (Miller and Kennedy, 1986; Fukunaga et al., 1993; Rich and Schulman, 1998; Lee et al., 2009). CaMKII's potential to remain active long after the postsynaptic calcium transient has subsided is why it is often thought of as a molecular switch capable of long-term memory storage. Indeed, increased threonine 286 phosphorylation of CaMKII can persist for at least 60 min (Barria et al., 1997b). However, one study reported that autonomous CaMKII activity returned to basal levels within minutes of LTP induction (Lengyel et al., 2004), suggesting that mechanisms other than threonine 286 phosphorylation control autonomous CaMKII activity. Although Thr286 autophosphorylated CaMKII may have roles independent of autonomous kinase activity during LTP maintenance, this study also highlights an uncertain role for CaMKII activity during LTP maintenance. Indeed, many laboratories have reported that intracellular perfusion of CaMKII inhibitor peptides after LTP induction does not disrupt maintenance (Malenka et al., 1989). In contrast, one recent study found that LTP maintenance can be disrupted by bath application of newly developed membrane-permeant CaMKII inhibitor following LTP induction (Sanhueza et al., 2007, 2011). It is possible that membrane-permeant peptides have better access to block CaMKII activity in the appropriate subcellular compartment (e.g., in the spines/PSD). Alternatively, the requirement for CaMKII activity may depend on the induction mechanism, or be evident only during certain phases of LTP maintenance. Interestingly, behavioral studies suggest that CaMKII activity in restricted time windows is required for memory consolidation (Wang et al., 2003, Wan et al., 2010).

Despite uncertainties about the role of CaMKII activity during LTP maintenance, a critical role of Thr286 phosphorylation was elegantly demonstrated by the creation of knock-in transgenic mice with a nonphosphorylated alanine residue replacing Thr286 in CaMKIIα. The resulting disruption of CaMKII phosphorylation at Thr286 leads to profound disruption of hippocampal LTP (Giese et al., 1998). Conversely, transgenic mice overexpressing CaMKIIα with a threonine 286 to aspartate mutation (CaMKIIα[T286D]), which mimics autophosphorylation, leads to a loss in LTP induction at lower frequencies and shifts the size and direction of synaptic change in favor of LTD induction (Mayford et al., 1995, 1996). Additionally overexpression of CaMKIIα[T286D] in the adult forebrain disrupts spatial memory and fear-conditioned memory (Mayford et al., 1996).

CaMKII activation shares many similarities with the induction of LTP. To determine if two forms of plasticity use a similar mechanism an occlusion test is performed. For example, LTP induction by HFS is prevented or occluded following potentiation induced by activated CaMKII, and vice versa, suggesting a common downstream mechanism (Pettit et al., 1994; Lledo et al., 1995). Addition of activated

CaMKII in the patch pipette can mimic LTP by enhancing the amplitude and frequency of spontaneous excitatory postsynaptic currents (sEPSCs, action potential dependent and independent currents that are recorded in a voltage clamped neuron in response to the release of synaptic glutamate) and decreasing the failure rate which is attributed with an increase in the probability of release (Lledo et al., 1995). These results suggest that CaMKII alone is sufficient to enhance synaptic transmission and that this enhancement shares common features underlying the mechanism seen following LTP induction.

The activation of CaMKII can strengthen synaptic transmission by multiple mechanisms. Active CaMKII can phosphorylate AMPAR GluR1 subunits at serine 831 and increase the conductance state of AMPARs (Barria et al., 1997a; Mammen et al., 1997; Derkach et al., 1999; Lee et al., 2000). Phosphorylation of GluR1^{Ser831} is increased following LTP induction and inhibiting CaMKII blocks phosphorylation at this site (Barria et al., 1997b). However, LTP can still occur when phosphorylation at GluR1^{Ser831} is blocked, suggesting that additional mechanisms underlie synaptic potentiation (Hayashi et al., 2000). Another important postsynaptic mechanism is the activity-dependent, CaMKII-dependent trafficking of AMPARs to the synapse. Hayashi et al. showed that constitutively activated CaMKII leads to the insertion of new AMPARs in the synapse; one potential mechanism driving the potentiation seen following LTP induction (Hayashi et al., 2000). A recent study suggested that CaMKII phosphorylation of the guanine–nucleotide exchange factor (GEF) kalirin-7 may mediate activity-dependent spine enlargement and enhanced AMPAR-mediated synaptic transmission under some conditions (Xie et al., 2007). However, despite the preponderance of data linking CaMKIIα to LTP induction in cortex and hippocampus, it is worth noting that CaMKIIα can also play a key role in the induction of LTD at parallel fiber–Purkinje cell synapses in the cerebellum (Hansel et al., 2006). As discussed later, CaMKII like in the hippocampus and cerebellum plays an important role in basal ganglia plasticity.

4.6   LONG-TERM SYNAPTIC DEPRESSION IN THE STRIATUM

Long-lasting neuroadaptations at glutamatergic synapses are not limited to the hippocampus, but include brain regions like the striatum as well. Indeed, plasticity at excitatory synapses on MSNs in the striatum has been clearly demonstrated. While plasticity at glutamatergic synapses in the striatum shares similarities with plasticity seen in the hippocampus, there are some notable exceptions, including a much more prominent neuromodulatory role for dopamine. Long-lasting changes in the strength of synaptic connections in the dorsal lateral striatum are likely to influence striatal control over motor activity and may play a role in neurodegenerative disease. Induction of LTP or LTD is dependent on the age of the animal and subregion within the dorsal striatum examined (Partridge et al., 2000). Unlike the hippocampus and cerebral cortex, *ex vivo* studies in adult rodent brain slices using HFS with or without concurrent depolarization typically leads to the induction of LTD rather than LTP at glutamatergic synapses in the dorsal lateral striatum (Calabresi et al., 1992b, 1996; Kerr and Wickens, 2001; Bonsi et al., 2003), and this form of LTD is independent of NMDAR activation (Calabresi et al., 1992b). Instead, induction of striatal HFS-LTD is dependent on membrane depolarization, activation of voltage-gated calcium channels, increases in postsynaptic calcium, coactivation of D1R and D2R signaling, and metabotropic glutamate receptors (mGluRs) (Calabresi et al., 1992b, 1996; Choi and Lovinger, 1997b; Sung et al., 2001). Indeed, either D1R or D2R antagonists are able to block HFS-LTD and mGluR antagonist significantly attenuates LTD magnitude (Calabresi et al., 1992b). Additionally, this form of LTD is absent in mice that lack the D2 receptor (Calabresi et al., 1997; Choi and Lovinger, 1997b) or the dopamine-regulated phosphatase regulator DARPP-32 (Calabresi et al., 2000). Alternatively, muscarinic acetylcholine receptor antagonists enhance the magnitude of LTD (Bonsi et al., 2008), while nicotinic acetylcholine receptor antagonists prevent LTD induction (Partridge et al., 2002).

At least some of these receptors/signaling molecules likely modulate, rather than mediate LTD, in the striatum. For example, Lovinger's group found that L-type calcium channel activation with modest depolarization and synaptic activity is sufficient for LTD, bypassing both D2Rs and mGluRs (Adermark and Lovinger, 2007a). Studies from this group further revealed that the mechanism underlying the expression of striatal LTD is a reduction in the probability of release as indicated by increases in the paired-pulse ratio (PPR, a measure of short-term plasticity thought to be presynaptically mediated and is inversely correlated with the probability of release) and coefficient of variation (CV, the SD of the EPSC amplitude normalized to the mean amplitude where the inverse square of CV is directly proportional to quantal content) of evoked excitatory responses following LTD induction (Choi and Lovinger, 1997a,b). Also supporting a presynaptic site of expression, LTD is associated with decreased mEPSC frequency, but not amplitude. This form of LTD is thought to be induced postsynaptically yet expressed presynaptically, evoking the need for a retrograde messenger. Common retrograde messengers include endocannabinoids (eCB) such as anandamide and 2-arachidonyl glycerol. Retrograde signaling via eCBs has become a prominent theme in synaptic plasticity throughout the brain. In the striatum, unlike other brain regions, depolarization alone is not sufficient to release eCBs to modulate glutamatergic transmission. Rather, there is an additional requirement for mGluR activation to induce eCB release (Kreitzer and Malenka, 2005). Additionally, striatal D2 receptor activation is known to mobilize endocannabinoid release as well (Giuffrida et al., 1999). The release of eCB postsynaptically can activate presynaptic CB1 receptors, which are Gi/o-coupled G-protein-coupled receptors, which suppress release at excitatory and inhibitory synapses (Szabo et al., 1998; Gerdeman and Lovinger, 2001; Huang et al., 2001). Studies using cannabinoid receptor (CB1R) agonists and antagonists and CB1R knockout mice support a role for postsynaptic eCB acting on presynaptic CB1Rs in a retrograde manner to induce LTD (Gerdeman et al., 2002; Kreitzer and Malenka, 2005). However, activation of CB1R alone is not enough to generate LTD; additional presynaptic stimulation is needed (Adermark and Lovinger, 2007b; Singla et al., 2007). This suggests that downstream signaling from CB1Rs synergize with depolarization-induced mechanisms, like calcium entry, to produce LTD. Indeed the striatum is enriched with CB1Rs. Within the striatum, CB1R mRNA is expressed with a gradient in the striatum with expression levels higher in the lateral striatum with a gradual decrease moving to the medial striatum with little expression in the ventral striatum (Matyas et al., 2006; Martin et al., 2008).

CB1R mRNA is expressed in cortical pyramidal neurons that project to the striatum (Tsou et al., 1998). While there is convincing functional data supporting the ability of CB1 receptors to suppress glutamatergic release (Gerdeman and Lovinger, 2001; Huang et al., 2001; Kofalvi et al., 2005), controversy exists over the ability of antibodies to recognize the expression of CB1 receptors on presynaptic glutamatergic receptor terminals (Matyas et al., 2006). Additional work and better antibodies are needed for a clearer understanding of eCB signaling in the striatum.

It is also unclear whether endocannabinoid-mediated LTD (eCB-LTD) exists on both direct and indirect populations of neurons or if eCB-LTD can only exist on indirect D2R-containing MSNs. Lovinger's and Surmeier's groups provided data consistent with a model in which both direct and indirect pathway MSNs support eCB-LTD via a cholinergic interneuron D2 receptor-dependent reduction in M1 receptor tone (Wang et al., 2006). The resulting reduction in M1 receptor tone promotes the opening of Cav1.3 calcium channels, which then enhances endocannabinoid production and CB1 receptor activation. However, data from the Malenka laboratory suggests that only indirect pathway MSNs can elicit eCB-LTD (Kreitzer and Malenka, 2007). These discrepancies could possibly be explained by the use of D1R-EGFP versus M4R-EGFP mice to label direct pathway MSNs. Additional conflicting data are reported by Calabresi's group showing that stimulation of M1 receptors can facilitate the induction of striatal LTP in ACSF lacking magnesium (see the following text) (Calabresi et al., 1999).

An additional form of LTD induced via LFS (5–10 min, 10–13 Hz) can also be detected in the dorsolateral striatum (Kreitzer and Malenka, 2005; Ronesi and Lovinger, 2005). Like HFS-LTD, LFS-LTD is blocked by D2R and CB1R antagonists or blunted by blockers of L-type calcium channels. However, LFS-LTD differs from HFS-LTD in that it is unaffected by postsynaptic depolarization, postsynaptic calcium chelation, mGluR antagonists, and an intracellular anandamide membrane transport inhibitor. The induction of LFS-LTD does not occlude the induction of HFS-LTD, suggesting both forms of LTD can occur at the same synapses. This presents a scenario where HFS-LTD is favored under strong bouts of cortical stimulation along with correlated postsynaptic depolarization, while LFS-LTD would be favored under moderate-frequency cortical stimulation uncorrelated with postsynaptic depolarization.

LTD is also the major form of plasticity observed in the ventral striatum. Typical stimulation protocols consist of 10–13 Hz stimulation applied for several minutes in the presence of $GABA_A$ receptor antagonists, leading to induction of a robust LTD of evoked excitatory transmission (Robbe et al., 2002c; Hoffman et al., 2003; Mato et al., 2004). This form of LTD, which is blocked by CB1R antagonists and in CB1RKO mice, is mediated by the postsynaptic release of eCB, which activates presynaptic CB1Rs on glutamatergic terminal afferents to the NAc (Robbe et al., 2001). The activation of $G_{i/o}$-coupled CB1Rs leads to the decrease in probability of glutamate release (Robbe et al., 2001). Consistent with this finding, eCB-mediated LTD is associated with a decrease in the sEPSC frequency, but not amplitude. Additionally, intracellular calcium chelation and antagonism of mGluR5 receptors blocks the induction of eCB-LTD, but NMDAR antagonists do not affect this form of LTD. The group1 mGluR agonist, DHPG, elicits LTD in the nucleus accumbens and 13 Hz eCB-LTD is occluded following DHPG-LTD, suggesting a similar shared mechanism. Interestingly, this form of LTD is blocked by prior chronic cocaine administration (Fourgeaud et al., 2004).

A second form of LTD is observed in the nucleus accumbens, mediated by presynaptic mGluR2/3 receptors and inhibition of P/Q-type calcium channels by the cAMP/PKA pathway (Robbe et al., 2002a). Both mGluR2/3 agonists and tetanic stimulation (three times for 1 s at 100 Hz, 20 s intervals) leads to mGluR2/3LTD in a majority of experiments. mGluR2/3 LTD, like eCB-LTD, leads to an increase in the PPR indicative of a decrease in the probability of release of glutamate. mGluR2/3 LTD induced by tetanic stimulation is also independent of NMDARs, postsynaptic intercellular calcium concentrations, and was occluded by prior mGluR2/3 agonist-induced chemical LTD.

Finally, a third form of LTD is observed in the nucleus accumbens. This form of LTD can be induced by three bursts of 5 Hz stimulation for 3 min paired with depolarization to –50 mV or 1 Hz for 3 min paired with depolarization (Thomas et al., 2000, 2001). This LFS paired with depolarization induces LTD, which unlike other forms is dependent on NMDARs (Thomas et al., 2001). This form of LTD is dependent on internal calcium, but independent of mGluRs and D1 and D2 receptor activation (Thomas et al., 2000).

4.7 LONG-TERM SYNAPTIC POTENTIATION IN THE STRIATUM

In addition to LTD, LTP can also occur at glutamate synapses on MSNs, but it is in general less well characterized. Interestingly, induction of striatal plasticity depends on the postnatal age of the animal and subregion of the dorsal striatum. In the dorsal lateral striatum, HFS stimulation in young animals (P12–14) leads to the induction of LTP, while in older animals (P15–34) it leads to LTD. Conversely, in dorsal medial striatum LTP is induced across both age ranges (Partridge et al., 2000). While the story is currently far from complete, striatal LTP may share some features with CA1 hippocampal LTP described earlier (Bliss and Collingridge, 1993), in particular that it is dependent on NMDARs for induction (Calabresi et al., 1997; Yamamoto et al., 1999; Partridge et al., 2000; Kerr and Wickens, 2001; Dang et al., 2006; Popescu et al., 2007), and has been observed both *in vitro* and *in vivo* (Charpier and Deniau, 1997; Charpier et al., 1999; Dos Santos Villar and Walsh, 1999). Interestingly, paired HFS and depolarization concurrently with pulsatile application (Wickens et al., 1996), but not bath application (Arbuthnott et al., 2000), of dopamine in normal aCSF (perhaps mimicking the natural pulsatile release of dopamine) leads to the induction of LTP instead of LTD in the dorsal striatum.

As mentioned earlier, less is known about striatal LTP relative to LTD, largely because of difficulty in identifying consistent means by which to induce and record the LTP at a single-cell level. For example, in many striatal LTP studies, aCSF lacking magnesium is used to both unblock the NMDA receptor and likely increase release probability (Calabresi et al., 1992c, 2000; Centonze et al., 1999). Potassium channel blockers have also been used as a means of facilitating striatal LTP (Wickens et al., 1998; Norman et al., 2005). Like LTD, striatal LTP appears to be heavily modulated by dopamine, apparently in a receptor-specific manner. D2 receptor knockout mice are reported to exhibit LTP with HFS in normal aCSF (Calabresi et al., 1997). Moreover, application of a D2 receptor antagonist can enhance LTP induced in magnesium-free aCSF in wildtype mice, suggesting that D2 receptors exert negative control on LTP induction (Calabresi et al., 1997). Pharmacological inhibition of D1 receptors, recordings in D1 receptor null mice, and postsynaptic inhibition of PKA/PKC activity block the induction of LTP (Akopian et al., 2000; Calabresi et al., 2000; Centonze et al., 2001, 2003; Kerr and Wickens, 2001; Ding and Perkel, 2004; Fino et al., 2005). DARRP-32 knockout mice, presumably via loss of PP1 inhibition, also prevent the induction of LTP in magnesium-free aCSF (Calabresi et al., 2000). Striatal LTP may also depend on the activation of M1 acetylcholine receptors (Calabresi et al., 1999; Lovinger et al., 2003) and mGluRs (Gubellini et al., 2003). In agreement with a role for M1 acetylcholine receptors in striatal LTP, blockade of M2 acetylcholine receptors enhances LTP (Calabresi et al., 1998).

With D1 receptor stimulation necessary for LTP induction and D2 receptor stimulation necessary for LTD induction, how can distinct populations of MSNs expressing only D1 or D2 receptors elicit both LTP and LTD? This question was addressed by Shen et al. using spike timing-dependent plasticity (STDP) in D1R- and D2R-EGFP BAC transgenic mice (Shen et al., 2008). STDP involves the precise timing of converging synaptic activity and backpropagating action potentials and has been studied in many systems after its initial discovery (Magee and Johnston, 1997; Markram et al., 1997). If an evoked excitatory postsynaptic potential (EPSP, a depolarization of a neuron induced by presynaptic stimulation increasing the likelihood of generating an action potential) precedes postsynaptic spiking by a few milliseconds then this continued pairing leads to LTP, whereas reversing the order induces LTD. Shen et al. report that disruption of adenosine A2a receptors—not D1R—blocked the induction of LTP on D2R MSNs. LTD in D2R MSNs induced with STDP, like conventional HFS-LTD, can be blocked by the single addition of either D2R antagonists, CB1R antagonists, or mGluR5 antagonists. In D1R MSNs, LTD was only induced in the presence of D1R antagonists, while D2R antagonists had no effect on LTD. So it seems that in D1 MSNs, D1 receptors and mGluR5 receptors have antagonistic actions while in D2 MSNs, it seems that adenosine A2a receptors substitute for D1 receptors to antagonize mGluR5. Thus, although this study showed that dopamine plays a key role in determining the directionality of plasticity, dopamine is not essential for the induction of striatal LTP/LTD under these conditions. Another study using STDP concluded that both LTP and LTD induction in the striatum is D1R dependent and NMDAR dependent with directionality determined by timing and order of EPSPs and backpropagating action potentials (Pawlak and Kerr, 2008). Like the hippocampus, this study also suggested that LTP involved a postsynaptic insertion of AMPA receptors via the unsilencing of synapses, but additional work is needed to confirm if a similar mechanism functions in the striatum. Recently, selective loss of DARPP-32 in D1R- or D2R-containing MSNs prevents the induction of STD-induced LTP in the striatum (Bateup et al., 2010). This is currently a rapidly evolving area of research.

As in the dorsal striatum, LTP is observed in the ventral striatum, but with some notable differences. HFS (100 Hz), in physiologically normal magnesium concentrations, elicits LTP in field recordings in the NAc (Pennartz et al., 1993; Kombian and Malenka, 1994; Mazzucchelli et al., 2002; Schramm et al., 2002;

Li and Kauer, 2004; Yao et al., 2004). This form of LTP is blocked by acute amphetamine, dopamine, or by dopamine receptor agonists (Li and Kauer, 2004). However, after chronic administration of amphetamine, the attenuation of LTP is lost. Another induction protocol uses pairings of depolarization via an intracellular patch pipette with LFS to induce LTP (Kombian and Malenka, 1994). Induction of LTP using this pairing approach, while sensitive to intracellular calcium concentration, is not blocked by acute amphetamine treatment (Li and Kauer, 2004). These results suggest that dopamine receptor activation normally decreases responses to HFS without disrupting excitatory synaptic transmission at low frequency. Indeed, D1R stimulation decreases glutamatergic transmission in the NAc response to both single and multiple stimuli (Pennartz et al., 1993; Harvey and Lacey, 1996; Nicola et al., 1996; Beurrier and Malenka, 2002). Additionally, NMDARs are required for this form of LTP induction in the NAc (Kombian and Malenka, 1994). Interestingly, the activation of glutamatergic basolateral amygdala afferents that innervate ventral striatum enhances LTP at cortical glutamatergic synapses on MSNs (Popescu et al., 2007).

In summary, it seems that repetitive stimulation of corticostriatal afferents can lead to the development of LTP or LTD depending on conditions. Therefore, it will be crucial to better understand and identify the factors that control the directionality, induction, and maintenance of plasticity in the striatum. Separating the role of direct and indirect projection neurons in plasticity and identifying the effects of local interneurons and local networks as well as determining how differing methodologies affect plasticity in the striatum will all be important in better understanding the mechanisms underlying function and disease.

4.8 GLUTAMATERGIC PLASTICITY AND MOTOR SKILL LEARNING

Long-lasting changes in synaptic efficacy lead to long-lasting changes at the cellular and network levels, which correlate with modifications in animal behavior and performance (Chapter 7). Deletion of the obligatory GluN1 subunit of the NMDAR in the striatum leads to a disruption of motor learning on the accelerating rotarod and a disruption in dorsal striatum LTP and ventral striatum LTD (Dang et al., 2006). Yin et al. (2009) were able to make a connection between plasticity at a cellular level and the following acquisition and consolidation of a long-lasting motor skill. Differing regions of the dorsal striatum subserve different forms of motor and procedural learning. The dorsal medial region of the striatum, or associative striatum, is preferentially involved in the rapid acquisition of action-outcome contingencies while the dorsal lateral, or sensorimotor, striatum is involved in the gradual acquisition of habit-based learning. Experiments recording the activity of MSNs in both the dorsal medial and dorsal lateral striatum *in vivo* in freely moving mice during a motor learning task found regional differences in neural activity at differing time points during learning (Yin et al., 2009). During early time points on the rotarod, the dorsal medial striatum showed the largest increase in activity, while at later time points when learning had reached a plateau dorsal lateral striatum activity peaked. In agreement with these data, excitotoxic lesions in the dorsal medial striatum only affected early motor skill learning, while dorsal lateral striatum excitotoxic lesions only affected late learning. To determine if these changes corresponded to changes in synaptic strength, saturation experiments were performed by inducing LTD. Yin et al. found that it was more difficult to saturate LTD in the dorsal medial striatum at early time points and found it more difficult to saturate LTD in the dorsal lateral striatum at late time points in motor skill acquisition. These data suggest that early in motor skill acquisition the dorsal medial striatum undergoes synaptic potentiation while at later time points in acquisition the dorsal lateral striatum shows potentiation. Additionally, using whole-cell voltage clamp they saw an increase in the sEPSC amplitude in the dorsal lateral striatum following extended training suggesting a postsynaptic increase in the function/number of AMPARs which is reflective of synaptic potentiation. Furthermore, recordings from D1- and D2-EGFP BAC mice during motor skill acquisition suggested that late in training the potentiation of excitatory transmission in the dorsal lateral striatum occurs predominantly in D2 receptor–expressing MSNs while becoming less dependent on the activation of D1 receptors (Yin et al., 2009). In all, these data show that LTP of glutamatergic transmission in the dorsal striatum is necessary for the acquisition of motor skill learning.

4.9 ALTERATIONS IN STRIATAL GLUTAMATERGIC SYNAPSES

4.9.1 FOLLOWING DOPAMINE DEPLETION

Unilateral lesion of the nigrostriatal dopaminergic pathway with 6-OHDA is one of the most widely used animal models of PD leading to profound biochemical, morphological, electrophysiological, and behavioral changes that in many cases mimic alterations described in patients with Parkinson's disease. Behaviorally, unilateral dopamine depletion leads to deficits in rotarod performance, locomotor performance, and increased limb-use asymmetry (Picconi et al., 2004a). Morphologically, the loss of dendritic spines from MSNs also is seen in both dopamine depletion models (Ingham et al., 1989, 1993) and in patients with Parkinson's disease (Anglade et al., 1996). While spine loss is occurring following dopamine depletion the remaining spines undergo changes as well. For example, an increase in the number of perforated synapses or synapses with bifurcated active zones—often seen after manipulations that enhance excitatory transmission (Edwards, 1995)—is observed following dopamine depletion (Ingham et al., 1998) and in PD patients (Anglade et al., 1996). Additional studies also point to dopamine depletion leading to an increase in glutamatergic transmission. Indeed, increased glutamate release from corticostriatal synapses (Lindefors and Ungerstedt, 1990) and increases in spontaneous glutamatergic transmission are seen following dopamine depletion (Galarraga et al., 1987; Calabresi et al., 1993; Tang et al., 2001; Gubellini et al., 2002; Picconi et al., 2004a). This increased sEPSC frequency following dopamine depletion can be renormalized by the addition of a D2 receptor agonist (Picconi et al., 2004a). This would suggest that dopamine tone normally exerts negative modulation on glutamatergic transmission. Conversely, work out of Surmeier's laboratory demonstrated that decreases, rather than increases, in spontaneous glutamatergic transmission—coinciding with spine loss—were only seen following dopamine depletion in indirect pathway MSNs

(Day et al., 2006). These differences could be due to differing experimental methodology like the inclusion of cesium in the patch pipette which blocks potassium channels and improves space clamp and voltage control at distal synapses. Another point for comparison was that this study only looked at sEPSC frequency shortly following reserpine treatment, a drug known to reversibly deplete all monoamines without dopamine cell and terminal loss. More recently, Warre et al. examined sEPSC and mEPSC frequency in indirect and direct pathway MSNs using adenosine A2a receptor-EGFP and D1R-EGFP mice (Warre et al., 2011). Using 6-OHDA lesions to deplete dopamine more persistently, they found the converse that frequency was reduced in the direct pathway cells rather than the indirect. Thus more work will be needed to address this important issue.

Calabresi's group amongst others has suggested that dopamine depletion leads to a global inability to induce striatal LTP in magnesium-free aCSF (Centonze et al., 1999; Kerr and Wickens, 2001; Picconi et al., 2003, 2004b) and prevents the induction of LTD *ex vivo* (Calabresi et al., 1992b). Alternatively, Malenka and colleagues have shown that LTD is absent only in indirect pathway MSNs in reserpine- and 6-OHDA-treated animals. This eCB-LTD, in addition to locomotor activity and catalepsy, could be rescued by a D2 receptor agonist or inhibitors of endocannabinoid degradation (Kreitzer and Malenka, 2007). This finding is intriguing since D2R-dependent expression of LTD has been reported in both D1R- and D2R-containing MSNs (Wang et al., 2006). Recently, Shen et al. showed that in animals lesioned with 6-OHDA, STDP leads to a flip in the polarity of plasticity so that LTP induction protocols induce LTD now in D1 MSNs (Shen et al., 2008) while in D2 MSNs LTP remained following dopamine depletion and could be rescued with a D2R agonist. However, *ex vivo* slice data contrast with *in vivo* LTD where alpha-methyl-para-tyrosine-induced dopamine depletion (Reynolds and Wickens, 2000) or blockade with D1R antagonists (Floresco et al., 2001) does not block LTD induction.

4.9.2 Following Subsequent Levodopa Administration

Administration of levodopa plus a peripheral dopa-decarboxylase inhibitor like benserazide is currently one of the most effective therapeutic options in Parkinson's disease. The therapeutic efficacy of levodopa is presumed to be due to its ability to counteract neuroadaptations that occur in the absence of dopamine innervation (Obeso et al., 2000). Experimentally, levodopa administration renormalizes many behavioral, biochemical, and electrophysiological deficits seen following dopamine depletion. However after chronic administration with dopamine replacement therapies the formation of abnormal involuntary movements or dyskinesias appears. Dyskinesias are observed in a subpopulation of rats and monkeys, independent of the degree of dopamine denervation, that undergo dopamine depletion followed by levodopa administration. Data suggest that early administration of levodopa (4 weeks postlesion) decreases the percentage of animals that eventually develop dyskinesias (Marin et al., 2009). Normal animals and patients that are given levodopa do not develop dyskinesias, suggesting that changes that take place following dopamine depletion are important in revealing dyskinesias. Behaviorally in nondyskinetic rats chronic levodopa administration rescues rotarod performance and limb-use asymmetry (Picconi et al., 2003, 2004b). The loss of striatal LTD seen following dopamine depletion can be rescued by the application of dopamine or a combination of D1 and D2 receptor agonists (Calabresi et al., 1992a,b). Also the *ex vivo* induction of striatal LTP is rescued following levodopa administration at dopamine depleted synapses whether or not the rats exhibit dyskinesias (Picconi et al., 2003). However, in animals that showed dyskinetic behaviors depotentiation—a reversal of LTP induced by LFS and mediated by phosphatases—could not be produced (Picconi et al., 2003). This same study showed that levodopa increased the phosphorylation of DARPP-32 at Thr34, to presumably inhibit PP1, in dyskinetic rats versus nondyskinetic rats. Additionally, in 6-OHDA-lesioned rats levodopa reverses the hypersensitivity of D2 receptors and renormalizes glutamatergic spontaneous EPSC frequency (Picconi et al., 2004a). Biochemical data implicate abnormalities in the subcellular localization of the NMDAR GluN2B subunit in animals undergoing levodopa-induced dyskinesias (Gardoni et al., 2006). Recently, inhibition of DARPP-32 only in D1R-containing MSNs significantly reduced the abnormal involuntary movements score (AIMS), a measure of dyskinetic-like behaviors, seen following repeated exposure to L-DOPA (Bateup et al., 2010). With these data, it is conceivable that pharmacological modulation of striatal glutamatergic synaptic plasticity might prove useful in the treatment of motor symptoms observed in PD.

4.10 CaMKII INHIBITION AS A THERAPEUTIC TARGET IN PARKINSON'S DISEASE

As mentioned earlier, CaMKII is a key signaling molecule in synaptic plasticity. Both the Colbran and Calabresi laboratories saw increases in the phosphorylation state of threonine 286 in CaMKIIα following dopamine depletion (Picconi et al., 2004b; Brown et al., 2005). In both studies, levodopa administration was able to restore the levels of phosphorylated CaMKIIα to normal. The increased Thr286 phosphorylation of CaMKIIα persisted for up to 20 months postlesion, but increased phosphorylation of GluR1^{Ser831}, a CaMKII substrate, was only detected 9–20 months after dopamine depletion. Additionally Picconi et al. showed that intrastriatal injection of two CaMKII inhibitors (KN93 or a membrane-permeant CaMKII inhibitor peptide) rescued LTP deficits following dopamine depletion, as well as limb-use asymmetry and rotarod performance. Together, these data suggest an important role for downstream signaling molecules like CaMKII in the aberrant plasticity and disrupted motor output following dopamine depletion.

4.11 EFFECTS OF DRUGS OF ABUSE ON PLASTICITY IN THE STRIATUM

Psychostimulants like cocaine and amphetamine, as well as other drugs of abuse, increase dopamine levels in the nucleus accumbens and in the dorsal striatum (Everitt et al., 2008). Behaviorally psychostimulant drugs of abuse lead to hyperlocomotion which is thought to be mediated via D1R signaling (Xu et al., 2000;

Bateup et al., 2010). Animals chronically treated with drugs of abuse show long-lasting modifications in excitatory transmission in the striatum. In the nucleus accumbens shell chronic administration of cocaine leads to the depression of glutamatergic synaptic strength reflected by decreases in the AMPA/NMDA current ratio (an index of the relative AMPA receptor- and NMDA receptor-mediated contribution to EPSCs), the amplitude of miniature EPSCs, and the magnitude of LTD (Thomas et al., 2001). This suggests that cocaine administration leads to the induction of LTD *in vivo*. Indeed, this form of NAc LTD was blocked by a peptide that disrupts clathrin-mediated endocytosis or by a GluA2-derived peptide that blocks regulated AMPA receptor endocytosis, a major mechanism underlying LTD. This GluA2-derived peptide also disrupted the development of behavioral sensitization, an enhancement in the locomotor-activating effects of cocaine with repeated administration, which has been shown to correlate with LTD (Brebner et al., 2005). eCB-LTD is abolished following single injection of cocaine and this effect is blocked in D1 receptor KO mice and when D1 receptor antagonists are administered with cocaine (Fourgeaud et al., 2004). Additionally, this form of LTD is blocked in morphine-withdrawn animals (Robbe et al., 2002b), during cocaine self-administration (Martin et al., 2006) and after the chronic treatment with cannabis derivatives (Hoffman et al., 2003). These data suggest that LTD is induced *in vivo* following chronic drug administration, therefore occluding subsequent LTD *ex vivo*. In addition to electrophysiological changes following psychostimulant exposure, long-lasting increases in spine density are observed following cocaine or amphetamine administration in the nucleus accumbens (Robinson and Kolb, 2004).

Recently, it has been demonstrated that extended withdrawal following chronic cocaine administration leads to synaptic potentiation in the NAc shell region and that subsequent re-exposure to cocaine reverses a synaptic potentiation to a synaptic depression (Kourrich et al., 2007). These data suggest that drug history determines the directionality of plasticity in the NAc shell. Additionally, exposure to cocaine leads to the development of silent synapses in adulthood, accompanied by the insertion of new GluN2B subunits of the NMDA receptor (Huang et al., 2009). Along with changes in synaptic transmission, drugs of abuse like cocaine and amphetamine can impart long-lasting changes in the intrinsic excitability of NAc MSNs. Chronic cocaine or amphetamine exposure leads to decreases in excitability in the nucleus accumbens shell region that starts within 1–3 days and persists for at least 2 weeks. The same drug regimen leads to an increase in intrinsic excitability in the NAc core region during early withdrawal (1–3 days), but returns to baseline after protracted withdrawal (2 weeks). These bidirectional changes in intrinsic excitability seem to be mediated by changes in the A-type potassium current (Kourrich and Thomas, 2009).

Interestingly, chronic cocaine administration leads to a downregulation of the PSD scaffolding protein PSD-95. Downregulation of PSD-95 correlates with synaptic potentiation and a PSD-95–targeted deletion enhances LTP and augments the locomotor-activating effects of cocaine (Yao et al., 2004). Cyclin-dependent kinase 5 (CDK5) is a downstream target gene of the transcription factor deltaFosB, which accumulates in striatal neurons following cocaine administration. Enhanced CDK5 expression in the NAc occurs following short access to self-administered cocaine

(Seiwell et al., 2007), while inhibition of CDK5 augments both the development and expression of cocaine sensitization and enhances the incentive-motivational effects of cocaine (Taylor et al., 2007).

In all, drugs of abuse have been shown to modulate both the induction and directionality of plasticity in the nucleus accumbens. Like in the dorsal striatum, the interactions between glutamatergic and dopaminergic systems are important for plasticity in this region. While many drugs have a common end point of elevating dopamine acutely in nucleus accumbens, chronic drug administration often leads to long-lasting modification of glutamatergic synapses, emphasizing the importance of drug history, including withdrawal, when interpreting whether LTP or LTD is expressed.

4.12   CONCLUSIONS

The striatum represents a major site of plasticity in the basal ganglia. Interactions between fast excitatory glutamatergic synaptic transmission and slower dopaminergic modulation are critical for plasticity in this region. This is a region of complex anatomy which exhibits a wealth of long-lasting synaptic modifications. Dopamine's importance is reflected by altered plasticity seen following dopamine depletion, subsequent dopamine replacement therapies, and chronic administration of drugs of abuse. We are just beginning to understand how plasticity in the striatum influences normal behaviors and its role in disease. The advent of BAC D1-EGFP- and D2-EGFP transgenic mice (Chapter 5) to separate direct and indirect pathway MSNs along with targeted whole-cell recordings of specific interneuron populations will continue to aid our understanding. New techniques utilizing channelrhodopsin or halorhodopsin, which can control firing rate of transfected cells with light *in vitro* or *in vivo*, will accelerate our knowledge of basal ganglia function and open the possibility of specific neural circuit control (Gradinaru et al., 2009). Gaining a more complete understanding of mechanisms underlying synaptic plasticity in the basal ganglia will hopefully allow for basic understanding of basal ganglia–associated behaviors as well as open new avenues for therapeutic intervention in disease.

REFERENCES

Abraham, W. C., Logan, B., Greenwood, J. M. and Dragunow, M. (2002) Induction and experience-dependent consolidation of stable long-term potentiation lasting months in the hippocampus. *J Neurosci*, 22, 9626–9634.

Adermark, L. and Lovinger, D. M. (2007a) Combined activation of L-type Ca^{2+} channels and synaptic transmission is sufficient to induce striatal long-term depression. *J Neurosci*, 27, 6781–6787.

Adermark, L. and Lovinger, D. M. (2007b) Retrograde endocannabinoid signaling at striatal synapses requires a regulated postsynaptic release step. *Proc Natl Acad Sci USA*, 104, 20564–20569.

Ahn, S., Ginty, D. D. and Linden, D. J. (1999) A late phase of cerebellar long-term depression requires activation of CaMKIV and CREB. *Neuron*, 23, 559–568.

Ahn, S. and Phillips, A. G. (2007) Dopamine efflux in the nucleus accumbens during within-session extinction, outcome-dependent, and habit-based instrumental responding for food reward. *Psychopharmacology (Berl)*, 191, 641–651.

Akopian, G., Musleh, W., Smith, R. and Walsh, J. P. (2000) Functional state of corticostriatal synapses determines their expression of short- and long-term plasticity. *Synapse*, 38, 271–280.

Anglade, P., Mouatt-Prigent, A., Agid, Y. and Hirsch, E. (1996) Synaptic plasticity in the caudate nucleus of patients with Parkinson's disease. *Neurodegeneration*, 5, 121–128.

Arbuthnott, G. W., Ingham, C. A. and Wickens, J. R. (2000) Dopamine and synaptic plasticity in the neostriatum. *J Anat*, 196 (Pt 4), 587–596.

Arbuthnott, G. W. and Wickens, J. (2007) Space, time and dopamine. *Trends Neurosci*, 30, 62–69.

Bamford, N. S., Robinson, S., Palmiter, R. D., Joyce, J. A., Moore, C. and Meshul, C. K. (2004b). Dopamine modulates release from corticostriatal terminals. *J Neurosci*, 24, 9541–9552.

Bamford, N. S., Zhang, H., Schmitz, Y., Wu, N. P., Cepeda, C., Levine, M. S., Schmauss, C., Zakharenko, S. S., Zablow, L. and Sulzer, D. (2004a) Heterosynaptic dopamine neurotransmission selects sets of corticostriatal terminals. *Neuron*, 42, 653–663.

Barria, A., Derkach, V. and Soderling, T. (1997a) Identification of the Ca^{2+}/calmodulin-dependent protein kinase II regulatory phosphorylation site in the alpha-amino-3-hydroxyl-5-methyl-4-isoxazole-propionate-type glutamate receptor. *J Biol Chem*, 272, 32727–32730.

Barria, A., Muller, D., Derkach, V., Griffith, L. C. and Soderling, T. R. (1997b) Regulatory phosphorylation of AMPA-type glutamate receptors by CaM-KII during long-term potentiation. *Science*, 276, 2042–2045.

Bateup, H. S., Santini, E., Shen, W., Birnbaum, S., Valjent, E., Surmeier, D. J., Fisone, G., Nestler, E. J. and Greengard, P. (2010) Distinct subclasses of medium spiny neurons differentially regulate striatal motor behaviors. *Proc Natl Acad Sci USA*, 107, 14845–14850.

Beattie, E. C., Carroll, R. C., Yu, X., Morishita, W., Yasuda, H., Von Zastrow, M. and Malenka, R. C. (2000) Regulation of AMPA receptor endocytosis by a signaling mechanism shared with LTD. *Nat Neurosci*, 3, 1291–1300.

Bergson, C., Mrzljak, L., Smiley, J. F., Pappy, M., Levenson, R. and Goldman-Rakic, P. S. (1995) Regional, cellular, and subcellular variations in the distribution of D1 and D5 dopamine receptors in primate brain. *J Neurosci*, 15, 7821–7836.

Beurrier, C. and Malenka, R. C. (2002) Enhanced inhibition of synaptic transmission by dopamine in the nucleus accumbens during behavioral sensitization to cocaine. *J Neurosci*, 22, 5817–5822.

Bliss, T. V. and Collingridge, G. L. (1993) A synaptic model of memory: Long-term potentiation in the hippocampus. *Nature*, 361, 31–39.

Bliss, T. V. and Lomo, T. (1973) Long-lasting potentiation of synaptic transmission in the dentate area of the anaesthetized rabbit following stimulation of the perforant path. *J Physiol*, 232, 331–356.

Bolshakov, V. Y. and Siegelbaum, S. A. (1994) Postsynaptic induction and presynaptic expression of hippocampal long-term depression. *Science*, 264, 1148–1152.

Bonsi, P., Martella, G., Cuomo, D., Platania, P., Sciamanna, G., Bernardi, G., Wess, J. and Pisani, A. (2008) Loss of muscarinic autoreceptor function impairs long-term depression but not long-term potentiation in the striatum. *J Neurosci*, 28, 6258–6263.

Bonsi, P., Pisani, A., Bernardi, G. and Calabresi, P. (2003) Stimulus frequency, calcium levels and striatal synaptic plasticity. *Neuroreport*, 14, 419–422.

Bordet, R., Ridray, S., Carboni, S., Diaz, J., Sokoloff, P. and Schwartz, J. C. (1997) Induction of dopamine D3 receptor expression as a mechanism of behavioral sensitization to levodopa. *Proc Natl Acad Sci USA*, 94, 3363–3367.

Brebner, K., Wong, T. P., Liu, L., Liu, Y., Campsall, P., Gray, S., Phelps, L., Phillips, A. G. and Wang, Y. T. (2005) Nucleus accumbens long-term depression and the expression of behavioral sensitization. *Science*, 310, 1340–1343.

Bredt, D. S. and Nicoll, R. A. (2003) AMPA receptor trafficking at excitatory synapses. *Neuron*, 40, 361–379.

Brown, A. M., Deutch, A. Y. and Colbran, R. J. (2005) Dopamine depletion alters phosphory-lation of striatal proteins in a model of Parkinsonism. *Eur J Neurosci*, 22, 247–256.

Calabresi, P., Centonze, D., Gubellini, P. and Bernardi, G. (1999) Activation of M1-like musca-rinic receptors is required for the induction of corticostriatal LTP. *Neuropharmacology*, 38, 323–326.

Calabresi, P., Centonze, D., Pisani, A., Sancesario, G., Gubellini, P., Marfia, G. A. and Bernardi, G. (1998) Striatal spiny neurons and cholinergic interneurons express differ-ential ionotropic glutamatergic responses and vulnerability: Implications for ischemia and Huntington's disease. *Ann Neurol*, 43, 586–597.

Calabresi, P., Gubellini, P., Centonze, D., Picconi, B., Bernardi, G., Chergui, K., Svenningsson, P., Fienberg, A. A. and Greengard, P. (2000) Dopamine and cAMP-regulated phospho-protein 32 kDa controls both striatal long-term depression and long-term potentiation, opposing forms of synaptic plasticity. *J Neurosci*, 20, 8443–8451.

Calabresi, P., Maj, R., Mercuri, N. B. and Bernardi, G. (1992a) Coactivation of D1 and D2 dopamine receptors is required for long-term synaptic depression in the striatum. *Neurosci Lett*, 142, 95–99.

Calabresi, P., Maj, R., Pisani, A., Mercuri, N. B. and Bernardi, G. (1992b) Long-term syn-aptic depression in the striatum: Physiological and pharmacological characterization. *J Neurosci*, 12, 4224–4233.

Calabresi, P., Mercuri, N. B., Sancesario, G. and Bernardi, G. (1993) Electrophysiology of dopamine-denervated striatal neurons. Implications for Parkinson's disease. *Brain*, 116 (Pt 2), 433–452.

Calabresi, P., Pisani, A., Mercuri, N. B. and Bernardi, G. (1992c) Long-term potentiation in the striatum is unmasked by removing the voltage-dependent magnesium block of NMDA receptor channels. *Eur J Neurosci*, 4, 929–935.

Calabresi, P., Pisani, A., Mercuri, N. B. and Bernardi, G. (1996) The corticostriatal projec-tion: From synaptic plasticity to dysfunctions of the basal ganglia. *Trends Neurosci*, 19, 19–24.

Calabresi, P., Saiardi, A., Pisani, A., Baik, J. H., Centonze, D., Mercuri, N. B., Bernardi, G. and Borrelli, E. (1997) Abnormal synaptic plasticity in the striatum of mice lacking dopamine D2 receptors. *J Neurosci*, 17, 4536–4544.

Carroll, R. C., Lissin, D. V., Von Zastrow, M., Nicoll, R. A. and Malenka, R. C. (1999) Rapid redistribution of glutamate receptors contributes to long-term depression in hippocam-pal cultures. *Nat Neurosci*, 2, 454–460.

Carter, A. G. and Sabatini, B. L. (2004) State-dependent calcium signaling in dendritic spines of striatal medium spiny neurons. *Neuron*, 44, 483–493.

Carter, A. G., Soler-Llavina, G. J. and Sabatini, B. L. (2007) Timing and location of synaptic inputs determine modes of subthreshold integration in striatal medium spiny neurons. *J Neurosci*, 27, 8967–8977.

Centonze, D., Grande, C., Saulle, E., Martin, A. B., Gubellini, P., Pavon, N., Pisani, A., Bernardi, G., Moratalla, R. and Calabresi, P. (2003) Distinct roles of D1 and D5 dopamine recep-tors in motor activity and striatal synaptic plasticity. *J Neurosci*, 23, 8506–8512.

Centonze, D., Gubellini, P., Picconi, B., Calabresi, P., Giacomini, P. and Bernardi, G. (1999) Unilateral dopamine denervation blocks corticostriatal LTP. *J Neurophysiol*, 82, 3575–3579.

Centonze, D., Picconi, B., Gubellini, P., Bernardi, G. and Calabresi, P. (2001) Dopaminergic control of synaptic plasticity in the dorsal striatum. *Eur J Neurosci*, 13, 1071–1077.

Chang, H. T., Wilson, C. J. and Kitai, S. T. (1982) A Golgi study of rat neostriatal neurons: Light microscopic analysis. *J Comp Neurol*, 208, 107–126.

Charpier, S. and Deniau, J. M. (1997) In vivo activity-dependent plasticity at cortico-striatal connections: Evidence for physiological long-term potentiation. *Proc Natl Acad Sci USA*, 94, 7036–7040.

Charpier, S., Mahon, S. and Deniau, J. M. (1999) In vivo induction of striatal long-term potentiation by low-frequency stimulation of the cerebral cortex. *Neuroscience*, 91, 1209–1222.

Choi, S. and Lovinger, D. M. (1997a) Decreased frequency but not amplitude of quantal synaptic responses associated with expression of corticostriatal long-term depression. *J Neurosci*, 17, 8613–8620.

Choi, S. and Lovinger, D. M. (1997b) Decreased probability of neurotransmitter release underlies striatal long-term depression and postnatal development of corticostriatal synapses. *Proc Natl Acad Sci USA*, 94, 2665–2670.

Chung, H. J., Steinberg, J. P., Huganir, R. L. and Linden, D. J. (2003) Requirement of AMPA receptor GluR2 phosphorylation for cerebellar long-term depression. *Science*, 300, 1751–1755.

Colbran, R. J. (2004) Protein phosphatases and calcium/calmodulin-dependent protein kinase II-dependent synaptic plasticity. *J Neurosci*, 24, 8404–8409.

Dang, M. T., Yokoi, F., Yin, H. H., Lovinger, D. M., Wang, Y. and Li, Y. (2006) Disrupted motor learning and long-term synaptic plasticity in mice lacking NMDAR1 in the striatum. *Proc Natl Acad Sci USA*, 103, 15254–15259.

Day, M., Wang, Z., Ding, J., An, X., Ingham, C. A., Shering, A. F., Wokosin, D., Ilijic, E., Sun, Z., Sampson, A. R., Mugnaini, E., Deutch, A. Y., Sesack, S. R., Arbuthnott, G. W. and Surmeier, D. J. (2006) Selective elimination of glutamatergic synapses on striatopallidal neurons in Parkinson disease models. *Nat Neurosci*, 9, 251–259.

Delong, M. and Wichmann, T. (2009) Update on models of basal ganglia function and dysfunction. *Parkinsonism Relat Disord*, 15 (Suppl 3), S237–S240.

Derkach, V., Barria, A. and Soderling, T. R. (1999) Ca^{2+}/calmodulin-kinase II enhances channel conductance of alpha-amino-3-hydroxy-5-methyl-4-isoxazolepropionate type glutamate receptors. *Proc Natl Acad Sci USA*, 96, 3269–3274.

Ding, J. B., Guzman, J. N., Peterson, J. D., Goldberg, J. A. and Surmeier, D. J. (2010) Thalamic gating of corticostriatal signaling by cholinergic interneurons. *Neuron*, 67, 294–307.

Ding, J., Peterson, J. D. and Surmeier, D. J. (2008) Corticostriatal and thalamostriatal synapses have distinctive properties. *J Neurosci*, 28, 6483–6492.

Ding, L. and Perkel, D. J. (2004) Long-term potentiation in an avian basal ganglia nucleus essential for vocal learning. *J Neurosci*, 24, 488–494.

Dos Santos Villar, F. and Walsh, J. P. (1999) Modulation of long-term synaptic plasticity at excitatory striatal synapses. *Neuroscience*, 90, 1031–1041.

Dudek, S. M. and Bear, M. F. (1992) Homosynaptic long-term depression in area CA1 of hippocampus and effects of N-methyl-D-aspartate receptor blockade. *Proc Natl Acad Sci USA*, 89, 4363–4367.

Edwards, F. A. (1995) Anatomy and electrophysiology of fast central synapses lead to a structural model for long-term potentiation. *Physiol Rev*, 75, 759–787.

Everitt, B. J., Belin, D., Economidou, D., Pelloux, Y., Dalley, J. W. and Robbins, T. W. (2008) Review. Neural mechanisms underlying the vulnerability to develop compulsive drug-seeking habits and addiction. *Philos Trans R Soc Lond B Biol Sci*, 363, 3125–3135.

Fino, E., Glowinski, J. and Venance, L. (2005) Bidirectional activity-dependent plasticity at corticostriatal synapses. *J Neurosci*, 25, 11279–11287.

Fisher, R. S., Levine, M. S., Sibley, D. R. and Ariano, M. A. (1994) D2 dopamine receptor protein location: Golgi impregnation-gold toned and ultrastructural analysis of the rat neostriatum. *J Neurosci Res*, 38, 551–564.

Fitzjohn, S. M., Kingston, A. E., Lodge, D. and Collingridge, G. L. (1999) DHPG-induced LTD in area CA1 of juvenile rat hippocampus; characterisation and sensitivity to novel mGlu receptor antagonists. *Neuropharmacology*, 38, 1577–1583.

Floresco, S. B., Blaha, C. D., Yang, C. R. and Phillips, A. G. (2001) Modulation of hippo-campal and amygdalar-evoked activity of nucleus accumbens neurons by dopamine: Cellular mechanisms of input selection. *J Neurosci*, 21, 2851–2860.

Fourgeaud, L., Mato, S., Bouchet, D., Hemar, A., Worley, P. F. and Manzoni, O. J. (2004) A single in vivo exposure to cocaine abolishes endocannabinoid-mediated long-term depression in the nucleus accumbens. *J Neurosci*, 24, 6939–6945.

Freund, T. F., Powell, J. F. and Smith, A. D. (1984) Tyrosine hydroxylase-immunoreactive boutons in synaptic contact with identified striatonigral neurons, with particular reference to dendritic spines. *Neuroscience*, 13, 1189–1215.

Fukazawa, Y., Saitoh, Y., Ozawa, F., Ohta, Y., Mizuno, K. and Inokuchi, K. (2003) Hippocampal LTP is accompanied by enhanced F-actin content within the dendritic spine that is essential for late LTP maintenance in vivo. *Neuron*, 38, 447–460.

Fukunaga, K., Stoppini, L., Miyamoto, E. and Muller, D. (1993) Long-term potentiation is associated with an increased activity of Ca^{2+}/calmodulin-dependent protein kinase II. *J Biol Chem*, 268, 7863–7867.

Galarraga, E., Bargas, J., Martinez-Fong, D. and Aceves, J. (1987) Spontaneous synaptic potentials in dopamine-denervated neostriatal neurons. *Neurosci Lett*, 81, 351–355.

Gallagher, S. M., Daly, C. A., Bear, M. F. and Huber, K. M. (2004) Extracellular signal-regulated protein kinase activation is required for metabotropic glutamate receptor-dependent long-term depression in hippocampal area CA1. *J Neurosci*, 24, 4859–4864.

Gardoni, F., Picconi, B., Ghiglieri, V., Polli, F., Bagetta, V., Bernardi, G., Cattabeni, F., Di Luca, M. and Calabresi, P. (2006) A critical interaction between NR2B and MAGUK in L-DOPA induced dyskinesia. *J Neurosci*, 26, 2914–2922.

Gerdeman, G. and Lovinger, D. M. (2001) CB1 cannabinoid receptor inhibits synaptic release of glutamate in rat dorsolateral striatum. *J Neurophysiol*, 85, 468–471.

Gerdeman, G. L., Ronesi, J. and Lovinger, D. M. (2002) Postsynaptic endocannabinoid release is critical to long-term depression in the striatum. *Nat Neurosci*, 5, 446–451.

Giese, K. P., Fedorov, N. B., Filipkowski, R. K. and Silva, A. J. (1998) Autophosphorylation at Thr286 of the alpha calcium-calmodulin kinase II in LTP and learning. *Science*, 279, 870–873.

Giuffrida, A., Parsons, L. H., Kerr, T. M., Rodriguez De Fonseca, F., Navarro, M. and Piomelli, D. (1999) Dopamine activation of endogenous cannabinoid signaling in dorsal striatum. *Nat Neurosci*, 2, 358–363.

Gladding, C. M., Collett, V. J., Jia, Z., Bashir, Z. I., Collingridge, G. L. and Molnar, E. (2009) Tyrosine dephosphorylation regulates AMPAR internalisation in mGluR-LTD. *Mol Cell Neurosci*, 40, 267–279.

Gradinaru, V., Mogri, M., Thompson, K. R., Henderson, J. M. and Deisseroth, K. (2009) Optical deconstruction of parkinsonian neural circuitry. *Science*, 324, 354–359.

Graybiel, A. M. (2008) Habits, rituals, and the evaluative brain. *Annu Rev Neurosci*, 31, 359–387.

Gubellini, P., Picconi, B., Bari, M., Battista, N., Calabresi, P., Centonze, D., Bernardi, G., Finazzi-Agro, A. and Maccarrone, M. (2002) Experimental parkinsonism alters endocannabinoid degradation: Implications for striatal glutamatergic transmission. *J Neurosci*, 22, 6900–6907.

Gubellini, P., Saulle, E., Centonze, D., Costa, C., Tropepi, D., Bernardi, G., Conquet, F. and Calabresi, P. (2003) Corticostriatal LTP requires combined mGluR1 and mGluR5 activation. *Neuropharmacology*, 44, 8–16.

Guigoni, C., Aubert, I., Li, Q., Gurevich, V. V., Benovic, J. L., Ferry, S., Mach, U., Stark, H., Leriche, L., Hakansson, K., Bioulac, B. H., Gross, C. E., Sokoloff, P., Fisone, G., Gurevich, E. V., Bloch, B. and Bezard, E. (2005) Pathogenesis of levodopa-induced dyskinesia: Focus on D1 and D3 dopamine receptors. *Parkinsonism Relat Disord*, 11 (Suppl 1), S25–S29.

Guillin, O., Griffon, N., Bezard, E., Leriche, L., Diaz, J., Gross, C. and Sokoloff, P. (2003) Brain-derived neurotrophic factor controls dopamine D3 receptor expression: Therapeutic implications in Parkinson's disease. *Eur J Pharmacol*, 480, 89–95.

Hansel, C., De Jeu, M., Belmeguenai, A., Houtman, S. H., Buitendijk, G. H., Andreev, D., DeZeeuw, C. I. and Elgersma, Y. (2006) AlphaCaMKII is essential for cerebellar LTD and motor learning. *Neuron*, 51, 835–843.

Harvey, J. and Lacey, M. G. (1996) Endogenous and exogenous dopamine depress EPSCs in rat nucleus accumbens in vitro via D1 receptors activation. *J Physiol*, 492 (Pt 1), 143–154.

Hayashi, Y., Shi, S. H., Esteban, J. A., Piccini, A., Poncer, J. C. and Malinow, R. (2000) Driving AMPA receptors into synapses by LTP and CaMKII: Requirement for GluR1 and PDZ domain interaction. *Science*, 287, 2262–2267.

Hersch, S. M., Ciliax, B. J., Gutekunst, C. A., Rees, H. D., Heilman, C. J., Yung, K. K., Bolam, J. P., Ince, E., Yi, H. and Levey, A. I. (1995) Electron microscopic analysis of D1 and D2 dopamine receptor proteins in the dorsal striatum and their synaptic relationships with motor corticostriatal afferents. *J Neurosci*, 15, 5222–5237.

Higley, M. J. and Sabatini, B. L. (2010) Competitive regulation of synaptic Ca^{2+} influx by D2 dopamine and A2A adenosine receptors. *Nat Neurosci*, 13, 958–966.

Hoffman, A. F., Oz, M., Caulder, T. and Lupica, C. R. (2003) Functional tolerance and blockade of long-term depression at synapses in the nucleus accumbens after chronic cannabinoid exposure. *J Neurosci*, 23, 4815–4820.

Huang, C. C. and Hsu, K. S. (2006) Sustained activation of metabotropic glutamate receptor 5 and protein tyrosine phosphatases mediate the expression of (S)-3,5-dihydroxyphenylglycine-induced long-term depression in the hippocampal CA1 region. *J Neurochem*, 96, 179–194.

Huang, Y. H., Lin, Y., Mu, P., Lee, B. R., Brown, T. E., Wayman, G., Marie, H., Liu, W., Yan, Z., Sorg, B. A., Schluter, O. M., Zukin, R. S. and Dong, Y. (2009) In vivo cocaine experience generates silent synapses. *Neuron*, 63, 40–47.

Huang, C. C., Lo, S. W. and Hsu, K. S. (2001) Presynaptic mechanisms underlying cannabinoid inhibition of excitatory synaptic transmission in rat striatal neurons. *J Physiol*, 532, 731–748.

Huber, K. M., Kayser, M. S. and Bear, M. F. (2000) Role for rapid dendritic protein synthesis in hippocampal mGluR-dependent long-term depression. *Science*, 288, 1254–1257.

Huber, K. M., Roder, J. C. and Bear, M. F. (2001) Chemical induction of mGluR5- and protein synthesis–dependent long-term depression in hippocampal area CA1. *J Neurophysiol*, 86, 321–325.

Ingham, C. A., Hood, S. H. and Arbuthnott, G. W. (1989) Spine density on neostriatal neurons changes with 6-hydroxydopamine lesions and with age. *Brain Res*, 503, 334–338.

Ingham, C. A., Hood, S. H., Taggart, P. and Arbuthnott, G. W. (1998) Plasticity of synapses in the rat neostriatum after unilateral lesion of the nigrostriatal dopaminergic pathway. *J Neurosci*, 18, 4732–4743.

Ingham, C. A., Hood, S. H., Van Maldegem, B., Weenink, A. and Arbuthnott, G. W. (1993) Morphological changes in the rat neostriatum after unilateral 6-hydroxydopamine injections into the nigrostriatal pathway. *Exp Brain Res*, 93, 17–27.

Ito, M., Sakurai, M. and Tongroach, P. (1982) Climbing fibre induced depression of both mossy fibre responsiveness and glutamate sensitivity of cerebellar Purkinje cells. *J Physiol*, 324, 113–134.

Johnston, J. G., Gerfen, C. R., Haber, S. N. and Van Der Kooy, D. (1990) Mechanisms of striatal pattern formation: Conservation of mammalian compartmentalization. *Brain Res Dev Brain Res*, 57, 93–102.

Jourdain, P., Fukunaga, K. and Muller, D. (2003) Calcium/calmodulin-dependent protein kinase II contributes to activity-dependent filopodia growth and spine formation. *J Neurosci*, 23, 10645–10649.

Kano, M. and Kato, M. (1987) Quisqualate receptors are specifically involved in cerebellar synaptic plasticity. *Nature*, 325, 276–279.

Kawasaki, H., Fujii, H., Gotoh, Y., Morooka, T., Shimohama, S., Nishida, E. and Hirano, T. (1999) Requirement for mitogen-activated protein kinase in cerebellar long term depression. *J Biol Chem*, 274, 13498–13502.

Kemp, J. M. and Powell, T. P. (1971) The synaptic organization of the caudate nucleus. *Philos Trans R Soc Lond B Biol Sci*, 262, 403–412.

Kerr, J. N. and Wickens, J. R. (2001) Dopamine D-1/D-5 receptor activation is required for long-term potentiation in the rat neostriatum in vitro. *J Neurophysiol*, 85, 117–124.

Kofalvi, A., Rodrigues, R. J., Ledent, C., Mackie, K., Vizi, E. S., Cunha, R. A. and Sperlagh, B. (2005) Involvement of cannabinoid receptors in the regulation of neurotransmitter release in the rodent striatum: A combined immunochemical and pharmacological analysis. *J Neurosci*, 25, 2874–2884.

Kombian, S. B. and Malenka, R. C. (1994) Simultaneous LTP of non-NMDA- and LTD of NMDA-receptor-mediated responses in the nucleus accumbens. *Nature*, 368, 242–246.

Kourrich, S., Rothwell, P. E., Klug, J. R. and Thomas, M. J. (2007) Cocaine experience controls bidirectional synaptic plasticity in the nucleus accumbens. *J Neurosci*, 27, 7921–7928.

Kourrich, S. and Thomas, M. J. (2009) Similar neurons, opposite adaptations: Psychostimulant experience differentially alters firing properties in accumbens core versus shell. *J Neurosci*, 29, 12275–12283.

Kravitz, A. V., Freeze, B. S., Parker, P. R., Kay, K., Thwin, M. T., Deisseroth, K. and Kreitzer, A. C. (2010) Regulation of parkinsonian motor behaviours by optogenetic control of basal ganglia circuitry. *Nature*, 466, 622–626.

Kreitzer, A. C. and Malenka, R. C. (2005) Dopamine modulation of state-dependent endocannabinoid release and long-term depression in the striatum. *J Neurosci*, 25, 10537–10545.

Kreitzer, A. C. and Malenka, R. C. (2007) Endocannabinoid-mediated rescue of striatal LTD and motor deficits in Parkinson's disease models. *Nature*, 445, 643–647.

Lee, H. K., Barbarosie, M., Kameyama, K., Bear, M. F. and Huganir, R. L. (2000) Regulation of distinct AMPA receptor phosphorylation sites during bidirectional synaptic plasticity. *Nature*, 405, 955–959.

Lee, S. J., Escobedo-Lozoya, Y., Szatmari, E. M. and Yasuda, R. (2009) Activation of CaMKII in single dendritic spines during long-term potentiation. *Nature*, 458, 299–304.

Lee, H. K., Kameyama, K., Huganir, R. L. and Bear, M. F. (1998) NMDA induces long-term synaptic depression and dephosphorylation of the GluR1 subunit of AMPA receptors in hippocampus. *Neuron*, 21, 1151–1162.

Lengyel, I., Voss, K., Cammarota, M., Bradshaw, K., Brent, V., Murphy, K. P., Giese, K. P., Rostas, J. A. and Bliss, T. V. (2004) Autonomous activity of CaMKII is only transiently increased following the induction of long-term potentiation in the rat hippocampus. *Eur J Neurosci*, 20, 3063–3072.

Li, Y. and Kauer, J. A. (2004) Repeated exposure to amphetamine disrupts dopaminergic modulation of excitatory synaptic plasticity and neurotransmission in nucleus accumbens. *Synapse*, 51, 1–10.

Liao, D., Hessler, N. A. and Malinow, R. (1995) Activation of postsynaptically silent synapses during pairing-induced LTP in CA1 region of hippocampal slice. *Nature*, 375, 400–404.

Liao, D., Jones, A. and Malinow, R. (1992) Direct measurement of quantal changes underlying long-term potentiation in CA1 hippocampus. *Neuron*, 9, 1089–1097.

Lindefors, N. and Ungerstedt, U. (1990) Bilateral regulation of glutamate tissue and extracellular levels in caudate-putamen by midbrain dopamine neurons. *Neurosci Lett*, 115, 248–252.

Linden, D. J. and Connor, J. A. (1991) Participation of postsynaptic PKC in cerebellar long-term depression in culture. *Science*, 254, 1656–1659.

Lisman, J. E. and Harris, K. M. (1993) Quantal analysis and synaptic anatomy—Integrating two views of hippocampal plasticity. *Trends Neurosci*, 16, 141–147.

Lisman, J., Schulman, H. and Cline, H. (2002) The molecular basis of CaMKII function in synaptic and behavioural memory. *Nat Rev Neurosci*, 3, 175–190.

Lisman, J. E. and Zhabotinsky, A. M. (2001) A model of synaptic memory: A CaMKII/PP1 switch that potentiates transmission by organizing an AMPA receptor anchoring assembly. *Neuron*, 31, 191–201.

Lledo, P. M., Hjelmstad, G. O., Mukherji, S., Soderling, T. R., Malenka, R. C. and Nicoll, R. A. (1995) Calcium/calmodulin-dependent kinase II and long-term potentiation enhance synaptic transmission by the same mechanism. *Proc Natl Acad Sci USA*, 92, 11175–11179.

Lovinger, D. M., Partridge, J. G. and Tang, K. C. (2003) Plastic control of striatal glutamatergic transmission by ensemble actions of several neurotransmitters and targets for drugs of abuse. *Ann N Y Acad Sci*, 1003, 226–240.

Lynch, G., Larson, J., Kelso, S., Barrionuevo, G. and Schottler, F. (1983) Intracellular injections of EGTA block induction of hippocampal long-term potentiation. *Nature*, 305, 719–721.

Madison, D. V., Malenka, R. C. and Nicoll, R. A. (1991) Mechanisms underlying long-term potentiation of synaptic transmission. *Annu Rev Neurosci*, 14, 379–397.

Magee, J. C. and Johnston, D. (1997) A synaptically controlled, associative signal for Hebbian plasticity in hippocampal neurons. *Science*, 275, 209–213.

Malenka, R. C., Kauer, J. A., Perkel, D. J., Mauk, M. D., Kelly, P. T., Nicoll, R. A. and Waxham, M. N. (1989) An essential role for postsynaptic calmodulin and protein kinase activity in long-term potentiation. *Nature*, 340, 554–557.

Malenka, R. C., Kauer, J. A., Zucker, R. S. and Nicoll, R. A. (1988) Postsynaptic calcium is sufficient for potentiation of hippocampal synaptic transmission. *Science*, 242, 81–84.

Malenka, R. C., Lancaster, B. and Zucker, R. S. (1992) Temporal limits on the rise in postsynaptic calcium required for the induction of long-term potentiation. *Neuron*, 9, 121–128.

Malenka, R. C. and Nicoll, R. A. (1999) Long-term potentiation—A decade of progress? *Science*, 285, 1870–1874.

Malinow, R. and Malenka, R. C. (2002) AMPA receptor trafficking and synaptic plasticity. *Annu Rev Neurosci*, 25, 103–126.

Malinow, R., Schulman, H. and Tsien, R. W. (1989) Inhibition of postsynaptic PKC or CaMKII blocks induction but not expression of LTP. *Science*, 245, 862–866.

Mameli, M., Balland, B., Lujan, R. and Luscher, C. (2007) Rapid synthesis and synaptic insertion of GluR2 for mGluR-LTD in the ventral tegmental area. *Science*, 317, 530–533.

Mammen, A. L., Kameyama, K., Roche, K. W. and Huganir, R. L. (1997) Phosphorylation of the alpha-amino-3-hydroxy-5-methylisoxazole4-propionic acid receptor GluR1 subunit by calcium/calmodulin-dependent kinase II. *J Biol Chem*, 272, 32528–32533.

Manabe, T., Renner, P. and Nicoll, R. A. (1992) Postsynaptic contribution to long-term potentiation revealed by the analysis of miniature synaptic currents. *Nature*, 355, 50–55.

Marin, C., Aguilar, E., Mengod, G., Cortes, R. and Obeso, J. A. (2009) Effects of early vs. late initiation of levodopa treatment in hemiparkinsonian rats. *Eur J Neurosci*, 30, 823–832.

Markram, H., Lubke, J., Frotscher, M. and Sakmann, B. (1997) Regulation of synaptic efficacy by coincidence of postsynaptic APs and EPSPs. *Science*, 275, 213–215.

Martin, M., Chen, B. T., Hopf, F. W., Bowers, M. S. and Bonci, A. (2006) Cocaine self-administration selectively abolishes LTD in the core of the nucleus accumbens. *Nat Neurosci*, 9, 868–869.

Martin, A. B., Fernandez-Espejo, E., Ferrer, B., Gorriti, M. A., Bilbao, A., Navarro, M., Rodriguez De Fonseca, F. and Moratalla, R. (2008) Expression and function of CB1 receptor in the rat striatum: Localization and effects on D1 and D2 dopamine receptor-mediated motor behaviors. *Neuropsychopharmacology*, 33, 1667–1679.

Mato, S., Chevaleyre, V., Robbe, D., Pazos, A., Castillo, P. E. and Manzoni, O. J. (2004) A single in-vivo exposure to delta 9THC blocks endocannabinoid-mediated synaptic plasticity. *Nat Neurosci*, 7, 585–586.

Matsuda, W., Furuta, T., Nakamura, K. C., Hioki, H., Fujiyama, F., Arai, R. and Kaneko, T. (2009) Single nigrostriatal dopaminergic neurons form widely spread and highly dense axonal arborizations in the neostriatum. *J Neurosci*, 29, 444–453.

Matsuzaki, M., Honkura, N., Ellis-Davies, G. C. and Kasai, H. (2004) Structural basis of long-term potentiation in single dendritic spines. *Nature*, 429, 761–766.

Matyas, F., Yanovsky, Y., Mackie, K., Kelsch, W., Misgeld, U. and Freund, T. F. (2006) Subcellular localization of type 1 cannabinoid receptors in the rat basal ganglia. *Neuroscience*, 137, 337–361.

Mayford, M., Bach, M. E., Huang, Y. Y., Wang, L., Hawkins, R. D. and Kandel, E. R. (1996) Control of memory formation through regulated expression of a CaMKII transgene. *Science*, 274, 1678–1683.

Mayford, M., Wang, J., Kandel, E. R. and O'dell, T. J. (1995) CaMKII regulates the frequency-response function of hippocampal synapses for the production of both LTD and LTP. *Cell*, 81, 891–904.

Mazzucchelli, C., Vantaggiato, C., Ciamei, A., Fasano, S., Pakhotin, P., Krezel, W., Welzl, H., Wolfer, D. P., Pages, G., Valverde, O., Marowsky, A., Porrazzo, A., Orban, P. C., Maldonado, R., Ehrengruber, M. U., Cestari, V., Lipp, H. P., Chapman, P. F., Pouyssegur, J. and Brambilla, R. (2002) Knockout of ERK1 MAP kinase enhances synaptic plasticity in the striatum and facilitates striatal-mediated learning and memory. *Neuron*, 34, 807–820.

Miller, S. G. and Kennedy, M. B. (1986) Regulation of brain type II Ca^{2+}/calmodulin-dependent protein kinase by autophosphorylation: A Ca^{2+}-triggered molecular switch. *Cell*, 44, 861–870.

Mirenowicz, J. and Schultz, W. (1994) Importance of unpredictability for reward responses in primate dopamine neurons. *J Neurophysiol*, 72, 1024–1027.

Mirenowicz, J. and Schultz, W. (1996) Preferential activation of midbrain dopamine neurons by appetitive rather than aversive stimuli. *Nature*, 379, 449–451.

Moult, P. R., Correa, S. A., Collingridge, G. L., Fitzjohn, S. M. and Bashir, Z. I. (2008) Co-activation of p38 mitogen-activated protein kinase and protein tyrosine phosphatase underlies metabotropic glutamate receptor-dependent long-term depression. *J Physiol*, 586, 2499–2510.

Moult, P. R., Gladding, C. M., Sanderson, T. M., Fitzjohn, S. M., Bashir, Z. I., Molnar, E. and Collingridge, G. L. (2006) Tyrosine phosphatases regulate AMPA receptor trafficking during metabotropic glutamate receptor-mediated long-term depression. *J Neurosci*, 26, 2544–2554.

Moult, P. R., Schnabel, R., Kilpatrick, I. C., Bashir, Z. I. and Collingridge, G. L. (2002) Tyrosine dephosphorylation underlies DHPG-induced LTD. *Neuropharmacology*, 43, 175–180.

Mulkey, R. M., Endo, S., Shenolikar, S. and Malenka, R. C. (1994) Involvement of a calcineurin/inhibitor-1 phosphatase cascade in hippocampal long-term depression. *Nature*, 369, 486–488.

Mulkey, R. M., Herron, C. E. and Malenka, R. C. (1993) An essential role for protein phosphatases in hippocampal long-term depression. *Science*, 261, 1051–1055.

Mulkey, R. M. and Malenka, R. C. (1992) Mechanisms underlying induction of homosynaptic long-term depression in area CA1 of the hippocampus. *Neuron*, 9, 967–975.

Nakazato, T. (2005) Striatal dopamine release in the rat during a cued lever-press task for food reward and the development of changes over time measured using high-speed voltammetry. *Exp Brain Res*, 166, 137–146.

Nicola, S. M., Kombian, S. B. and Malenka, R. C. (1996) Psychostimulants depress excitatory synaptic transmission in the nucleus accumbens via presynaptic D1-like dopamine receptors. *J Neurosci*, 16, 1591–1604.

Norman, E. D., Egli, R. E., Colbran, R. J. and Winder, D. G. (2005) A potassium channel blocker induces a long-lasting enhancement of corticostriatal responses. *Neuropharmacology*, 48, 311–321.

Obeso, J. A., Olanow, C. W. and Nutt, J. G. (2000) Levodopa motor complications in Parkinson's disease. *Trends Neurosci*, 23, S2–S7.

Oliet, S. H., Malenka, R. C. and Nicoll, R. A. (1997) Two distinct forms of long-term depression coexist in CA1 hippocampal pyramidal cells. *Neuron*, 18, 969–982.

Otmakhov, N., Griffith, L. C. and Lisman, J. E. (1997) Postsynaptic inhibitors of calcium/calmodulin-dependent protein kinase type II block induction but not maintenance of pairing-induced long-term potentiation. *J Neurosci*, 17, 5357–5365.

Partridge, J. G., Apparsundaram, S., Gerhardt, G. A., Ronesi, J. and Lovinger, D. M. (2002) Nicotinic acetylcholine receptors interact with dopamine in induction of striatal long-term depression. *J Neurosci*, 22, 2541–2549.

Partridge, J. G., Tang, K. C. and Lovinger, D. M. (2000) Regional and postnatal heterogeneity of activity-dependent long-term changes in synaptic efficacy in the dorsal striatum. *J Neurophysiol*, 84, 1422–1429.

Pawlak, V. and Kerr, J. N. (2008) Dopamine receptor activation is required for corticostriatal spike-timing-dependent plasticity. *J Neurosci*, 28, 2435–2446.

Pennartz, C. M., Ameerun, R. F., Groenewegen, H. J. and Lopes Da Silva, F. H. (1993) Synaptic plasticity in an in vitro slice preparation of the rat nucleus accumbens. *Eur J Neurosci*, 5, 107–117.

Pettit, D. L., Perlman, S. and Malinow, R. (1994) Potentiated transmission and prevention of further LTP by increased CaMKII activity in postsynaptic hippocampal slice neurons. *Science*, 266, 1881–1885.

Picconi, B., Centonze, D., Hakansson, K., Bernardi, G., Greengard, P., Fisone, G., Cenci, M. A. and Calabresi, P. (2003) Loss of bidirectional striatal synaptic plasticity in L-DOPA-induced dyskinesia. *Nat Neurosci*, 6, 501–506.

Picconi, B., Centonze, D., Rossi, S., Bernardi, G. and Calabresi, P. (2004a) Therapeutic doses of L-dopa reverse hypersensitivity of corticostriatal D2-dopamine receptors and glutamatergic overactivity in experimental parkinsonism. *Brain*, 127, 1661–1669.

Picconi, B., Gardoni, F., Centonze, D., Mauceri, D., Cenci, M. A., Bernardi, G., Calabresi, P. and Di Luca, M. (2004b) Abnormal Ca^{2+}-calmodulin-dependent protein kinase II function mediates synaptic and motor deficits in experimental parkinsonism. *J Neurosci*, 24, 5283–5291.

Plenz, D. and Kitai, S. T. (1998) Up and down states in striatal medium spiny neurons simultaneously recorded with spontaneous activity in fast-spiking interneurons studied in cortex-striatum-substantia nigra organotypic cultures. *J Neurosci*, 18, 266–283.

Popescu, A. T., Saghyan, A. A. and Pare, D. (2007) NMDA-dependent facilitation of corticostriatal plasticity by the amygdala. *Proc Natl Acad Sci USA*, 104, 341–346.

Reynolds, J. N. and Wickens, J. R. (2000) Substantia nigra dopamine regulates synaptic plasticity and membrane potential fluctuations in the rat neostriatum, in vivo. *Neuroscience*, 99, 199–203.

Rich, R. C. and Schulman, H. (1998) Substrate-directed function of calmodulin in autophosphorylation of Ca^{2+}/calmodulin-dependent protein kinase II. *J Biol Chem*, 273, 28424–28429.

Robbe, D., Alonso, G., Chaumont, S., Bockaert, J. and Manzoni, O. J. (2002a) Role of p/q-Ca^{2+} channels in metabotropic glutamate receptor 2/3-dependent presynaptic long-term depression at nucleus accumbens synapses. *J Neurosci*, 22, 4346–4356.

Robbe, D., Alonso, G., Duchamp, F., Bockaert, J. and Manzoni, O. J. (2001) Localization and mechanisms of action of cannabinoid receptors at the glutamatergic synapses of the mouse nucleus accumbens. *J Neurosci*, 21, 109–116.

Robbe, D., Bockaert, J. and Manzoni, O. J. (2002b) Metabotropic glutamate receptor 2/3-dependent long-term depression in the nucleus accumbens is blocked in morphine withdrawn mice. *Eur J Neurosci*, 16, 2231–2235.

Robbe, D., Kopf, M., Remaury, A., Bockaert, J. and Manzoni, O. J. (2002c) Endogenous cannabinoids mediate long-term synaptic depression in the nucleus accumbens. *Proc Natl Acad Sci USA*, 99, 8384–8388.

Robbins, T. W., Ersche, K. D. and Everitt, B. J. (2008) Drug addiction and the memory systems of the brain. *Ann N Y Acad Sci*, 1141, 1–21.

Robinson, T. E. and Kolb, B. (2004) Structural plasticity associated with exposure to drugs of abuse. *Neuropharmacology*, 47 (Suppl 1), 33–46.

Ronesi, J. A. and Huber, K. M. (2008) Homer interactions are necessary for metabotropic glutamate receptor-induced long-term depression and translational activation. *J Neurosci*, 28, 543–547.

Ronesi, J. and Lovinger, D. M. (2005) Induction of striatal long-term synaptic depression by moderate frequency activation of cortical afferents in rat. *J Physiol*, 562, 245–256.

Sadikot, A. F., Parent, A., Smith, Y. and Bolam, J. P. (1992) Efferent connections of the centromedian and parafascicular thalamic nuclei in the squirrel monkey: A light and electron microscopic study of the thalamostriatal projection in relation to striatal heterogeneity. *J Comp Neurol*, 320, 228–242.

Sanhueza, M., Fernandez-Villalobos, G., Stein, I. S., Kasumova, G., Zhang, P., Bayer, K. U., Otmakhov, N., Hell, J. W. and Lisman, J. (2011) Role of the CaMKII/NMDA receptor complex in the maintenance of synaptic strength. *J Neurosci*, 31, 9170–9178.

Sanhueza, M., Mcintyre, C. C. and Lisman, J. E. (2007) Reversal of synaptic memory by Ca^{2+}/calmodulin-dependent protein kinase II inhibitor. *J Neurosci*, 27, 5190–5199.

Schramm, N. L., Egli, R. E. and Winder, D. G. (2002) LTP in the mouse nucleus accumbens is developmentally regulated. *Synapse*, 45, 213–219.

Schultz, W. (1997) Dopamine neurons and their role in reward mechanisms. *Curr Opin Neurobiol*, 7, 191–197.

Schultz, W. (1998) Predictive reward signal of dopamine neurons. *J Neurophysiol*, 80, 1–27.

Seiwell, A. P., Reveron, M. E. and Duvauchelle, C. L. (2007) Increased accumbens Cdk5 expression in rats after short-access to self-administered cocaine, but not after long-access sessions. *Neurosci Lett*, 417, 100–105.

Sesack, S. R., Aoki, C. and Pickel, V. M. (1994) Ultrastructural localization of D2 receptor-like immunoreactivity in midbrain dopamine neurons and their striatal targets. *J Neurosci*, 14, 88–106.

Shen, W., Flajolet, M., Greengard, P. and Surmeier, D. J. (2008) Dichotomous dopaminergic control of striatal synaptic plasticity. *Science*, 321, 848–851.

Sidibe, M. and Smith, Y. (1996) Differential synaptic innervation of striatofugal neurones projecting to the internal or external segments of the globus pallidus by thalamic afferents in the squirrel monkey. *J Comp Neurol*, 365, 445–465.

Singla, S., Kreitzer, A. C. and Malenka, R. C. (2007) Mechanisms for synapse specificity during striatal long-term depression. *J Neurosci*, 27, 5260–5264.

Smeal, R. M., Gaspar, R. C., Keefe, K. A. and Wilcox, K. S. (2007) A rat brain slice preparation for characterizing both thalamostriatal and corticostriatal afferents. *J Neurosci Methods*, 159, 224–235.

Smith, Y., Bennett, B. D., Bolam, J. P., Parent, A. and Sadikot, A. F. (1994) Synaptic relationships between dopaminergic afferents and cortical or thalamic input in the sensorimotor territory of the striatum in monkey. *J Comp Neurol*, 344, 1–19.

Smith, A. D. and Bolam, J. P. (1990) The neural network of the basal ganglia as revealed by the study of synaptic connections of identified neurones. *Trends Neurosci*, 13, 259–265.

Smith, Y. and Kieval, J. Z. (2000) Anatomy of the dopamine system in the basal ganglia. *Trends Neurosci*, 23, S28–S33.

Smith, Y., Raju, D. V., Pare, J. F. and Sidibe, M. (2004) The thalamostriatal system: A highly specific network of the basal ganglia circuitry. *Trends Neurosci*, 27, 520–527.

Song, I. and Huganir, R. L. (2002) Regulation of AMPA receptors during synaptic plasticity. *Trends Neurosci*, 25, 578–588.

Squire, L. R. (2003) *Fundamental Neuroscience*, San Diego, CA, Academic.

Sung, K. W., Choi, S. and Lovinger, D. M. (2001) Activation of group I mGluRs is necessary for induction of long-term depression at striatal synapses. *J Neurophysiol*, 86, 2405–2412.

Szabo, B., Dorner, L., Pfreundtner, C., Norenberg, W. and Starke, K. (1998) Inhibition of GABAergic inhibitory postsynaptic currents by cannabinoids in rat corpus striatum. *Neuroscience*, 85, 395–403.

Tang, K., Low, M. J., Grandy, D. K. and Lovinger, D. M. (2001) Dopamine-dependent synaptic plasticity in striatum during in vivo development. *Proc Natl Acad Sci USA*, 98, 1255–1260.

Taylor, J. R., Lynch, W. J., Sanchez, H., Olausson, P., Nestler, E. J. and Bibb, J. A. (2007) Inhibition of Cdk5 in the nucleus accumbens enhances the locomotor-activating and incentive-motivational effects of cocaine. *Proc Natl Acad Sci USA*, 104, 4147–4152.

Tepper, J. M., Abercrombie, E. D. and Bolam, J. P. (2007) Basal ganglia macrocircuits. *Prog Brain Res*, 160, 3–7.

Thomas, M. J., Beurrier, C., Bonci, A. and Malenka, R. C. (2001) Long-term depression in the nucleus accumbens: A neural correlate of behavioral sensitization to cocaine. *Nat Neurosci*, 4, 1217–1223.

Thomas, G. M. and Huganir, R. L. (2004) MAPK cascade signalling and synaptic plasticity. *Nat Rev Neurosci*, 5, 173–183.

Thomas, M. J., Malenka, R. C. and Bonci, A. (2000) Modulation of long-term depression by dopamine in the mesolimbic system. *J Neurosci*, 20, 5581–5586.

Tokumitsu, H., Chijiwa, T., Hagiwara, M., Mizutani, A., Terasawa, M. and Hidaka, H. (1990) KN-62, 1-[N,O-bis(5-isoquinolinesulfonyl)-N-methyl-L-tyrosyl]-4-phenylpiperazine, a specific inhibitor of Ca^{2+}/calmodulin-dependent protein kinase II. *J Biol Chem*, 265, 4315–4320.

Tsou, K., Brown, S., Sanudo-Pena, M. C., Mackie, K. and Walker, J. M. (1998) Immunohistochemical distribution of cannabinoid CB1 receptors in the rat central nervous system. *Neuroscience*, 83, 393–411.

Volk, L. J., Daly, C. A. and Huber, K. M. (2006) Differential roles for group 1 mGluR subtypes in induction and expression of chemically induced hippocampal long-term depression. *J Neurophysiol*, 95, 2427–2438.

Wan, H., Mackay, B., Iqbal, H., Naskar, S. and Kemenes, G. (2010) Delayed intrinsic activation of an NMDA-independent CaM-kinase II in a critical time window is necessary for late consolidation of an associative memory. *J Neurosci*, 30, 56–63.

Wang, Q., Chang, L., Rowan, M. J. and Anwyl, R. (2007) Developmental dependence, the role of the kinases p38 MAPK and PKC, and the involvement of tumor necrosis factor-R1 in the induction of mGlu-5 LTD in the dentate gyrus. *Neuroscience*, 144, 110–118.

Wang, Z., Kai, L., Day, M., Ronesi, J., Yin, H. H., Ding, J., Tkatch, T., Lovinger, D. M. and Surmeier, D. J. (2006) Dopaminergic control of corticostriatal long-term synaptic depression in medium spiny neurons is mediated by cholinergic interneurons. *Neuron*, 50, 443–452.

Wang, Y. T. and Linden, D. J. (2000) Expression of cerebellar long-term depression requires postsynaptic clathrin-mediated endocytosis. *Neuron*, 25, 635–647.

Wang, H. and Pickel, V. M. (2002) Dopamine D2 receptors are present in prefrontal cortical afferents and their targets in patches of the rat caudate-putamen nucleus. *J Comp Neurol*, 442, 392–404.

Wang, H., Shimizu, E., Tang, Y. P., Cho, M., Kyin, M., Zuo, W., Robinson, D. A., Alaimo, P. J., Zhang, C., Morimoto, H., Zhuo, M., Feng, R., Shokat, K. M. and Tsien, J. Z. (2003) Inducible protein knockout reveals temporal requirement of CaMKII reactivation for memory consolidation in the brain. *Proc Natl Acad Sci USA*, 100, 4287–4292.

Warre, R., Thiele, S., Talwar, S., Kamal, M., Johnson, T. H., Wang, S., Lam, D., Lo, C., Khademullah, C. S., Perera, G., Reyes, G., Sun, X. S., Brotchie, J. M. and Nash, J. E. (2011) Altered function of glutamatergic cortico-striatal synapses causes output pathway abnormalities in a chronic model of parkinsonism. *Neurobiol Dis*, 41, 591–604.

Wickens, J. R., Begg, A. J. and Arbuthnott, G. W. (1996) Dopamine reverses the depression of rat corticostriatal synapses which normally follows high-frequency stimulation of cortex in vitro. *Neuroscience*, 70, 1–5.

Wickens, J. R., Mckenzie, D., Costanzo, E. and Arbuthnott, G. W. (1998) Effects of potassium channel blockers on synaptic plasticity in the corticostriatal pathway. *Neuropharmacology*, 37, 523–533.

Wilson, C. J. and Kawaguchi, Y. (1996) The origins of two-state spontaneous membrane potential fluctuations of neostriatal spiny neurons. *J Neurosci*, 16, 2397–2410.

Wise, R. A. (2004) Dopamine and food reward: Back to the elements. *Am J Physiol Regul Integr Comp Physiol*, 286, R13.

Wu, Y., Richard, S. and Parent, A. (2000) The organization of the striatal output system: A single-cell juxtacellular labeling study in the rat. *Neurosci Res*, 38, 49–62.

Xie, Z., Srivastava, D. P., Photowala, H., Kai, L., Cahill, M. E., Woolfrey, K. M., Shum, C. Y., Surmeier, D. J. and Penzes, P. (2007) Kalirin-7 controls activity-dependent structural and functional plasticity of dendritic spines. *Neuron*, 56, 640–656.

Xu, M., Guo, Y., Vorhees, C. V. and Zhang, J. (2000) Behavioral responses to cocaine and amphetamine administration in mice lacking the dopamine D1 receptor. *Brain Res*, 852, 198–207.

Yamagata, Y., Kobayashi, S., Umeda, T., Inoue, A., Sakagami, H., Fukaya, M., Watanabe, M., Hatanaka, N., Totsuka, M., Yagi, T., Obata, K., Imoto, K., Yanagawa, Y., Manabe, T. and Okabe, S. (2009) Kinase-dead knock-in mouse reveals an essential role of kinase activity of Ca^{2+}/calmodulin-dependent protein kinase IIalpha in dendritic spine enlargement, long-term potentiation, and learning. *J Neurosci*, 29, 7607–7618.

Yamamoto, Y., Nakanishi, H., Takai, N., Shimazoe, T., Watanabe, S. and Kita, H. (1999) Expression of N-methyl-D-aspartate receptor-dependent long-term potentiation in the neostriatal neurons in an in vitro slice after ethanol withdrawal of the rat. *Neuroscience*, 91, 59–68.

Yao, W. D., Gainetdinov, R. R., Arbuckle, M. I., Sotnikova, T. D., Cyr, M., Beaulieu, J. M., Torres, G. E., Grant, S. G. and Caron, M. G. (2004) Identification of PSD-95 as a regulator of dopamine-mediated synaptic and behavioral plasticity. *Neuron*, 41, 625–638.

Yin, H. H. and Knowlton, B. J. (2006) The role of the basal ganglia in habit formation. *Nat Rev Neurosci*, 7, 464–476.

Yin, H. H. and Lovinger, D. M. (2006) Frequency-specific and D2 receptor-mediated inhibition of glutamate release by retrograde endocannabinoid signaling. *Proc Natl Acad Sci USA*, 103, 8251–8256.

Yin, H. H., Mulcare, S. P., Hilario, M. R., Clouse, E., Holloway, T., Davis, M. I., Hansson, A. C., Lovinger, D. M. and Costa, R. M. (2009) Dynamic reorganization of striatal circuits during the acquisition and consolidation of a skill. *Nat Neurosci*, 12, 333–341.

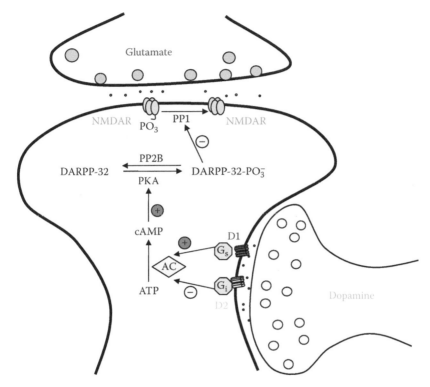

FIGURE 3.1 Cartoon representation of dopamine receptor modulation of NMDA receptors. In this classical pathway, D1 receptors couple to G_s G-proteins, raising cAMP and resulting in protein kinase A (PKA) phosphorylation of DARPP-32 (the dopamine- and cyclic AMP-regulated phosphoprotein of Mr 32000) and inhibition of protein phosphatase-1 (PP1). The calcium- and calmodulin-dependent phosphatase, calcineurin (PP2B), dephosphorylates DARPP-32. NMDA receptor phosphorylation may increase the NMDA response, by directly increasing channel open probability or by stabilizing receptors at the cell surface. D2 receptor stimulation would be expected to give the opposite effect compared to D1 activation and the mechanism should be blocked by inhibitors of G-protein signaling or PKA. Because PKA can also directly phosphorylate the NMDA receptor on the GluN1 C-terminal, the balance of effect on NMDA responses is complicated to predict.

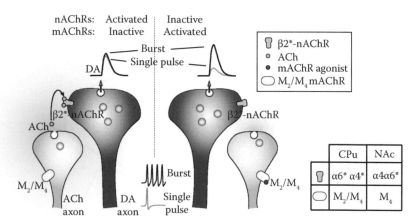

FIGURE 6.2 Scheme illustrating cholinergic regulation of dopamine release during burst and nonburst activity in dopamine axons and nAChR/mAChR subtypes responsible in CPu and NAc. Left, When mAChRs are inactive under control conditions, endogenous acetylcholine (ACh) released from tonically active striatal cholinergic interneurons (ChIs) maintains ACh tone at β2*-nAChRs on dopaminergic axons. This tonic β2*-nAChR activity ensures that dopamine (DA) release has a high probability of occurring in response to a single stimulus pulse. Short-term synaptic depression follows such DA release and limits re-release by successive pulses within bursts, that is, DA release is insensitive to frequency. Right, Activation of mAChR autoreceptors on ChIs (e.g., with mAChR agonist Oxo-M) reduces ACh tone at β2*-nAChRs on DA axon terminals, thereby reducing DA release probability by a single stimulus, relieving short-term depression and increasing the relative probability of release at subsequent action potentials. After reduction of ACh tone at β2*-nAChRs by activation of mAChR autoreceptors on ChIs, or blockade/desensitization of nAChRs, DA release becomes more sensitive to frequency of presynaptic activity (see Exley and Cragg 2008; Rice and Cragg 2004). In CPu, M_2 and M_4 populations of mAChRs and α6*- and α4*-nAChRs appear to modulate DA release. In NAc, only M_4-mAChRs and α4α6*-nAChRs appear to modulate DA release.

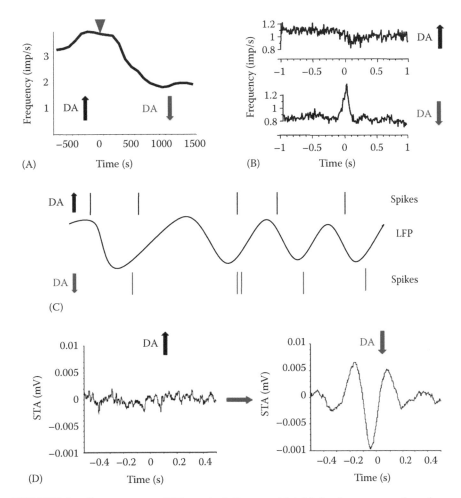

FIGURE 9.1 Consequences of DA manipulations on striatal firing frequency and synchrony. (A) Acute DA depletion in DDD mice results in a rapid decrease in the average activity of striatal neurons. The population activity in the striatum of a DDD mouse before and after DA depletion is depicted (time 0). (B) Rapid DA depletion results in an increase in synchrony in the activity of striatal MSNs. Example of cross-correlation of firing between the same pair of striatal MSNs before (top) and after (bottom) DA depletion in DDD mice. (C) Scheme depicting increased entrainment of MSN firing to the striatal LFP oscillations after DA depletion. (D) Increased entrainment of a striatal MSN to the LFP oscillations after acute DA receptor blockade. (Adapted from Costa, et al., 2006a; Burkhardt, J.M., et al., *Front Integr. Neurosci.*, 3, 28, 2009.) The spike-triggered average (STA) of the LFP oscillations measures if spikes occur preferentially at a particular phase of the LFP oscillations. Significant fluctuations of the STA of the LFP around time zero indicates that spikes occur at a particular phase of the LFP oscillations while a flat STA of the LFP around time zero indicates no relation between the time of the spike and the LFP oscillations.

5 Dopaminergic Modulation of Glutamatergic Synaptic Plasticity in Striatal Circuits
New Insights from BAC-Transgenic Mice

D. James Surmeier, Weixing Shen,
Tracy S. Gertler, and C. Savio Chan

CONTENTS

5.1 INTRODUCTION

The largest of the basal ganglia nuclei, the dorsal striatum, integrates information about sensory and motor state conveyed by cortical and thalamic neurons, facilitating the selection of actions that achieve desirable outcomes, like reward, and avoid undesirable ones. Current models of how this happens have been built upon the notion that reward prediction errors signaled by mesencephalic dopaminergic neurons innervating the striatum provide a means by which experience shapes the strength of glutamatergic corticostriatal synapses and, in so doing, shapes action selection (Yin and Knowlton, 2006; Schultz, 2007; Cohen and Frank, 2009). One of the most compelling pieces of evidence for this view comes from the difficulty of Parkinson's disease (PD) patients, who have lost their striatal dopaminergic innervation, to readily choose context appropriate actions (Dujardin and Laurent, 2003).

Although there is strong support for the basic tenets of these models, precisely how dopamine modulates the strength of corticostriatal synapses has been the subject of continuing debate. One of the experimental obstacles that have slowed physiological study is the cellular heterogeneity of the striatum and the seemingly random anatomical distribution of cell types within it (Chapter 4). The principal neurons of the striatum are medium spiny neurons (MSNs), constituting roughly 90% of all striatal neurons in most mammals (Kawaguchi, 1997). MSNs can be divided into at least two groups based upon their dopamine receptor expression and axonal projection site: striatopallidal MSNs send their principal axonal arbor to the globus pallidus (external GP in primates) and express high levels of the D_2 dopamine receptor whereas striatonigral MSNs send their principal axonal arbor to the substantia nigra (and internal GP in primates) and express high levels of the D_1 dopamine receptor (Gerfen et al., 1990). In physiological studies performed either *in vitro* or *in vivo*, these two types of MSNs have been virtually impossible to tell apart, clouding the interpretation of plasticity studies exploring the role of dopamine. The recent development of bacteria artificial chromosome (BAC)-transgenic mice in which the expression of D_1 or D_2 receptors is reported by the expression of red or green fluorescent protein (Gong et al., 2003) has eliminated this problem and led to a flurry of discoveries about striatal synaptic plasticity—providing the grist for this chapter.

5.2 DOPAMINERGIC MODULATION OF STRIATAL LTD

Long-term depression (LTD) at MSN glutamatergic synapses is the easiest form of synaptic plasticity to see in the dorsal striatum and, as a consequence, has been studied the most thoroughly. Unlike the situation at many other synapses, striatal LTD induction typically requires pairing of postsynaptic depolarization with moderate-to-high-frequency (not low-frequency, but see the following text) afferent stimulation at physiological temperatures (Kreitzer and Malenka, 2005). Typically for the induction to be successful, postsynaptic L-type Ca^{2+} channels and G_q-linked mGluR5 receptors need to be co-activated (Lovinger et al., 1993; Kreitzer and Malenka, 2005). Both L-type Ca^{2+} channels and mGluR5 receptors are appropriately positioned at glutamatergic synapses on MSN spines (Testa et al., 1994; Carter and Sabatini, 2004; Day et al., 2006; Carter et al., 2007). What is less clear is the nature of the interaction between these two membrane proteins in the process of induction. A clue has come from recent work showing that prolonging the opening of L-type channels with an allosteric modulator eliminates the need to stimulate mGluR5 receptors (Adermark and Lovinger, 2007), pointing to shared regulation of dendritic Ca^{2+} concentration, elevation of which is required for LTD induction. However, there is an asymmetry, as increasing mGluR5 activation by bath application of agonists does not eliminate the need for L-type Ca^{2+} channel opening (Ronesi et al., 2004; Kreitzer and Malenka, 2005). This might reflect a requirement for Ca^{2+}-induced Ca^{2+} release (CICR) from intracellular stores in LTD induction. In many cell types, CICR depends upon Ca^{2+} influx through L-type channels (Nakamura et al., 2000). Activation of mGluR5 and the production of inositol-1,4,5-triphosphate (IP3) could serve to prime these dendritic Ca^{2+} stores, boosting CICR evoked by activity-dependent Ca^{2+} entry through L-type Ca^{2+} channels and thus promoting LTD induction (Berridge, 1998; Wang et al., 2000; Taufiq Ur et al., 2009).

Although the induction of LTD is postsynaptic, its expression is presynaptic. The activity-induced elevation in dendritic Ca^{2+} concentration triggers the production of an endocannabinoid (EC) that diffuses to presynaptically located CB1 receptors. The combination of presynaptic CB1 receptor activation, spiking, and altered gene expression in the presynaptic cell leads to a long-term reduction in glutamate release (Lovinger, 2008). Having both pre- and postsynaptic determinants confers synaptic specificity on LTD expression (Singla et al., 2007). The molecular identity of the metabolic pathway leading to EC production in MSNs is still uncertain. This is important for a variety of reasons, not the least of which is knowledge of the critical signaling event responsible for triggering plasticity. There are two abundant striatal ECs: anandamide and 2-arachidonylglycerol (2-AG). Although previous studies have underscored the neural regulation of anandamide synthesis in the striatum (Giuffrida et al., 1999), collateral support for it as the obligate signaling molecule has been scant; but recent work has provided additional support for a role of anandamide, rather than 2-AG, in striatal synaptic plasticity (Ade and Lovinger, 2007). A dependence on anandamide would help to explain the ability of Ca^{2+} alone to induce EC-dependent LTD, as phospholipase D (one of key enzymes in the anandamide cascade) is activated by high intracellular Ca^{2+} (Brenowitz et al., 2006).

5.3   STRIATAL LTD: INSIGHT FROM DOPAMINE RECEPTOR BAC-TRANSGENIC MICE

One still unresolved question about the induction of striatal LTD is whether the activation of D_2 receptors is necessary. Activation of D_2 receptors is a potent stimulus for anandamide production (Giuffrida et al., 1999). However, recent work showing the sufficiency of L-type channel opening in EC-dependent LTD (Adermark and Lovinger, 2007) makes it clear that D_2 receptors play a modulatory—not obligatory—role. The real issue is the role of D_2 receptors in LTD induction using synaptic stimulation.

Attempts to address this question using BAC mice have consistently found that in D_2 receptor–expressing striatopallidal MSNs, D_2 receptor activation seems to be necessary (Wang et al., 2006; Kreitzer and Malenka, 2007; Shen et al., 2008). This could be due to the need to suppress A2a adenosine receptor signaling that prevents efficient EC synthesis (see the following text) (Fuxe et al., 2007b). The issue is whether EC-dependent LTD is inducible in the other major population of MSNs that do not express D_2 receptors—the D_1 receptor–dominated striatonigral MSNs. Kreitzer and Malenka (Kreitzer and Malenka, 2007) reported that LTD was not inducible in these MSNs using a minimal local stimulation. This result was confirmed subsequently (Shen et al., 2008). However, using macroelectrode stimulation, EC-dependent LTD is readily inducible in identified D_1 MSNs (Wang et al., 2006), consistent with the high probability of MSN induction seen in previous work (Calabresi et al., 2007). Thus, the stimulation paradigm seems critical to LTD induction in D_1 MSNs. The problem with these induction protocols is that the type of axon and cell activated by the electrical stimulus is poorly controlled. With intrastriatal stimulation or with nominal white matter stimulation in coronal brain slices, glutamatergic afferent fibers, dopaminergic fibers, and fibers intrinsic to the striatum are all activated, producing a mixture of neuromodulators that makes the interpretation of results less than straightforward. In Kreitzer and Malenka's case, minimal local stimulation of both dopaminergic and glutamatergic fibers appears to be critical to the LTD induction failure, as blocking D_1 receptors unmasked a robust EC-dependent LTD in D_1 MSNs (Shen et al., 2008). This kind of complication also appears to be responsible for the apparent D_2 receptor dependence of LTD induction in D_1 MSNs using macroelectrodes that more effectively activate cholinergic interneuron axons (Wang et al., 2006).

The neuromodulator mixture created by nonspecific electrical stimulation could also be a factor in slice studies implicating nitric oxide (NO) signaling in LTD induction. First, it must be acknowledged that this form of LTD might not be EC dependent, in spite of the fact that its induction occludes conventional HFS stimulation-induced LTD (Calabresi et al., 1999). Because activation of striatal interneurons containing nitric oxide synthase (NOS) depends upon NMDA and D_1/D_5 dopamine receptors (Ondracek et al., 2008), this form of LTD should be sensitive to antagonism of either. Interestingly, recent work using field potential recording has identified an LFS-induced form of LTD in dorsal striatum with these properties (Kung et al., 2007).

The lack of specificity in activating inputs to MSNs during the induction of plasticity also raises questions about the type of glutamatergic synapse being affected by EC-dependent LTD. Studies using nominal white matter or cortical stimulation in coronal brain slices typically assume that the glutamatergic fibers being stimulated are of cortical origin, but very few of these fibers are left intact in this preparation (Kawaguchi et al., 1989). The thalamic glutamatergic innervation of MSNs is similar in magnitude to that of the cerebral cortex, perhaps constituting as much as 40% of the total glutamatergic input to MSNs, terminating on both shafts and spines (Wilson, 2004). As a consequence, it is not really known whether EC-dependent LTD is present at corticostriatal or thalamostriatal synapses or both. The localization of CB1 receptors on corticostriatal terminals, but not thalamostriatal terminals (Uchigashima et al., 2007), is consistent with the hypothesis that LTD is a corticostriatal phenomenon, but more definitive studies are needed. Cutting brain slices in planes that preserve cortical and/or thalamic connectivity is one way to sort this out (Kawaguchi et al., 1989; Smeal et al., 2007; Ding et al., 2008). But these approaches have limitations, given the highly convergent nature of the glutamatergic input to

MSNs (Wilson, 2004). Optogenetic approaches offer a powerful alternative strategy (Zhang et al., 2006) that would allow glutamatergic inputs from various cortical and thalamic regions to be dissected.

5.4 LONG-TERM POTENTIATION AT GLUTAMATERGIC SYNAPSES ON MSNs

The mechanisms controlling induction and expression of long-term potentiation (LTP) at glutamatergic synapses are considerably more mysterious. Most of the work describing LTP at glutamatergic synapses has been done with sharp electrodes (either *in vivo* or *in vitro*), not with patch clamp electrodes in brain slices that afford greater experimental control and definition of the cellular and molecular determinants of induction. However, there have been a number of studies using these approaches in the last few years that have shed new light on LTP mechanisms. Previous studies have argued that LTP induced in MSNs by pairing HFS of glutamatergic inputs and postsynaptic depolarization depends upon co-activation of D_1 dopamine and NMDA receptors (Calabresi et al., 2007). The involvement of NMDA receptors in LTP induction is not controversial. What is controversial is the involvement of D_1 receptors. Robust expression of these receptors is only found in striatonigral MSNs, roughly half of the MSN population, making it difficult to understand how LTP induction could be universally dependent upon them unless some rather complicated, indirect mechanism was involved.

5.5 STRIATAL LTP: INSIGHT FROM DOPAMINE RECEPTOR BAC-TRANSGENIC MICE

Again, the advent of BAC-transgenic mice has provided a tool to sort this issue out. Using perforated patch recording to preserve the intracellular milieu controlling the induction of synaptic plasticity, our group found that the induction of LTP at glutamatergic synapses was dependent upon D_1 dopamine receptors only in striatonigral MSNs, not in D_2 receptor–expressing striatopallidal MSNs (Flajolet et al., 2008; Shen et al., 2008). In these MSNs, LTP induction required activation of A2a adenosine receptors. These receptors are robustly expressed in striatopallidal MSNs and have a very similar intracellular signaling linkage to that of D_1 receptors; that is, they positively couple to adenylyl cyclase and protein kinase A (PKA). A2a receptors also negatively couple to D_2 dopamine receptors linked to the induction of LTD (Fuxe et al., 2007a). Acting through PKA, D_1 and A2a receptor activation leads to the phosphorylation of DARPP-32 and a variety of other signaling molecules, including MAPKs, linked to synaptic plasticity (Sweatt, 2004).

The nature of the cooperativity between NMDA receptors and D_1/A2a receptor signaling in the induction of LTP remains to be resolved. Certainly, postsynaptic Ca^{2+} will figure prominently in this equation, but it is hard to map the simple model derived from the study of hippocampal synapses onto the striatum, given the apparent necessity of high intracellular Ca^{2+} concentration for the production of ECs and LTD induction. The site of Ca^{2+} entry will undoubtedly prove to be important, as this will contain information about the timing and magnitude of pre- and postsynaptic activity. For example, Ca^{2+} entry through NMDA receptors occurs only when glutamate is released at a time when there is sufficient postsynaptic depolarization to relieve Mg^{2+} block; scaffolding signaling molecules, like CaMKII, near the cytoplasmic mouth of these receptors could lead to their selective activation during conditions favoring LTP induction, and not during conditions favoring LTD induction, in spite of a very similar spatially averaged Ca^{2+} signal.

Neuromodulator signaling should also be a factor governing the effects of an activity-induced elevation in intracellular Ca^{2+} concentration. A concrete example of the role of neuromodulators in regulating the sign of synaptic plasticity can be found in our recent study of spike timing–dependent plasticity (STDP) using BAC-transgenic mice. In both types of MSN, STDP plasticity was Hebbian in the sense that when synaptic activation was followed by postsynaptic spiking, LTP was induced; whereas when the order of stimulation was reversed, LTD was induced. This rule for STDP in MSNs was also reported by Pawlak and Kerr (2008). The LTP and LTD induced in these protocols had all the features of plasticity induced by more conventional protocols, as outlined earlier, arguing that the core mechanisms were the same. However, the Hebbian character of the plasticity was malleable. For example, the sign and timing dependence of plasticity depended upon dopamine signaling. In striatopallidal MSNs, D_2 receptor signaling was necessary for STDP LTD induction when postsynaptic spiking preceded synaptic stimulation; when D_2 receptors were blocked and A2a receptors were stimulated, this pairing led to LTP induction. In contrast, in striatonigral MSNs, pairing postsynaptic spiking with a trailing presynaptic volley only produced LTD in the absence of D_1 receptor stimulation, suggesting that PKA signaling could abrogate LTD induction. Reversing the order of stimulation gave LTP only when D_1 receptors were stimulated and yielded LTD otherwise, arguing that PKA signaling not only could shut down LTD induction, but was also necessary for LTP induction. Conceptually similar results have been reported in other cell types (Seol et al., 2007; Tzounopoulos et al., 2007), leading to the notion that LTD and LTP induction is governed by "opponent processes" that interact at synaptic sites to determine the sign of synaptic plasticity. Altered activation of these processes could be responsible for "anti-Hebbian" plasticity reported in the striatum (Fino et al., 2005). How these opponent processes interact with one another and the cellular mechanisms underlying changes in synaptic strength remains to be determined. Molecules like RCS (regulator of calmodulin signaling), whose affinity for calmodulin and negative regulation of Ca^{2+} signaling is dramatically elevated by PKA phosphorylation, could be mediators of the opponent interaction (Xia and Storm, 2005).

The nature of this interaction also has implications for the distal reward problem (Sutton and Barto, 1981). The change in dopamine release produced by the consequences of action selection occurs later in time than the pre- and postsynaptic activity that produced the action. In theoretical treatments of this issue, there are two strategies for dealing with this temporal delay or distal reward. One way is to have temporally coincident pre- and postsynaptic activity create an eligibility trace that subsequently can be acted on by an outcome-dependent signal, in this case dopamine. However, if dopamine receptor signaling changes the impact of patterned synaptic stimulation on intracellular signaling cascades controlling the induction of plasticity, it is difficult to see how this could work. An alternative approach is to have the memory of the action selection be held not as a biochemical trace but as a network activity. Fictive activity in the network could be maintained for some seconds until the outcome of the action is determined and the change in dopamine brought about in the striatum. The looping structure of the basal ganglia–thalamus–cortex network is well suited to this kind of reverberating activity (Houk et al., 2007). Of course, this would not work for actions where the outcome is delayed for a greater period of time. It would seem that in this sort of scenario, the action leading to the outcome would have to be remembered and in the process of remembering the key corticostriatal circuits, underlying the action selection would have to reactivated, allowing dopamine to correctly modify synaptic strength. This reactivation of the corticostriatal circuitry would have to be rapid to coincide with the postsynaptic effects of dopamine.

5.6 WHAT TYPE OF STRIATAL ACTIVITY NORMALLY TRIGGERS THE INDUCTION OF SYNAPTIC PLASTICITY?

Although most of the induction protocols for synaptic plasticity that have been used to study striatal plasticity are decidedly unphysiological, involving sustained, strong depolarization, and/or high-frequency synaptic stimulation that induces dendritic depolarization, they do make the necessity of postsynaptic depolarization clear. In a physiological setting, what types of depolarization are likely to gate induction?

5.6.1 SPIKE TIMING–DEPENDENT PLASTICITY

One possibility is that spikes generated in the axon initial segment (AIS) propagate into dendritic regions where synapses are formed. Recent work has shown that STDP is present in MSNs (Fino et al., 2005; Pawlak and Kerr, 2008; Shen et al., 2008). But there are reasons to believe that this type of plasticity is relevant for only a subset of the synapses formed on MSNs. MSN dendrites are several 100 μm long, thin, and modestly branched. Their initial 20–30 μm are largely devoid of spines and glutamatergic synapses. Glutamatergic synapse and spine density peaks near 50 μm from the soma and then modestly declines with distance (Wilson, 2004). Because of their geometry and ion channel expression, AIS-generated spikes rapidly decline in amplitude as they invade MSN dendrites (as judged by their ability to open voltage-dependent Ca^{2+} channels), producing only a modest depolarization 80–100 μm from the soma. This is less than half the way to the dendritic tips (Day et al., 2008), arguing that a large portion of the synaptic surface area is not normally accessible to somatic feedback about the outcome of aggregate synaptic activity. High-frequency, repetitive somatic spiking improves dendritic invasion, but distal (>100 μm) synapses remain relatively inaccessible.

5.6.2 SYNAPTIC CONVERGENCE

What controls the induction of plasticity in the more distal dendritic regions, if not backpropagating action potentials? The situation in MSNs might be very similar to that found in hippocampal pyramidal neurons where somatically generated bAPs do not invade the apical dendritic tuft (Golding et al., 2002). In this region, convergent synaptic stimulation is capable of producing a local Ca^{2+} spike or plateau potential that produces a strong enough depolarization to open L-type Ca^{2+} channels, to unblock NMDA receptors, and to promote plasticity. *In vivo*, convergent synaptic inputs to MSNs can trigger plateau potentials called up-states (Wilson and Kawaguchi, 1996). Although transitions from the resting down-state to the up-state have all the hallmarks of an active, regenerative process (e.g., stereotyped transition kinetics, a narrow range of up-state potentials), transitions are very difficult to manipulate with a sharp electrode impaling the somatic region (Wilson and Kawaguchi, 1996). This suggests that the site of up-state generation is in distal dendritic regions that cannot be easily manipulated. If this were the case, distal dendrites should have ionic conductances that could support a plateau. Ca^{2+} imaging using two-photon laser scanning microscopy (2PLSM) has shown that there is robust expression of both low-threshold Cav3 and Cav1 channels in MSN dendrites (Carter and Sabatini, 2004; Day et al., 2006; Carter et al., 2007), a result that has been confirmed using cell type-specific gene profiling (Day et al., 2006) (unpublished observations). The rich investment of MSN dendrites with strongly rectifying Kir2 K^+ channels also creates a favorable biophysical condition for plateau potential generation.

The question is how the plateaus or up-states are normally generated. Based upon the sparse connectivity between individual cortical axons and MSNs (Kincaid et al., 1998; Wilson, 2004), modeling studies have concluded that several hundred pyramidal neurons need to be near simultaneously active for a sufficient amount of current to be injected into dendrites for an up-state to be generated (Stern et al., 1997; Wilson, 2004). These studies have assumed that MSN dendrites are passive. However, if dendrites are not passive but active, then the convergence requirements could be dramatically different. Although glutamate uncaging experiments at proximal spines have not revealed regenerative processes (Carter et al., 2007), the situation could be different at more distal locations. If this is the case, spatial convergence of glutamatergic inputs onto a distal dendrite could induce a local plateau potential capable of pulling the rest of the cell into the up-state, fundamentally altering the impact of synaptic input on other dendrites. This is a way in which spatially convergent excitatory input to one dendrite could gate synaptic input to another. The lack of temporal correlation between up-state transitions and EPSP-driven spike generation is consistent with a scenario like this one (Stern et al., 1998). If this were how MSNs operated, it would fundamentally change our models of striatal information processing. The problem is how to test it. Because glutamatergic connections are sparse, it is virtually impossible to reliably stimulate a collection of synapses onto a particular MSN dendrite with an electrode in a brain slice. Optogenetic techniques might provide a feasible alternative strategy. Another strategy would be to employ two-photon laser uncaging (2PLU) of glutamate at visualized synaptic sites (Carter and Sabatini, 2004). These tools are becoming more widely available and should allow the regenerative capacity of MSN dendrites to be tested soon. If it turns out to be the case that up-states are locally generated in dendrites, then it also becomes feasible to characterize their role in the induction of synaptic plasticity. Up-states could be sufficient, as in the apical tuft of pyramidal neurons, or they could simply be necessary by promoting backpropagation of spikes into the distal dendrites (Kerr and Plenz, 2002).

5.7   HOMEOSTATIC PLASTICITY IN PARKINSON'S DISEASE MODELS

Sorting out how dopamine regulates synaptic plasticity in striatal MSNs has obvious implications for disease states that are triggered by alterations in the function of dopaminergic neurons. Second in prominence among dopamine-dependent disorders only to drug abuse, PD is a common neurodegenerative disorder whose motor symptoms are attributable largely to the loss of dopaminergic neurons innervating the dorsal striatum. In the prevailing model, the excitability of the two major populations of MSNs shift in opposite directions following dopamine-depleting lesions, creating an "imbalance" in the regulation of the motor thalamus favoring suppression of movement (Albin et al., 1989). In particular, D_2 receptor–expressing striatopallidal MSNs spike more, whereas D_1 receptor–expressing striatonigral MSNs spike less in the PD state. The mechanisms underlying this shift were not known at the time the model was formulated, but have widely been assumed to reflect changes in intrinsic excitability that accompanied the loss of inhibitory D_2 receptor signaling and excitatory D_1 receptor signaling. Indeed, studies by our group and others have found electrophysiological support for this view (Mallet et al., 2006; Surmeier et al., 2007).

What about synaptic remodeling? Several studies have suggested that in the absence of dopamine, synaptic plasticity is lost, essentially "freezing" the striatal circuit in its pre-depleted state (Calabresi et al., 2007; Kreitzer and Malenka, 2007). However, recent studies of defined MSN populations have shown that although dopamine is necessary for plasticity to be bidirectional and Hebbian, it is not necessary for the induction of plasticity *per se* (Shen et al., 2008). Following DA depletion, pairing pre- and postsynaptic activity—regardless of which came first—induced LTP in D_2 MSNs and LTD in D_1 MSNs. This result adds a new dimension to the prevailing model by showing that activity-dependent changes in synaptic strength parallel to those of intrinsic excitability following DA depletion. Work *in vivo* examining the responsiveness of antidromically identified MSNs to cortical stimulation following unilateral lesions of the striatal dopaminergic innervation is consistent with this broader model (Mallet et al., 2006).

But this poses a problem. Neurons are homeostatic; sustained perturbations in synaptic or intrinsic properties that make neurons spike more or less than their set-point engage homeostatic mechanisms that attempt to bring activity back to the desired level (Turrigiano, 1999; Marder and Goaillard, 2006). One of the most common mechanisms of homeostatic plasticity is to alter the synaptic strength or to scale synapses. In striatopallidal MSNs, the elevation in activity following dopamine depletion triggers a dramatic downregulation of glutamatergic synapses formed on spines (Day et al., 2006). This can be viewed as a form of homeostatic plasticity. Like scaling seen in other cell types, the synaptic modification depends upon Ca^{2+} entry through voltage-dependent L-type channels that presumably report activity levels. There are other recently described network adaptations relevant to homeostatic plasticity in PD models. For example, although feed-forward inhibition through fast-spiking GABAergic interneurons does not appear to be directly altered, low-threshold GABAergic interneurons do elevate their input to at least a subset of MSNs in PD models (Mallet et al., 2005; Dehorter et al., 2009). Recurrent collateral inhibition between MSNs, which is normally strongest between D_2 MSNs, is almost abolished following dopamine depletion (Taverna et al., 2008). These adaptations in conjunction with enhanced striatopallidal MSN excitability are likely to contribute to the transmission of beta band activity from the cortex through the striatum to the globus pallidus (Murer et al., 2002). That said, a major gap in the existing literature is a description of the intrinsic changes in MSN excitability following prolonged DA depletion. All of the work with identified cell types has relied upon short-term (~<1 week) DA depletions (e.g., Day et al., 2006; Kreitzer and Malenka, 2007), but there clearly are slower adaptations that take 3–4 weeks to stabilize. Given the robust differences in the anatomy and intrinsic physiology of striatonigral and striatopallidal MSNs that exists in the normal striatum (Gertler et al., 2008), it is easy to conjecture that these resting differences are due to differential regulation of basal excitability by DA. If that were true, losing DA could trigger homeostatic processes that make MSNs much more alike.

5.8 CONCLUDING REMARKS

In the last few years, our understanding of the mechanisms controlling synaptic plasticity in the corticostriatal circuits underlying action selection has significantly deepened. Dopamine remains an important player in the induction of plasticity at corticostriatal synapses on principal MSNs, but it is not the only player and its effects are dictated by the type of dopamine receptor expressed. In large measure, this advance has been made possible by the development of BAC-transgenic mice that make the cellular heterogeneity of the striatum tractable. How the relatively sparse but functionally important interneuron populations contribute to plasticity remains to be clearly defined. The development transgenic mice expressing Cre in select neuronal populations (and the growing stable of mice with floxed genes) should propel this effort forward and allow a molecular dissection of these mechanisms in coming years. The growing application of optical approaches, like 2PLSM and 2PLU, also promises to yield insights into synaptic integration and plasticity not achievable with any other approach. Coupling these new tools with optogenetic strategies for activating microcircuits relevant to action selection should prove to be a watershed for basal ganglia and motor systems research. These approaches should allow us to gain a better grasp of basal ganglia pathophysiology in disease states—like PD, Huntington's disease, and drug abuse—and in so doing develop a new generation of therapeutics.

ACKNOWLEDGMENTS

This work was supported by NS34696 to DJS.

REFERENCES

Ade KK, Lovinger DM (2007) Anandamide regulates postnatal development of long-term synaptic plasticity in the rat dorsolateral striatum. *J Neurosci* 27:2403–2409.
Adermark L, Lovinger DM (2007) Combined activation of L-type Ca^{2+} channels and synaptic transmission is sufficient to induce striatal long-term depression. *J Neurosci* 27:6781–6787.

Albin RL, Young AB, Penney JB (1989) The functional anatomy of basal ganglia disorders. *Trends Neurosci* 12:366–375.

Berridge MJ (1998) Neuronal calcium signaling. *Neuron* 21:13–26.

Brenowitz SD, Best AR, Regehr WG (2006) Sustained elevation of dendritic calcium evokes widespread endocannabinoid release and suppression of synapses onto cerebellar Purkinje cells. *J Neurosci* 26:6841–6850.

Calabresi P, Gubellini P, Centonze D, Sancesario G, Morello M, Giorgi M, Pisani A, Bernardi G (1999) A critical role of the nitric oxide/cGMP pathway in corticostriatal long-term depression. *J Neurosci* 19:2489–2499.

Calabresi P, Picconi B, Tozzi A, Di Filippo M (2007) Dopamine-mediated regulation of corticostriatal synaptic plasticity. *Trends Neurosci* 30:211–219.

Carter AG, Sabatini BL (2004) State-dependent calcium signaling in dendritic spines of striatal medium spiny neurons. *Neuron* 44:483–493.

Carter AG, Soler-Llavina GJ, Sabatini BL (2007) Timing and location of synaptic inputs determine modes of subthreshold integration in striatal medium spiny neurons. *J Neurosci* 27:8967–8977.

Cohen MX, Frank MJ (2009) Neurocomputational models of basal ganglia function in learning, memory and choice. *Behav Brain Res* 199:141–156.

Day M, Wang Z, Ding J, An X, Ingham CA, Shering AF, Wokosin D, Ilijic E, Sun Z, Sampson AR, Mugnaini E, Deutch AY, Sesack SR, Arbuthnott GW, Surmeier DJ (2006) Selective elimination of glutamatergic synapses on striatopallidal neurons in Parkinson's disease models. *Nat Neurosci* 9:251–259.

Day M, Wokosin D, Plotkin JL, Tian X, Surmeier DJ (2008) Differential excitability and modulation of striatal medium spiny neuron dendrites. *J Neurosci* 28:11603–11614.

Dehorter N, Guigoni C, Lopez C, Hirsch J, Eusebio A, Ben-Ari Y, Hammond C (2009) Dopamine-deprived striatal gabaergic interneurons burst and generate repetitive gigantic IPSCs in medium spiny neurons. *J Neurosci* 29:7776–7787.

Ding J, Peterson JD, Surmeier DJ (2008) Corticostriatal and thalamostriatal synapses have distinctive properties. *J Neurosci* 28:6483–6492.

Dujardin K, Laurent B (2003) Dysfunction of the human memory systems: Role of the dopaminergic transmission. *Curr Opin Neurol* 16 (Suppl 2):S11–S16.

Fino E, Glowinski J, Venance L (2005) Bidirectional activity-dependent plasticity at corticostriatal synapses. *J Neurosci* 25:11279–11287.

Flajolet M, Wang Z, Futter M, Shen W, Nuangchamnong N, Bendor J, Wallach I, Nairn AC, Surmeier DJ, Greengard P (2008) FGF acts as a co-transmitter through adenosine A(2A) receptor to regulate synaptic plasticity. *Nat Neurosci* 11:1402–1409.

Fuxe K, Ferre S, Genedani S, Franco R, Agnati LF (2007b) Adenosine receptor-dopamine receptor interactions in the basal ganglia and their relevance for brain function. *Physiol Behav* 92:210–217.

Fuxe K, Marcellino D, Genedani S, Agnati L (2007a) Adenosine A(2A) receptors, dopamine D(2) receptors and their interactions in Parkinson's disease. *Mov Disord* 22:1990–2017.

Gerfen CR, Engber TM, Mahan LC, Susel Z, Chase TN, Monsma FJ, Jr., Sibley DR (1990) D1 and D2 dopamine receptor-regulated gene expression of striatonigral and striatopallidal neurons. *Science* 250:1429–1432.

Gertler TS, Chan CS, Surmeier DJ (2008) Dichotomous anatomical properties of adult striatal medium spiny neurons. *J Neurosci* 28:10814–10824.

Giuffrida A, Parsons LH, Kerr TM, Rodriguez de Fonseca F, Navarro M, Piomelli D (1999) Dopamine activation of endogenous cannabinoid signaling in dorsal striatum. *Nat Neurosci* 2:358–363.

Golding NL, Staff NP, Spruston N (2002) Dendritic spikes as a mechanism for cooperative long-term potentiation. *Nature* 418:326–331.

Gong S, Zheng C, Doughty ML, Losos K, Didkovsky N, Schambra UB, Nowak NJ, Joyner A, Leblanc G, Hatten ME, Heintz N (2003) A gene expression atlas of the central nervous system based on bacterial artificial chromosomes. *Nature* 425:917–925.

Houk JC, Bastianen C, Fansler D, Fishbach A, Fraser D, Reber PJ, Roy SA, Simo LS (2007) Action selection and refinement in subcortical loops through basal ganglia and cerebellum. *Philos Trans R Soc Lond B Biol Sci* 362:1573–1583.

Kawaguchi Y (1997) Neostriatal cell subtypes and their functional roles. *Neurosci Res* 27:1–8.

Kawaguchi Y, Wilson CJ, Emson PC (1989) Intracellular recording of identified neostriatal patch and matrix spiny cells in a slice preparation preserving cortical inputs. *J Neurophysiol* 62:1052–1068.

Kerr JN, Plenz D (2002) Dendritic calcium encodes striatal neuron output during up-states. *J Neurosci* 22:1499–1512.

Kincaid AE, Zheng T, Wilson CJ (1998) Connectivity and convergence of single corticostriatal axons. *J Neurosci* 18:4722–4731.

Kreitzer AC, Malenka RC (2005) Dopamine modulation of state-dependent endocannabinoid release and long-term depression in the striatum. *J Neurosci* 25:10537–10545.

Kreitzer AC, Malenka RC (2007) Endocannabinoid-mediated rescue of striatal LTD and motor deficits in Parkinson's disease models. *Nature* 445:643–647.

Kung VW, Hassam R, Morton AJ, Jones S (2007) Dopamine-dependent long term potentiation in the dorsal striatum is reduced in the R6/2 mouse model of Huntington's disease. *Neuroscience* 146:1571–1580.

Lovinger DM (2008) Presynaptic modulation by endocannabinoids. *Handb Exp Pharmacol* 184:435–477.

Lovinger DM, Tyler EC, Merritt A (1993) Short- and long-term synaptic depression in rat neostriatum. *J Neurophysiol* 70:1937–1949.

Mallet N, Ballion B, Le Moine C, Gonon F (2006) Cortical inputs and GABA interneurons imbalance projection neurons in the striatum of parkinsonian rats. *J Neurosci* 26:3875–3884.

Mallet N, Le Moine C, Charpier S, Gonon F (2005) Feedforward inhibition of projection neurons by fast-spiking GABA interneurons in the rat striatum in vivo. *J Neurosci* 25:3857–3869.

Marder E, Goaillard JM (2006) Variability, compensation and homeostasis in neuron and network function. *Nat Rev Neurosci* 7:563–574.

Murer MG, Tseng KY, Kasanetz F, Belluscio M, Riquelme LA (2002) Brain oscillations, medium spiny neurons, and dopamine. *Cell Mol Neurobiol* 22:611–632.

Nakamura T, Nakamura K, Lasser-Ross N, Barbara JG, Sandler VM, Ross WN (2000) Inositol 1,4,5-trisphosphate (IP3)-mediated Ca^{2+} release evoked by metabotropic agonists and backpropagating action potentials in hippocampal CA1 pyramidal neurons. *J Neurosci* 20:8365–8376.

Ondracek JM, Dec A, Hoque KE, Lim SA, Rasouli G, Indorkar RP, Linardakis J, Klika B, Mukherji SJ, Burnazi M, Threlfell S, Sammut S, West AR (2008) Feed-forward excitation of striatal neuron activity by frontal cortical activation of nitric oxide signaling in vivo. *Eur J Neurosci* 27:1739–1754.

Pawlak V, Kerr JN (2008) Dopamine receptor activation is required for corticostriatal spike-timing-dependent plasticity. *J Neurosci* 28:2435–2446.

Ronesi J, Gerdeman GL, Lovinger DM (2004) Disruption of endocannabinoid release and striatal long-term depression by postsynaptic blockade of endocannabinoid membrane transport. *J Neurosci* 24:1673–1679.

Schultz W (2007) Multiple dopamine functions at different time courses. *Annu Rev Neurosci* 30:259–288.

Seol GH, Ziburkus J, Huang S, Song L, Kim IT, Takamiya K, Huganir RL, Lee HK, Kirkwood A (2007) Neuromodulators control the polarity of spike-timing-dependent synaptic plasticity. *Neuron* 55:919–929.

Shen W, Flajolet M, Greengard P, Surmeier DJ (2008) Dichotomous dopaminergic control of striatal synaptic plasticity. *Science* 321:848–851.

Singla S, Kreitzer AC, Malenka RC (2007) Mechanisms for synapse specificity during striatal long-term depression. *J Neurosci* 27:5260–5264.

Smeal RM, Gaspar RC, Keefe KA, Wilcox KS (2007) A rat brain slice preparation for characterizing both thalamostriatal and corticostriatal afferents. *J Neurosci Methods* 159:224–235.

Stern EA, Jaeger D, Wilson CJ (1998) Membrane potential synchrony of simultaneously recorded striatal spiny neurons in vivo. *Nature* 394:475–478.

Stern EA, Kincaid AE, Wilson CJ (1997) Spontaneous subthreshold membrane potential fluctuations and action potential variability of rat corticostriatal and striatal neurons in vivo. *J Neurophysiol* 77:1697–1715.

Surmeier DJ, Ding J, Day M, Wang Z, Shen W (2007) D1 and D2 dopamine-receptor modulation of striatal glutamatergic signaling in striatal medium spiny neurons. *Trends Neurosci* 30:228–235.

Sutton RS, Barto AG (1981) Toward a modern theory of adaptive networks: Expectation and prediction. *Psychol Rev* 88:135–170.

Sweatt JD (2004) Mitogen-activated protein kinases in synaptic plasticity and memory. *Curr Opin Neurobiol* 14:311–317.

Taufiq Ur R, Skupin A, Falcke M, Taylor CW (2009) Clustering of InsP3 receptors by InsP3 retunes their regulation by InsP3 and Ca^{2+}. *Nature* 458:655–659.

Taverna S, Ilijic E, Surmeier DJ (2008) Recurrent collateral connections of striatal medium spiny neurons are disrupted in models of Parkinson's disease. *J Neurosci* 28:5504–5512.

Testa CM, Standaert DG, Young AB, Penney JB, Jr. (1994) Metabotropic glutamate receptor mRNA expression in the basal ganglia of the rat. *J Neurosci* 14:3005–3018.

Turrigiano GG (1999) Homeostatic plasticity in neuronal networks: The more things change, the more they stay the same. *Trends Neurosci* 22:221–227.

Tzounopoulos T, Rubio ME, Keen JE, Trussell LO (2007) Coactivation of pre- and postsynaptic signaling mechanisms determines cell-specific spike-timing-dependent plasticity. *Neuron* 54:291–301.

Uchigashima M, Narushima M, Fukaya M, Katona I, Kano M, Watanabe M (2007) Subcellular arrangement of molecules for 2-arachidonoyl-glycerol-mediated retrograde signaling and its physiological contribution to synaptic modulation in the striatum. *J Neurosci* 27:3663–3676.

Wang SS, Denk W, Hausser M (2000) Coincidence detection in single dendritic spines mediated by calcium release. *Nat Neurosci* 3:1266–1273.

Wang Z, Kai L, Day M, Ronesi J, Yin HH, Ding J, Tkatch T, Lovinger DM, Surmeier DJ (2006) Dopaminergic control of corticostriatal long-term synaptic depression in medium spiny neurons is mediated by cholinergic interneurons. *Neuron* 50:443–452.

Wilson CJ (2004) Basal ganglia. In: *The Synaptic Organization of The Brain* (Shepherd GM, ed.), pp. 361–414. Oxford, U.K.: Oxford University Press.

Wilson CJ, Kawaguchi Y (1996) The origins of two-state spontaneous membrane potential fluctuations of neostriatal spiny neurons. *J Neurosci* 16:2397–2410.

Xia Z, Storm DR (2005) The role of calmodulin as a signal integrator for synaptic plasticity. *Nat Rev Neurosci* 6:267–276.

Yin HH, Knowlton BJ (2006) The role of the basal ganglia in habit formation. *Nat Rev Neurosci* 7:464–476.

Zhang F, Wang LP, Boyden ES, Deisseroth K (2006) Channelrhodopsin-2 and optical control of excitable cells. *Nat Methods* 3:785–792.

6 Striatal Acetylcholine–Dopamine Crosstalk and the Dorsal–Ventral Divide

Sarah Threlfell and Stephanie J. Cragg

CONTENTS

6.1 INTRODUCTION

The striatum is a large subcortical nucleus involved in motor coordination, cognition, as well as disorders such as Parkinson's disease, Tourette's syndrome, Huntington's disease, schizophrenia, and drug addiction (Wilson 2004). The principal neurons of the striatum are the medium spiny neurons (MSNs) which constitute ~90% of the striatal neuron population and form the striatal output

(Bolam et al. 2000). The remaining striatal neurons comprise at least three types of interneuron, including the large, aspiny, tonically active cholinergic interneuron (Kawaguchi 1993). As the input nucleus of the basal ganglia, the striatum receives and gates massive convergent innervation via the MSNs to generate appropriate behaviors. Among this convergent innervation, striatal projection neurons receive excitatory inputs from both cortex and thalamus and dopaminergic inputs from the midbrain. Nigrostriatal dopaminergic afferents and corticostriatal/thalamostriatal glutamatergic afferents commonly synapse onto the same MSN dendritic spine (Moss and Bolam 2008), and this synaptic triad is likely to be of great importance within the striatum helping to powerfully shape basal ganglia network activity. At the level of this synaptic triad, dopamine is able to modulate corticostriatal transmission via presynaptic D_2 receptors (Bamford et al. 2004) and is also able to regulate corticostriatal plasticity (Calabresi et al. 2007; Surmeier et al. 2007) as well as directly regulating the excitability of MSNs (Albin et al. 1995; Day et al. 2008; Nicola et al. 2000). The intimacy of the dopamine–glutamate striatal interaction has recently been highlighted further with the discovery that dopamine and glutamate may be co-released within ventral but not dorsal striatum (Stuber et al. 2010; Tecuapetla et al. 2010).

To fully appreciate the ability of dopamine to interact with glutamate at the level of the MSNs, it is important to recognize that other neurotransmitters present within the striatum are able to powerfully regulate the release of striatal dopamine. Dopamine and acetylcholine particularly have a potent reciprocal relationship within the striatum. Mesostriatal dopamine neurons and striatal cholinergic interneurons participate in signaling the motivational significance of environmental stimuli and regulate striatal plasticity (Calabresi et al. 2000; Morris et al. 2004; Partridge et al. 2002; Schultz 1998, 2002). Interactions between these neurotransmitters in the striatum occur at postsynaptic and presynaptic levels, through synchronous changes in parent neuron activities and reciprocal presynaptic regulation of release (Calabresi et al. 2000; Centonze et al. 2003; Cragg 2006; Morris et al. 2004; Pisani et al. 2003; Zhou et al. 2002). Furthermore, it is also essential to consider that the striatum is not a homogeneous nucleus, but rather it is organized into different albeit overlapping territories, anatomically, biochemically, functionally, and also neurochemically in terms of neurotransmitter interactions (Belin et al. 2009; Cragg 2003; Haber et al. 2000; Nakano et al. 2000; Voorn et al. 2004).

A major interest of our group is characterizing how acetylcholine is able to regulate striatal dopamine neurotransmission across different striatal territories and under conditions representing an array of activity patterns of dopamine neurons. To study cholinergic regulation of dopamine transmission, we use the real-time electrochemical detection technique, fast-scan cyclic voltammetry at carbon fiber microelectrodes. This chapter summarizes our current understanding of how acetylcholine, via actions at both nicotinic and muscarinic receptors (mAChRs), is able to powerfully modulate dopamine transmission in a variable manner dependent on DA neuron activity, and moreover, differentially between sensorimotor- versus limbic-associated striatum.

6.2   DOPAMINE AND ACETYLCHOLINE SYSTEMS IN THE STRIATUM

6.2.1   STRIATAL SUBTERRITORIES

Collectively, the striatum participates in a wide range of motivational, associative, and sensorimotor-related brain functions (Albin et al. 1995; Everitt and Robbins 2005; Gerdeman et al. 2003; Haber et al. 2000; Schultz 2006; Voorn et al. 2004). The striatum is a highly heterogeneous structure where the dominant neurons, the MSNs, are intermingled in an irregular fashion with at least three different types of interneuron as well as a variety of inputs from cortex, thalamus, and midbrain. Within this heterogeneous structure, subterritories exist which differ according to their functional roles as well as their afferent inputs from and projections to other nuclei. For example, dorsal striatal regions are associated with sensorimotor function with glutamatergic afferents from motor cortex and dopaminergic afferents from substantia nigra pars compacta (SNc) (Haber et al. 2000; McGeorge and Faull 1989; Reep et al. 2003), whereas ventral striatal regions are associated with limbic function with dopaminergic afferents from ventral tegmental area (VTA) and glutamatergic afferents from prefrontal cortex, amygdala, and hippocampus (Brog et al. 1993; Haber et al. 2000; Kelley and Domesick 1982; McGeorge and Faull 1989; O'Donnell and Grace 1995; Pennartz et al. 1994; Sesack et al. 1989). Increasingly, it is becoming accepted that these subterritories, despite being distinct regions, are functionally related and interact (Belin and Everitt 2008; Belin et al. 2009; Haber et al. 2000). Therefore, factors that have regulatory functions within one striatal region may have important outcomes on function in other regions.

6.2.2   DOPAMINERGIC AND CHOLINERGIC INNERVATIONS IN STRIATUM

Dopamine inputs to the striatum arise from the "A9"–"A10" dopamine neurons in the midbrain. Neurons in the VTA and SNc project in a topographic pattern to differentially innervate the ventral striatum (nucleus accumbens, NAc) and the dorsal striatum (caudate-putamen, CPu), respectively (Bjorklund and Lindvall 1984; Gerfen et al. 1987; McFarland and Haber 2000; Voorn et al. 2004). Mesostriatal dopamine neurons exhibit two well-defined firing modes *in vivo*—tonic firing (~1–10 Hz) and burst firing of three to five spikes at a frequency of ~15–100 Hz (Grace and Bunney 1984a,b; Hyland et al. 2002). The switch in firing mode of the mesostriatal DA neurons, from tonic to burst firing, is thought to encode information about the prediction and receipt of rewards or behaviorally salient stimuli (Matsumoto and Hikosaka 2009; Morris et al. 2004; Schultz 1986, 2002). Despite dopamine neurons forming a relatively small population of neurons, a single dopamine neuron projecting to the striatum forms a very dense arbor, occupying up to 5.7% of the volume of the striatum, with a high density of axonal varicosities forming a synapse approximately every 10–20 μm^3 (Arbuthnott and Wickens 2007; Descarries et al. 1996; Matsuda et al. 2009; Moss and Bolam 2010). Furthermore, an individual DA neuron is thought to make several hundred thousand synapses—at least an order of magnitude higher than most other CNS neurons (Matsuda et al. 2009). Although dopamine neurons make a huge number of structurally defined synapses, dopamine receptors and uptake transporters exist extrasynaptically (Nirenberg et al. 1996, 1997; Pickel 2000), and dopamine is able to spillover from the synapse to participate in extrasynaptic or "volume" transmission (Cragg and Rice 2004; Fuxe and Agnati 1991; Garris et al. 1994; Gonon 1997; Rice and Cragg 2008). Dopamine is therefore extremely well positioned to modulate striatal function: dopamine synapses are situated on the necks of spines of MSNs adjacent to corticostriatal or thalamostriatal glutamatergic inputs (Freund et al. 1984; Groves et al. 1994; Moss and Bolam 2008; Smith and Bolam 1990), and D1-like and D2-like receptors are present throughout the striatum on MSNs, interneurons, as well as dopamine axons (Alcantara et al. 2003; Gerfen 1992; Hersch et al. 1995; Sesack et al. 1994; Surmeier et al. 1996).

In the striatum, cholinergic innervation arises solely from cholinergic interneurons (Calabresi et al. 2000; Contant et al. 1996; Woolf 1991), which, despite occupying only 1%–2% of all the striatal neurons, provide an extensive axonal arborization throughout the striatal subterritories in a manner similar to dopaminergic axons. Quantitative electron microscopy indicates approximately 2×10^8 acetylcholine varicosities/mm^3 in the striatum, and each striatal cholinergic interneuron has 500,000 axon varicosities. Similar to dopaminergic axon terminals, a 10 µm radius sphere of striatal neuropil contains about 400 cholinergic axon terminals (Contant et al. 1996; Descarries and Mechawar 2000). Cholinergic interneurons, like dopaminergic neurons, are also critically involved in signaling learning associated with events of unexpected high salience (Berridge and Robinson 2003; Calabresi et al. 2000; Centonze et al. 2003; Schultz 2002; Wickens et al. 2003; Wise 2004). These cholinergic interneurons are spontaneously active in the striatum (Bennett and Wilson 1999; Zhou et al. 2001; Zhou et al. 2003) and as a result supply endogenous acetylcholine tone in the striatum.

A high density of dopaminergic and cholinergic terminals exists in the striatum, and they are often only about 1 µm apart (Descarries et al. 1997). Both dopamine and acetylcholine are able to participate in extrasynaptic signaling (i.e., volume transmission) (Descarries et al. 1996, 1997; Zoli et al. 1998), but the close proximity of striatal dopaminergic and cholinergic terminals ensures interactions between the dopaminergic and cholinergic systems. Indeed, the cooperative interactions of the dopaminergic and cholinergic systems are very important for the proper functioning of the striatum (Calabresi et al. 2000; Centonze et al. 2003; Cragg et al. 2005; Rice and Cragg 2004; Zhang and Sulzer 2004). There is a longstanding hypothesis of an antagonistic balance between dopamine and acetylcholine in normal striatal function (Calabresi et al. 2000; Centonze et al. 2003; Pisani et al. 2003; Zhou et al. 2002). This hypothesis arose from studies which found that both pro-dopaminergic treatments and anti-cholinergic treatments were able to alleviate the debilitating motor symptoms of Parkinson's disease (Barbeau 1962; Pisani et al. 2003). In support of this hypothesis, at the postsynaptic cellular level, dopamine and acetylcholine can have opposing effects on the excitability of striatal output neurons and on corticostriatal plasticity (Calabresi et al. 2000, 2007; Centonze et al. 2003; Morris et al. 2004; Pisani et al. 2003).

Striatal acetylcholine targets two classes of cholinergic receptors in the mammalian brain, nicotinic acetylcholine receptors (nAChRs), and muscarinic acetylcholine receptors (mAChRs). Nicotinic receptors are ligand-gated ion channels composed of five subunits, arranged symmetrically around a central pore. mAChRs, however, are seven transmembrane domain, G-protein-coupled receptors. Acetylcholine via actions at both mAChRs and nAChRs is able to powerfully regulate striatal dopamine availability in a manner which differs according to the striatal subregion and dopamine neuron activity. In addition, the mAChR subtypes and nAChR subunits responsible differ between dorsal and ventral striatal regions.

6.3   ACTIVITY-DEPENDENT REGULATION OF DOPAMINE RELEASE PROBABILITY BY PRESYNAPTIC NICOTINIC RECEPTORS (nAChRs)

Striatal dopamine release does not correlate directly with the activity of dopamine neurons due to use-dependent changes or "plasticity" of release probability (Chergui et al. 1994; Cragg 2003; Montague et al. 2004). Within the striatum, dopamine release probability following a single action potential is relatively high, with subsequent short-term depression limiting further release by subsequent action potentials within a burst of action potentials (Rice and Cragg 2004; Zhou et al. 2001).

nAChR ligands have long been known to have a powerful control over striatal dopamine release (Dajas-Bailador and Wonnacott 2004; Di Chiara and Imperato 1988). However, we now appreciate that the control of dopamine release probability by endogenous acetylcholine and presynaptic nAChRs is dynamic and complex depending on the activity of dopamine neurons (Drenan et al. 2010; Exley et al. 2008; Rice and Cragg 2004; Threlfell et al. 2010; Zhang and Sulzer 2004; Zhou et al. 2001). Typically, striatal dopamine release is associated with a use-dependent, short-term depression of release probability at short interpulse intervals (Abeliovich et al. 2000; Cragg 2003; Montague et al. 2004). Studies using real-time electrochemical detection of dopamine in striatal slices indicate that acetylcholine released from tonically active cholinergic interneurons (Bennett and Wilson 1999; Zhou et al. 2001, 2003) acts at β2-subunit-containing (denoted by *)-nAChRs on striatal dopamine axons to maintain a high probability of dopamine release evoked by a single pulse (Rice and Cragg 2004; Zhou et al. 2001). Reducing cholinergic actions at nAChRs both via desensitization of nAChRs with nicotine, or by application of nAChR antagonists, reduces initial dopamine release probability and subsequently relieves short-term depression (Figure 6.1) (Rice and Cragg 2004; Zhang and Sulzer 2004). Acetylcholine therefore contributes to subsequent short-term depression of dopamine release.

This alteration of presynaptic release probability by nAChRs is also dependent on the frequency of dopamine neuron firing (Rice and Cragg 2004; Zhang and Sulzer 2004), the shorter the interpulse interval, that is, the higher the dopamine neuron firing frequency, the greater the relief from short-term depression. Alternatively, it can be viewed as nAChR blockade suppressing initial dopamine release but permitting a high-frequency pass filtering that facilitates bursts of dopamine release (Cragg 2006; Rice and Cragg 2004). Consequently, reduced nAChR activity can (a) *enhance* dopamine release following *bursts* of dopamine neuron activity (>20 Hz) such as those that accompany the presentation of unexpected rewards or conditioned salient

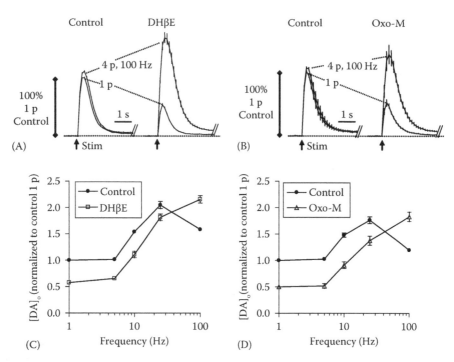

FIGURE 6.1 Striatal nAChRs and mAChRs modify activity dependence of dopamine release. (A, B) Profiles of mean [DA]$_o$ ± SEM versus time in NAc after stimuli (arrows) of either a single-pulse (1 p) or a high-frequency burst (4 p/100 Hz) in control or drug conditions (DHβE, 1 μM; Oxo-M, 10 μM). Data are normalized to peak [DA]$_o$ released by 1 p in controls. Either β2*-nAChR antagonist (DHβE) or mAChR agonist (Oxo-M) reduce release by a single pulse and enhance release by a burst, thereby increasing the contrast between dopamine released by burst and nonburst activity. (C, D) Mean peak [DA]$_o$ ± SEM versus frequency during four pulse trains (1–100 Hz) in control (filled circles) or DHβE (squares) or Oxo-M (triangles) normalized to [DA]$_o$ released by a single pulse in control conditions. Blockade of nAChRs with DHβE or activation of mAChRs with Oxo-M results in an increased frequency dependence of dopamine release as described previously (see Exley et al. 2008; Rice and Cragg 2004; Threlfell et al. 2010).

stimuli (Hyland et al. 2002; Mirenowicz and Schultz 1996) and (b) *reduce* dopamine release following a reduction in dopamine neuron activity such as that seen following omission of a expected reward (Tobler et al. 2003).

This ability of acetylcholine via nAChRs to regulate dopamine availability in a manner that is dependent on activity in dopamine neurons has obvious implications in physiological circumstances whereby dopamine neurons switch from firing at tonic frequencies to bursts of high-frequency activity, such as during unexpected reward delivery or other events of unusual salience. This nAChR filtering mechanism will increase the contrast in dopamine signaling during different modes of activity of dopamine neurons and hence ultimately encode this filtering at the level of the MSN by varying dopamine concentrations able to act at either D1 or D2 receptors on discrete MSN populations.

6.3.1 nAChR Subunits on Striatal Dopamine Axons

Dopamine neurons express a range of nAChR subunits, and several different stoichiometric configurations are proposed to exist in dopamine axons and regulate striatal dopamine transmission. Currently, 14 mammalian nAChR subunits α1–α10 and β1–β4 have been identified. These subunits are organized into subfamilies I–IV, according to gene sequence and structure (Corringer et al. 2000; Le Novere et al. 2002). To date, only nine nAChR subunits from subfamilies II (α7) and III (α2–α6, β2–β4) have been identified in mammalian brain (Corringer et al. 2000; Le Novere et al. 2002). These subunits can form homomeric pentamers (all of one type of subunit, e.g., α7) or heteromeric pentamers, consisting of combinations of various α- and β-type subunits. In rodents, dopamine neurons in VTA and SNc express mRNAs for the nAChR subunits α4, α5, α6, β2, and β3, as well as lower levels of α3, α7, and β4 (Azam et al. 2002). This diversity of subunit expression has the potential to give rise to multiple types of pentameric receptors in somatodendritic and axon terminal regions of dopamine neurons with a corresponding multitude of potential functions.

Differences in the nAChR subunits present in axon terminals versus somatodendritic regions of dopamine neurons exist. Whereas α7 and β4 subunits are present in VTA and SNc, they do not get transported to dopamine axon terminals in striatum (Champtiaux et al. 2003; Quik et al. 2005). Furthermore, despite existing in the primate striatum, there is no α3 subunit in rodent striatum (Champtiaux et al. 2003; Gotti et al. 2005; Quik et al. 2005; Wonnacott et al. 2000; Zoli et al. 2002). The α4, α5, α6, β2, and β3 subunits are however found at high density in dopamine axon terminals in rodent striatum, where they can assemble as functional heteromeric nicotinic receptors. As heteromeric receptors, nicotinic subunits arrange to form two α/β pairs, and a fifth subunit, α or β. The boundary of each α/β pair is one of two acetylcholine-binding sites at each nAChR (Gotti and Clementi 2004) which must be occupied for the receptor to function. In striatal dopamine axons, the two α/β pairs are the α4/β2 and/or α6/β2 and/or α4/β4 (Luetje 2004; Quik et al. 2005; Salminen et al. 2004). The fifth subunit in the nAChR pentamer may consist of any other subunit, including α5 or β3.

All nAChRs on striatal dopamine axon terminals are thought to contain the β2 subunit (Champtiaux et al. 2003; Salminen et al. 2004). Deduction of the other subunits present within these β2*-nAChRs has been aided by use of an antagonist α-conotoxin MII (α-CtxMII) which binds selectively to α6-/α3-*nAChRs (Cartier et al. 1996; McIntosh et al. 2004; Whiteaker et al. 2000). The majority of striatal nAChRs located on DA terminals can therefore be classed into two groups according to their sensitivity to α-CtxMII. Studies using α-CtxMII to determine which nAChR subunits are responsible for nicotine-evoked striatal dopamine release from synaptosomes have revealed that up to 60% of nicotine-evoked dopamine is not prevented by α-CtxMII and therefore results from non-α6*-nAChRs (Kaiser et al. 1998; Kulak et al. 1997; Salminen et al. 2007). Therefore both α6*-nAChRs (comprising α6α4β2β3 and α6/β2*) and non-α6*-nAChRs (comprising α4β2 and α4α5β2) exist on striatal dopamine axons (Champtiaux et al. 2003; Exley and Cragg 2008; Gotti et al. 2009; Jennings et al. 2009; Quik et al. 2005; Salminen et al. 2004; Zoli et al. 2002). However, as emerging data are now beginning to show, expression does not necessarily indicate function.

6.3.2 nAChR Subunits Responsible for Regulation of Striatal Dopamine Transmission in Dorsal versus Ventral Striatum

Previous studies identifying the subunits responsible for nicotinic regulation of striatal dopamine release from synaptosomes have not typically differentiated between striatal subterritories. It is important to distinguish factors (such as nAChR subunits) regulating dopamine availability across different striatal subregions due to different the behaviors associated with distinct striatal subterritories. For example, dopamine neurons innervating more dorsal striatal regions are those most vulnerable in Parkinson's disease; therefore, understanding mechanisms regulating availability of dopamine within such territories will help to unveil future therapeutic targets/ neuroprotective strategies in such diseases.

Recent work from our own laboratory using the $\alpha6$-selective nAChR antagonist α-CtxMII has revealed that $\alpha6^*$-nAChRs dominate activity-dependent regulation of dopamine transmission in ventral striatum, compared with dorsal striatum where the $\alpha6^*$-nAChRs seem to dominate less (Exley and Cragg 2008; Exley et al. 2008). Due to a lack of effective and subunit-specific pharmacological nAChR ligands, further delineation of the nAChR subunits necessary for activity-dependent regulation of dopamine transmission requires the use of subunit-null and transgenic mice expressing mutant nAChR subunits. A recent study using mice expressing a nonendogenous hypersensitive $\alpha6$ subunit ($\alpha6'$) revealed activity-dependent dopamine release in central CPu in the absence of any nAChR ligand. This activity dependence was absent in $\alpha6'$ mice lacking $\alpha4$ subunits ($\alpha6'\alpha4KO$), which the authors suggested might indicate a role for $\alpha6\alpha4\beta2^*$ nAChRs in the regulation of dopamine release in the central CPu region (Drenan et al. 2010). However, this role remains to be established for endogenous α subunits, and for other noncentral CPu regions, and moreover, unpublished data from our own laboratory in knockout mice continues to suggest that $\alpha6$-containing nAChRs in dorsal striatum play only a limited role compared to $\alpha4$-subunit-containing receptors in dorsal striatum (Exley et al. 2008), whereas $\alpha4\alpha6$-nAChRs may be necessary in NAc (Exley et al. 2008).

6.4 ACTIVITY-DEPENDENT REGULATION OF DOPAMINE RELEASE PROBABILITY BY mAChRs ON CHOLINERGIC INTERNEURONS

Like nAChRs, mAChRs are also able to powerfully regulate striatal dopamine availability. Until recently, data within the literature were conflicting, with some reports suggesting that mAChRs enhance dopamine release (De Klippel et al. 1993; Grilli et al. 2008; Lehmann and Langer 1982; Raiteri et al. 1984; Schoffelmeer et al. 1986; Xu et al. 1989; Zhang et al. 2002), and others reporting the opposite— that mAChRs suppress dopamine availability (De Klippel et al. 1993; Kemel et al. 1989; Kudernatsch and Sutor 1994; Schoffelmeer et al. 1986; Tzavara et al. 2004; Xu et al. 1989; Zhang et al. 2002). Recent data from our laboratory have revealed that mAChRs can regulate dopamine bidirectionally depending on the activity of dopamine neurons (Threlfell et al. 2010). This activity-dependent regulation of dopamine by mAChRs is reminiscent of that shown by nicotinic receptors (Exley et al. 2008;

Rice and Cragg 2004; Zhang and Sulzer 2004). Moreover, the mAChR subtypes responsible for this regulation differ between striatal subterritories (Threlfell et al. 2010).

Activation of striatal mAChRs inhibits dopamine release by single pulses or low, tonic-like frequencies of presynaptic activity, but enhances the sensitivity of dopamine release to frequency, increasing dopamine released by higher, burst-like frequencies (Threlfell et al. 2010), in a manner similar to the activity-dependent outcome on dopamine release probability following the inhibition of striatal nAChRs (Exley et al. 2008; Rice and Cragg 2004; Zhang and Sulzer 2004). Thus, either activation of mAChRs or inhibition/desensitization of nicotinic receptors results in dopamine release probability becoming proportional to frequency and as such restores release to exhibit classical, dynamic probability of neurotransmitter release as seen at other synapses (Thomson 2000a,b). However, unlike for striatal nAChRs, there is no anatomical evidence that striatal mAChRs are present on dopaminergic axon terminals to influence DA release directly (Jones et al. 2001; Zhang et al. 2002b; Zhou et al. 2003). In contrast, many other striatal neurons express and are regulated by mAChRs, including striatal cholinergic interneurons (Alcantara et al. 2001; Bernard et al. 1992, 1998; Bonsi et al. 2008; Yan and Surmeier 1996), GABAergic interneurons (Koos and Tepper 2002), MSNs (Calabresi et al. 2000; Levey et al. 1991; Weiner et al. 1990; Yan et al. 2001), and glutamatergic afferents (Calabresi et al. 2000; Pakhotin and Bracci 2007; Sugita et al. 1991). Previous studies have identified the regulation of dopamine release by mAChRs on a variety of striatal neurons/ inputs. For example, mAChR regulation of GABA acting at $GABA_A$ receptors has been implicated in the control of dopamine in protocols using high $[K^+]$ to evoke [3H] dopamine release (Zhang et al. 2002b), but the dynamic mAChR modulation of endogenous dopamine release we described during discrete electrical stimuli persists in the presence of synaptic blockers for GABA/glutamate. The dynamic regulation of striatal dopamine by mAChRs described here requires cholinergic tone at nAChRs on dopamine axons, since blockade of acetylcholine input from cholinergic interneurons to nAChRs precludes the effects of mAChR activation and vice versa. Striatal cholinergic interneurons express somatodendritic and axonal mAChRs (Alcantara et al. 2001; Bernard et al. 1992, 1998; Zhang et al. 2002a), which are autoreceptors. When these autoreceptors are activated, cholinergic interneurons become silenced and acetylcholine release is inhibited (via inhibition of Ca_v2-type Ca^{2+} conductances and/or K_v^+ channel opening) (Bonsi et al. 2008; Calabresi et al. 1998, 2000; Ding et al. 2006; Raiteri et al. 1984; Schoffelmeer et al. 1986; Yan and Surmeier 1996; Zhang et al. 2002a; Zhou et al. 2003). The reduction in acetylcholine release following mAChR autoreceptor activation consequently deactivates nAChRs on dopaminergic axons, and in turn, increases the sensitivity of DA release to presynaptic depolarization frequency as seen following the inhibition/desensitization of nicotinic receptors (see Figure 6.1).

These data revise our understanding of striatal mAChR–dopamine interactions in several ways. First, striatal mAChRs offer variable, bidirectional control of dopamine release probability depending on presynaptic activity. This reconciles previous contradictory findings since striatal mAChRs do not simply suppress or enhance dopamine release but can do both depending on the frequency of depolarization.

Second, this variable mAChR control of dopamine release is not via multiple striatal neuron types but via the control of acetylcholine release from striatal cholinergic interneurons. These mAChRs, by modifying acetylcholine release, powerfully gate the nAChR regulation of dopamine release. Finally, we also show important differences in the mAChR subtypes regulating dopamine (and cholinergic interneurons) in sensorimotor- versus limbic-associated striatum.

6.4.1 mAChR Subtypes and Locations within Striatum

The G-protein-coupled mAChRs are present throughout the striatum on a variety of neurons and axon terminals. Five different types of mAChRs exist in the striatum, M_1–M_5. Like dopamine receptors, these mAChRs are commonly divided into two families—M_1-like (M_1, M_3, and M_5) and M_2-like (M_2 and M_4). M_1-like receptors are coupled to the G_s class of G_α proteins, whereas M_2-like receptors are coupled to G_i proteins, thereby modulating different intracellular signaling pathways. The mAChR subtypes M_1, M_2, and M_4 appear to be dominant in the striatum (Zhang et al. 2002b; Zhou et al. 2003). There is now considerable evidence that discrete expression and function of M_1, M_2, and M_4 receptors can be partitioned to different striatal neurons and neurotransmitter interactions. For example, MSNs express primarily M_1 and M_4, with M_4 receptors dominant on striatonigral MSNs with very low or undetectable levels of M_2, M_3, and M_5 (Levey et al. 1991; Santiago and Potter 2001; Shen et al. 2005; Wang et al. 2006; Weiner et al. 1990; Yan et al. 2001). Striatal cholinergic interneurons by contrast have dominant expression and function of M_2- and M_4-mAChRs (Alcantara et al. 2001; Bernard et al. 1998; Ding et al. 2006; Yan and Surmeier 1996). M_5 receptors may exist on dopamine neurons in the midbrain, but there is no evidence to suggest that they exist on dopamine terminals in the striatum (Weiner et al. 1990).

6.4.2 mAChR Subtypes Responsible for Regulation of Striatal Dopamine in Dorsal versus Ventral Striatum

The mAChRs regulating striatal dopamine release differ between dorsal and ventral striatal subterritories. Due to a lack of pharmacological ligands able to act selectively at distinct mAChR subtypes (ligands are subtype preferring but not selective), we have made use of transgenic mice lacking individual mAChR subtypes (Wess et al. 2003) to study which subtypes are responsible for the regulation of dopamine in specific striatal subregions. Given that mAChRs on cholinergic interneurons are responsible for modulation of striatal dopamine, we studied mice lacking M_2- and M_4-mAChRs (cholinergic interneuron autoreceptors), as well as M_5-mAChR-null mice should there be any mAChRs present on dopamine axon terminals.

In dorsal striatum, a region classically associated with motor function, both M_2- and M_4-mAChRs are necessary for muscarinic regulation of dopamine release. Deletion of either M_2- or M_4-mAChRs prevents mAChR modulation of dopamine (Threlfell et al. 2010). By contrast, in more limbic striatal regions, in the NAc core or shell, only the M_4-mAChR is necessary for mAChR control (Threlfell et al. 2010). Deletion of M_2-mAChRs in the NAc core or shell does not prevent muscarinic control of dopamine; deletion of only the M_4-mAChR eliminates muscarinic control.

In both dorsal striatum and NAc, elimination of the M_5-mAChR has no effect on the activity-dependent regulation of dopamine by mAChRs. This suggests that if M_5 receptors do exist presynaptically on striatal dopamine terminals, they do not participate in activity-dependent regulation of dopamine by mAChR agonists. Muscarinic-mediated suppression of dopamine release following single-pulse stimulation in mice lacking the M_5R appears to be more pronounced in both dorsal and ventral striatum, suggesting that if presynaptic M_5Rs on dopamine terminals exist they function normally to enhance dopamine release, as suggested by others (Bendor et al. 2010; Zhang et al. 2002b).

At present, it is unclear why such differences in mAChR control would exist between dorsal and ventral striatum. It is currently unknown whether these differences are attributable to different expression/role of M_4Rs versus M_2Rs in cholinergic interneurons that supply each territory, or a different basal acetylcholine tone at nAChRs, or whether there are other intrinsic regional differences in cholinergic interneurons including compensatory adaptations that can accommodate loss of M_2Rs in NAc but not in dorsal striatum. There is no comparative study of expression/function of mAChRs in the regulation of cholinergic activity in each region and no evidence for a difference in the levels of acetylcholine reaching nAChRs (Exley et al. 2008). There are, however, reports that mAChRs are found at higher levels in ventral than dorsal striatum in rats (Tayebati et al. 2004) and that cholinergic interneurons in limbic/prefrontal versus sensorimotor territories can be differentiated by different expression levels of other types of receptors (μ-opioid) (Jabourian et al. 2005). Thus, distinct muscarinic mechanisms/receptors in subpopulations of cholinergic interneurons may explain the differing mAChR control of dopamine release in dorsal striatum and NAc. These regional differences in M_2/M_4 function could ultimately be exploited for discrete modulation of dopamine/acetylcholine. For example, acetylcholine/dopamine function might be modified selectively in NAc by activation of M_4Rs, and in dorsal striatum by M_2R inhibition. It is possible that regional differences may exist due to differential intracellular coupling of M_2- and M_4-mAChRs on cholinergic interneurons in different striatal regions, or perhaps differential expression of M_2- and M_4-mAChRs on cholinergic interneurons in these two regions. Further electrophysiological characterization of cholinergic interneurons in mice lacking M_2- or M_4-mAChRs may shed light on potential differences between dorsal and ventral striatum and is currently underway.

6.5   SUMMARY AND PERSPECTIVE

Striatal nAChRs and mAChRs powerfully modulate dopamine transmission in an activity-dependent manner. This regulation will strongly influence how dopamine synapses convey the discrete changes or bursts in neuronal activity that signals events of motivational salience. Interestingly, cholinergic modulation of dopamine transmission within striatum is via a different profile of nicotinic (Exley et al. 2008) and mAChRs (Threlfell et al. 2010) in CPu compared to NAc (Figure 6.2), which may ultimately enable region-specific targeting of striatal function. For example, acetylcholine/dopamine function might be modified selectively in NAc by the activation of M_4Rs, and in dorsal striatum by M_2R inhibition. Or a specific nAChR such

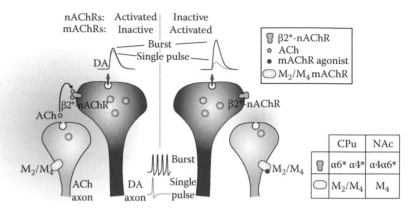

FIGURE 6.2 **(See color insert.)** Scheme illustrating cholinergic regulation of dopamine release during burst and nonburst activity in dopamine axons and nAChR/mAChR subtypes responsible in CPu and NAc. Left, When mAChRs are inactive under control conditions, endogenous acetylcholine (ACh) released from tonically active striatal cholinergic interneurons (ChIs) maintains ACh tone at β2*-nAChRs on dopaminergic axons. This tonic β2*-nAChR activity ensures that dopamine (DA) release has a high probability of occurring in response to a single stimulus pulse. Short-term synaptic depression follows such DA release and limits re-release by successive pulses within bursts, that is, DA release is insensitive to frequency. Right, Activation of mAChR autoreceptors on ChIs (e.g., with mAChR agonist Oxo-M) reduces ACh tone at β2*-nAChRs on DA axon terminals, thereby reducing DA release probability by a single stimulus, relieving short-term depression and increasing the relative probability of release at subsequent action potentials. After reduction of ACh tone at β2*-nAChRs by activation of mAChR autoreceptors on ChIs, or blockade/desensitization of nAChRs, DA release becomes more sensitive to frequency of presynaptic activity (see Exley and Cragg 2008; Rice and Cragg 2004). In CPu, M_2 and M_4 populations of mAChRs and α6*- and α4*-nAChRs appear to modulate DA release. In NAc, only M_4-mAChRs and α4α6*-nAChRs appear to modulate DA release.

as the α4α6β2*-nAChR may be exploited in NAc in the treatment of nicotine addiction. More generally, these data emphasize that regulation of neurotransmission by a given neuromodulator is not fixed, or unidirectional; it can be variable and bidirectional in an activity-dependent manner, a principle which may apply to many other neuromodulators and neurotransmitters.

    We now have an ever increasing availability of new molecular biological tools to facilitate nucleus-specific and cell-specific elimination/expression of proteins. We are therefore in a great position to advance our appreciation of how receptors are able to regulate striatal circuitry. Dissecting the complex interplay of the multitude of neurotransmitters and neuromodulators within the highly heterogeneous striatum across a wide array of behaviors will undoubtedly be facilitated by the use of such tools. It is imperative to assess the role played by all neurotransmitters and neuromodulators when considering how striatal output changes during different behaviors and as such the acetylcholine–dopamine interaction is one powerful relationship that should be taken into account. In particular, glutamate–dopamine interactions must not be considered in isolation from other neuromodulators such as acetylcholine, which have important parallel roles in shaping striatal output.

ACKNOWLEDGMENTS

The authors thank their sponsors Parkinson's UK (Grants G0808, G1103, G4078, and the Discovery Award), the Medical Research Council (Grant G0700932), and the BBSRC.

REFERENCES

Abeliovich, A., Y. Schmitz, I. Farinas, D. Choi-Lundberg, W. H. Ho, P. E. Castillo, N. Shinsky, J. M. Verdugo, M. Armanini, A. Ryan, M. Hynes, H. Phillips, D. Sulzer, and A. Rosenthal. 2000. Mice lacking alpha-synuclein display functional deficits in the nigrostriatal dopamine system. *Neuron* 25 (1):239–252.

Albin, R. L., A. B. Young, and J. B. Penney. 1995. The functional anatomy of disorders of the basal ganglia. *Trends Neurosci* 18 (2):63–64.

Alcantara, A. A., V. Chen, B. E. Herring, J. M. Mendenhall, and M. L. Berlanga. 2003. Localization of dopamine D2 receptors on cholinergic interneurons of the dorsal striatum and nucleus accumbens of the rat. *Brain Res* 986 (1–2):22–29.

Alcantara, A. A., L. Mrzljak, R. L. Jakab, A. I. Levey, S. M. Hersch, and P. S. Goldman-Rakic. 2001. Muscarinic m1 and m2 receptor proteins in local circuit and projection neurons of the primate striatum: Anatomical evidence for cholinergic modulation of glutamatergic prefronto-striatal pathways. *J Comp Neurol* 434 (4):445–460.

Arbuthnott, G. W. and J. Wickens. 2007. Space, time and dopamine. *Trends Neurosci* 30 (2):62–69.

Azam, L., U. H. Winzer-Serhan, Y. Chen, and F. M. Leslie. 2002. Expression of neuronal nicotinic acetylcholine receptor subunit mRNAs within midbrain dopamine neurons. *J Comp Neurol* 444 (3):260–274.

Bamford, N. S., H. Zhang, Y. Schmitz, N. P. Wu, C. Cepeda, M. S. Levine, C. Schmauss, S. S. Zakharenko, L. Zablow, and D. Sulzer. 2004. Heterosynaptic dopamine neurotransmission selects sets of corticostriatal terminals. *Neuron* 42 (4):653–663.

Barbeau, A. 1962. The pathogenesis of Parkinson's disease: A new hypothesis. *Can Med Assoc J* 87:802–807.

Belin, D. and B. J. Everitt. 2008. Cocaine seeking habits depend upon dopamine-dependent serial connectivity linking the ventral with the dorsal striatum. *Neuron* 57 (3):432–441.

Belin, D., S. Jonkman, A. Dickinson, T. W. Robbins, and B. J. Everitt. 2009. Parallel and interactive learning processes within the basal ganglia: Relevance for the understanding of addiction. *Behav Brain Res* 199 (1):89–102.

Bendor, J., J. E. Lizardi-Ortiz, R. I. Westphalen, M. Brandstetter, H. C. Hemmings, Jr., D. Sulzer, M. Flajolet, and P. Greengard. 2010. AGAP1/AP-3-dependent endocytic recycling of M(5) muscarinic receptors promotes dopamine release. *Embo J* 29 (16):2813–2826.

Bennett, B. D. and C. J. Wilson. 1999. Spontaneous activity of neostriatal cholinergic interneurons in vitro. *J Neurosci* 19 (13):5586–5596.

Bernard, V., O. Laribi, A. I. Levey, and B. Bloch. 1998. Subcellular redistribution of m2 muscarinic acetylcholine receptors in striatal interneurons in vivo after acute cholinergic stimulation. *J Neurosci* 18 (23):10207–10218.

Bernard, V., E. Normand, and B. Bloch. 1992. Phenotypical characterization of the rat striatal neurons expressing muscarinic receptor genes. *J Neurosci* 12 (9):3591–3600.

Berridge, K. C. and T. E. Robinson. 2003. Parsing reward. *Trends Neurosci* 26 (9):507–513.

Bjorklund, A. and O. Lindvall. 1984. Dopamine-containing systems in the CNS, in ed. H. T. Bjorklund, *A Handbook of Chemical Neuroanatomy*. New York: Elsevier.

Bolam, J. P., J. J. Hanley, P. A. Booth, and M. D. Bevan. 2000. Synaptic organisation of the basal ganglia. *J Anat* 196 (Pt 4):527–542.

Bonsi, P., G. Martella, D. Cuomo, P. Platania, G. Sciamanna, G. Bernardi, J. Wess, and
     A. Pisani. 2008. Loss of muscarinic autoreceptor function impairs long-term depression
     but not long-term potentiation in the striatum. *J Neurosci* 28 (24):6258–6263.
Brog, J. S., A. Salyapongse, A. Y. Deutch, and D. S. Zahm. 1993. The patterns of afferent
     innervation of the core and shell in the "accumbens" part of the rat ventral striatum:
     Immunohistochemical detection of retrogradely transported fluoro-gold. *J Comp Neurol*
     338 (2):255–278.
Calabresi, P., D. Centonze, P. Gubellini, A. Pisani, and G. Bernardi. 2000. Acetylcholine-
     mediated modulation of striatal function. *Trends Neurosci* 23 (3):120–126.
Calabresi, P., D. Centonze, A. Pisani, G. Sancesario, R. A. North, and G. Bernardi. 1998.
     Muscarinic IPSPs in rat striatal cholinergic interneurones. *J Physiol* 510 (Pt 2):421–427.
Calabresi, P., B. Picconi, A. Tozzi, and M. Di Filippo. 2007. Dopamine-mediated regulation of
     corticostriatal synaptic plasticity. *Trends Neurosci* 30 (5):211–219.
Cartier, G. E., D. Yoshikami, W. R. Gray, S. Luo, B. M. Olivera, and J. M. McIntosh. 1996.
     A new alpha-conotoxin which targets alpha3beta2 nicotinic acetylcholine receptors.
     J Biol Chem 271 (13):7522–7528.
Centonze, D., P. Gubellini, A. Pisani, G. Bernardi, and P. Calabresi. 2003. Dopamine, acetyl-
     choline and nitric oxide systems interact to induce corticostriatal synaptic plasticity. *Rev
     Neurosci* 14 (3):207–216.
Champtiaux, N., C. Gotti, M. Cordero-Erausquin, D. J. David, C. Przybylski, C. Lena,
     F. Clementi, M. Moretti, F. M. Rossi, N. Le Novere, J. M. McIntosh, A. M. Gardier, and
     J. P. Changeux. 2003. Subunit composition of functional nicotinic receptors in dopami-
     nergic neurons investigated with knock-out mice. *J Neurosci* 23 (21):7820–7829.
Chergui, K., M. F. Suaud-Chagny, and F. Gonon. 1994. Nonlinear relationship between impulse
     flow, dopamine release and dopamine elimination in the rat brain in vivo. *Neuroscience*
     62 (3):641–645.
Contant, C., D. Umbriaco, S. Garcia, K. C. Watkins, and L. Descarries. 1996. Ultrastructural
     characterization of the acetylcholine innervation in adult rat neostriatum. *Neuroscience*
     71 (4):937–947.
Corringer, P. J., N. Le Novere, and J. P. Changeux. 2000. Nicotinic receptors at the amino acid
     level. *Annu Rev Pharmacol Toxicol* 40:431–458.
Cragg, S. J. 2003. Variable dopamine release probability and short-term plasticity between
     functional domains of the primate striatum. *J Neurosci* 23 (10):4378–4385.
Cragg S. J. 2006. Meaningful silences: How dopamine listens to the ACh pause. *Trends
     Neurosci* 29 (3):125–131.
Cragg, S. J., R. Exley, and M. A. Clements. 2005. Striatal acetylcholine control of reward-
     related dopamine signalling, in eds. J. P. Bolam, C. A. Ingham, and P. J. Magill,
     The Basal Ganglia VIII. New York: Springer.
Cragg, S. J. and M. E. Rice. 2004. Dancing past the DAT at a DA synapse. *Trends Neurosci*
     27 (5):270–277.
Dajas-Bailador, F. and S. Wonnacott. 2004. Nicotinic acetylcholine receptors and the regula-
     tion of neuronal signalling. *Trends Pharmacol Sci* 25 (6):317–324.
Day, M., D. Wokosin, J. L. Plotkin, X. Tian, and D. J. Surmeier. 2008. Differential excitability and
     modulation of striatal medium spiny neuron dendrites. *J Neurosci* 28 (45):11603–11614.
De Klippel, N., S. Sarre, G. Ebinger, and Y. Michotte. 1993. Effect of M1- and M2-muscarinic
     drugs on striatal dopamine release and metabolism: An in vivo microdialysis study com-
     paring normal and 6-hydroxydopamine-lesioned rats. *Brain Res* 630 (1–2):57–64.
Descarries, L., V. Gisiger, and M. Steriade. 1997. Diffuse transmission by acetylcholine in the
     CNS. *Prog Neurobiol* 53 (5):603–625.
Descarries, L. and N. Mechawar. 2000. Ultrastructural evidence for diffuse transmission by
     monoamine and acetylcholine neurons of the central nervous system. *Prog Brain Res*
     125:27–47.

Descarries, L., K. C. Watkins, S. Garcia, O. Bosler, and G. Doucet. 1996. Dual character, asynaptic and synaptic, of the dopamine innervation in adult rat neostriatum: A quantitative autoradiographic and immunocytochemical analysis. *J Comp Neurol* 375 (2):167–186.

Di Chiara, G. and A. Imperato. 1988. Drugs abused by humans preferentially increase synaptic dopamine concentrations in the mesolimbic system of freely moving rats. *Proc Natl Acad Sci USA* 85 (14):5274–5278.

Ding, J., J. N. Guzman, T. Tkatch, S. Chen, J. A. Goldberg, P. J. Ebert, P. Levitt, C. J. Wilson, H. E. Hamm, and D. J. Surmeier. 2006. RGS4-dependent attenuation of M4 autoreceptor function in striatal cholinergic interneurons following dopamine depletion. *Nat Neurosci* 9 (6):832–842.

Drenan, R. M., S. R. Grady, A. D. Steele, S. McKinney, N. E. Patzlaff, J. M. McIntosh, M. J. Marks, J. M. Miwa, and H. A. Lester. 2010. Cholinergic modulation of locomotion and striatal dopamine release is mediated by alpha6alpha4* nicotinic acetylcholine receptors. *J Neurosci* 30 (29):9877–9889.

Everitt, B. J. and T. W. Robbins. 2005. Neural systems of reinforcement for drug addiction: From actions to habits to compulsion. *Nat Neurosci* 8 (11):1481–1489.

Exley, R., M. A. Clements, H. Hartung, J. M. McIntosh, and S. J. Cragg. 2008. Alpha6-containing nicotinic acetylcholine receptors dominate the nicotine control of dopamine neurotransmission in nucleus accumbens. *Neuropsychopharmacology* 33 (9):2158–2166.

Exley, R. and S. J. Cragg. 2008. Presynaptic nicotinic receptors: A dynamic and diverse cholinergic filter of striatal dopamine neurotransmission. *Br J Pharmacol* 153 (1):S283–S297.

Freund, T. F., L. F. Powell, and A. D. Smith. 1984. Tyrosine hydroxylase-immunoreactive boutons in synaptic contact with identified striatonigral neurons, with particular reference to dendritic spines. *Neuroscience* 13 (4):1189–1215.

Fuxe, K. and L.F. Agnati. 1991. *Volume Transmission in the Brain*. New York: Raven Press.

Garris, P. A., E. L. Ciolkowski, P. Pastore, and R. M. Wightman. 1994. Efflux of dopamine from the synaptic cleft in the nucleus accumbens of the rat brain. *J Neurosci* 14 (10):6084–6093.

Gerdeman, G. L., J. G. Partridge, C. R. Lupica, and D. M. Lovinger. 2003. It could be habit forming: Drugs of abuse and striatal synaptic plasticity. *Trends Neurosci* 26 (4):184–192.

Gerfen, C. R. 1992. The neostriatal mosaic: Multiple levels of compartmental organization. *Trends Neurosci* 15 (4):133–139.

Gerfen, C. R., M. Herkenham, and J. Thibault. 1987. The neostriatal mosaic: II. Patch- and matrix-directed mesostriatal dopaminergic and non-dopaminergic systems. *J Neurosci* 7 (12):3915–3934.

Gonon, F. 1997. Prolonged and extrasynaptic excitatory action of dopamine mediated by D1 receptors in the rat striatum in vivo. *J Neurosci* 17 (15):5972–5978.

Gotti, C. and F. Clementi. 2004. Neuronal nicotinic receptors: From structure to pathology. *Prog Neurobiol* 74 (6):363–396.

Gotti, C., F. Clementi, A. Fornari, A. Gaimarri, S. Guiducci, I. Manfredi, M. Moretti, P. Pedrazzi, L. Pucci, and M. Zoli. 2009. Structural and functional diversity of native brain neuronal nicotinic receptors. *Biochem Pharmacol* 78 (7):703–711.

Gotti, C., M. Moretti, F. Clementi, L. Riganti, J. M. McIntosh, A. C. Collins, M. J. Marks, and P. Whiteaker. 2005. Expression of nigrostriatal alpha 6-containing nicotinic acetylcholine receptors is selectively reduced, but not eliminated, by beta 3 subunit gene deletion. *Mol Pharmacol* 67 (6):2007–2015.

Grace, A. A. and B. S. Bunney. 1984a. The control of firing pattern in nigral dopamine neurons: Burst firing. *J Neurosci* 4 (11):2877–2890.

Grace, A. A. and B. S. Bunney. 1984b. The control of firing pattern in nigral dopamine neurons: Single spike firing. *J Neurosci* 4 (11):2866–2876.

Grilli, M., L. Patti, F. Robino, S. Zappettini, M. Raiteri, and M. Marchi. 2008. Release-enhancing pre-synaptic muscarinic and nicotinic receptors co-exist and interact on dopaminergic nerve endings of rat nucleus accumbens. *J Neurochem* 105 (6):2205–2213.

Groves, P. M., J. C. Linder, and S. J. Young. 1994. 5-hydroxydopamine-labeled dopaminergic axons: Three-dimensional reconstructions of axons, synapses and postsynaptic targets in rat neostriatum. *Neuroscience* 58 (3):593–604.

Haber, S. N., J. L. Fudge, and N. R. McFarland. 2000. Striatonigrostriatal pathways in primates form an ascending spiral from the shell to the dorsolateral striatum. *J Neurosci* 20 (6):2369–2382.

Hersch, S. M., B. J. Ciliax, C. A. Gutekunst, H. D. Rees, C. J. Heilman, K. K. Yung, J. P. Bolam, E. Ince, H. Yi, and A. I. Levey. 1995. Electron microscopic analysis of D1 and D2 dopamine receptor proteins in the dorsal striatum and their synaptic relationships with motor corticostriatal afferents. *J Neurosci* 15 (7 Pt 2):5222–5237.

Hyland, B. I., J. N. Reynolds, J. Hay, C. G. Perk, and R. Miller. 2002. Firing modes of midbrain dopamine cells in the freely moving rat. *Neuroscience* 114 (2):475–492.

Jabourian, M., L. Venance, S. Bourgoin, S. Ozon, S. Perez, G. Godeheu, J. Glowinski, and M. L. Kemel. 2005. Functional mu opioid receptors are expressed in cholinergic interneurons of the rat dorsal striatum: Territorial specificity and diurnal variation. *Eur J Neurosci* 21 (12):3301–3309.

Jennings, K. A., S. Threlfell, R. Exley, and S. J. Cragg. 2009. Unmasking the role of nicotine receptors in nicotine addiction: Recent advances in understanding nicotine action on dopamine systems. *Cell Sci Rev* 5 (4):1–31.

Jones, I. W., J. P. Bolam, and S. Wonnacott. 2001. Presynaptic localisation of the nicotinic acetylcholine receptor beta2 subunit immunoreactivity in rat nigrostriatal dopaminergic neurones. *J Comp Neurol* 439 (2):235–247.

Kaiser, S. A., L. Soliakov, S. C. Harvey, C. W. Luetje, and S. Wonnacott. 1998. Differential inhibition by alpha-conotoxin-MII of the nicotinic stimulation of [3H]dopamine release from rat striatal synaptosomes and slices. *J Neurochem* 70 (3):1069–1076.

Kawaguchi, Y. 1993. Physiological, morphological, and histochemical characterization of three classes of interneurons in rat neostriatum. *J Neurosci* 13 (11):4908–4923.

Kelley, A. E. and V. B. Domesick. 1982. The distribution of the projection from the hippocampal formation to the nucleus accumbens in the rat: An anterograde- and retrograde-horseradish peroxidase study. *Neuroscience* 7 (10):2321–2335.

Kemel, M. L., M. Desban, J. Glowinski, and C. Gauchy. 1989. Distinct presynaptic control of dopamine release in striosomal and matrix areas of the cat caudate nucleus. *Proc Natl Acad Sci USA* 86 (22):9006–9010.

Koos, T. and J. M. Tepper. 2002. Dual cholinergic control of fast-spiking interneurons in the neostriatum. *J Neurosci* 22 (2):529–535.

Kudernatsch, M. and B. Sutor. 1994. Cholinergic modulation of dopamine overflow in the rat neostriatum: A fast cyclic voltammetric study in vitro. *Neurosci Lett* 181 (1–2):107–112.

Kulak, J. M., T. A. Nguyen, B. M. Olivera, and J. M. McIntosh. 1997. Alpha-conotoxin MII blocks nicotine-stimulated dopamine release in rat striatal synaptosomes. *J Neurosci* 17 (14):5263–5270.

Le Novere, N., P. J. Corringer, and J. P. Changeux. 2002. The diversity of subunit composition in nAChRs: Evolutionary origins, physiologic and pharmacologic consequences. *J Neurobiol* 53 (4):447–456.

Lehmann, J. and S. Z. Langer. 1982. Muscarinic receptors on dopamine terminals in the cat caudate nucleus: Neuromodulation of [3H]dopamine release in vitro by endogenous acetylcholine. *Brain Res* 248 (1):61–69.

Levey, A. I., C. A. Kitt, W. F. Simonds, D. L. Price, and M. R. Brann. 1991. Identification and localization of muscarinic acetylcholine receptor proteins in brain with subtype-specific antibodies. *J Neurosci* 11 (10):3218–3226.

Luetje, C. W. 2004. Getting past the asterisk: The subunit composition of presynaptic nicotinic receptors that modulate striatal dopamine release. *Mol Pharmacol* 65 (6):1333–1335.

Matsuda, W., T. Furuta, K. C. Nakamura, H. Hioki, F. Fujiyama, R. Arai, and T. Kaneko. 2009. Single nigrostriatal dopaminergic neurons form widely spread and highly dense axonal arborizations in the neostriatum. *J Neurosci* 29 (2):444–453.

Matsumoto, M. and O. Hikosaka. 2009. Two types of dopamine neuron distinctly convey positive and negative motivational signals. *Nature* 459 (7248):837–841.

McFarland, N. R. and S. N. Haber. 2000. Convergent inputs from thalamic motor nuclei and frontal cortical areas to the dorsal striatum in the primate. *J Neurosci* 20 (10):3798–3813.

McGeorge, A. J. and R. L. Faull. 1989. The organization of the projection from the cerebral cortex to the striatum in the rat. *Neuroscience* 29 (3):503–537.

McIntosh, J. M., L. Azam, S. Staheli, C. Dowell, J. M. Lindstrom, A. Kuryatov, J. E. Garrett, M. J. Marks, and P. Whiteaker. 2004. Analogs of alpha-conotoxin MII are selective for alpha6-containing nicotinic acetylcholine receptors. *Mol Pharmacol* 65 (4):944–952.

Mirenowicz, J. and W. Schultz. 1996. Preferential activation of midbrain dopamine neurons by appetitive rather than aversive stimuli. *Nature* 379 (6564):449–451.

Montague, P. R., S. M. McClure, P. R. Baldwin, P. E. Phillips, E. A. Budygin, G. D. Stuber, M. R. Kilpatrick, and R. M. Wightman. 2004. Dynamic gain control of dopamine delivery in freely moving animals. *J Neurosci* 24 (7):1754–1759.

Morris, G., D. Arkadir, A. Nevet, E. Vaadia, and H. Bergman. 2004. Coincident but distinct messages of midbrain dopamine and striatal tonically active neurons. *Neuron* 43 (1):133–143.

Moss, J. and J. P. Bolam. 2008. A dopaminergic axon lattice in the striatum and its relationship with cortical and thalamic terminals. *J Neurosci* 28 (44):11221–11230.

Moss, J. and J. P. Bolam. 2010. The relationship between dopaminergic axons and glutamatergic synapses in the straitum: Structural considerations, in eds. L. L. Iversen, S. D. Iversen, S. B. Dunnett, and A. Bjorklund, *Dopamine Handbook*. Oxford, U.K.: Oxford University Press.

Nakano, K., T. Kayahara, T. Tsutsumi, and H. Ushiro. 2000. Neural circuits and functional organization of the striatum. *J Neurol* 247 (5):V1–V15.

Nicola, S. M., J. Surmeier, and R. C. Malenka. 2000. Dopaminergic modulation of neuronal excitability in the striatum and nucleus accumbens. *Annu Rev Neurosci* 23:185–215.

Nirenberg, M. J., J. Chan, R. A. Vaughan, G. R. Uhl, M. J. Kuhar, and V. M. Pickel. 1997. Immunogold localization of the dopamine transporter: An ultrastructural study of the rat ventral tegmental area. *J Neurosci* 17 (11):4037–4044.

Nirenberg, M. J., R. A. Vaughan, G. R. Uhl, M. J. Kuhar, and V. M. Pickel. 1996. The dopamine transporter is localized to dendritic and axonal plasma membranes of nigrostriatal dopaminergic neurons. *J Neurosci* 16 (2):436–447.

O'Donnell, P. and A. A. Grace. 1995. Synaptic interactions among excitatory afferents to nucleus accumbens neurons: Hippocampal gating of prefrontal cortical input. *J Neurosci* 15 (5 Pt 1):3622–3639.

Pakhotin, P. and E. Bracci. 2007. Cholinergic interneurons control the excitatory input to the striatum. *J Neurosci* 27 (2):391–400.

Partridge, J. G., S. Apparsundaram, G. A. Gerhardt, J. Ronesi, and D. M. Lovinger. 2002. Nicotinic acetylcholine receptors interact with dopamine in induction of striatal long-term depression. *J Neurosci* 22 (7):2541–2549.

Pennartz, C. M., H. J. Groenewegen, and F. H. Lopes da Silva. 1994. The nucleus accumbens as a complex of functionally distinct neuronal ensembles: An integration of behavioural, electrophysiological and anatomical data. *Prog Neurobiol* 42 (6):719–761.

Pickel, V. M. 2000. Extrasynaptic distribution of monoamine transporters and receptors. *Prog Brain Res* 125:267–276.

Pisani, A., P. Bonsi, D. Centonze, P. Gubellini, G. Bernardi, and P. Calabresi. 2003. Targeting striatal cholinergic interneurons in Parkinson's disease: Focus on metabotropic glutamate receptors. *Neuropharmacology* 45 (1):45–56.

Quik, M., S. Vailati, T. Bordia, J. M. Kulak, H. Fan, J. M. McIntosh, F. Clementi, and C. Gotti. 2005. Subunit composition of nicotinic receptors in monkey striatum: Effect of treatments with 1-methyl-4-phenyl-1,2,3,6-tetrahydropyridine or L-DOPA. *Mol Pharmacol* 67 (1):32–41.

Raiteri, M., R. Leardi, and M. Marchi. 1984. Heterogeneity of presynaptic muscarinic receptors regulating neurotransmitter release in the rat brain. *J Pharmacol Exp Ther* 228 (1):209–214.

Reep, R. L., J. L. Cheatwood, and J. V. Corwin. 2003. The associative striatum: Organization of cortical projections to the dorsocentral striatum in rats. *J Comp Neurol* 467 (3):271–292.

Rice, M. E. and S. J. Cragg. 2004. Nicotine amplifies reward-related dopamine signals in striatum. *Nat Neurosci* 7 (6):583–584.

Rice, M. E. and S. J. Cragg. 2008. Dopamine spillover after quantal release: Rethinking dopamine transmission in the nigrostriatal pathway. *Brain Res Rev* 58 (2):303–313.

Salminen, O., J. A. Drapeau, J. M. McIntosh, A. C. Collins, M. J. Marks, and S. R. Grady. 2007. Pharmacology of alpha-conotoxin MII-sensitive subtypes of nicotinic acetylcholine receptors isolated by breeding of null mutant mice. *Mol Pharmacol* 71 (6):1563–1571.

Salminen, O., K. L. Murphy, J. M. McIntosh, J. Drago, M. J. Marks, A. C. Collins, and S. R. Grady. 2004. Subunit composition and pharmacology of two classes of striatal presynaptic nicotinic acetylcholine receptors mediating dopamine release in mice. *Mol Pharmacol* 65 (6):1526–1535.

Santiago, M. P. and L. T. Potter. 2001. Biotinylated m4-toxin demonstrates more M4 muscarinic receptor protein on direct than indirect striatal projection neurons. *Brain Res* 894 (1):12–20.

Schoffelmeer, A. N., B. J. Van Vliet, G. Wardeh, and A. H. Mulder. 1986. Muscarine receptor-mediated modulation of [3H]dopamine and [14C]acetylcholine release from rat neostriatal slices: Selective antagonism by gallamine but not pirenzepine. *Eur J Pharmacol* 128 (3):291–294.

Schultz, W. 1986. Responses of midbrain dopamine neurons to behavioral trigger stimuli in the monkey. *J Neurophysiol* 56 (5):1439–1461.

Schultz, W. 1998. Predictive reward signal of dopamine neurons. *J Neurophysiol* 80 (1):1–27.

Schultz, W. 2002. Getting formal with dopamine and reward. *Neuron* 36 (2):241–263.

Schultz, W. 2006. Behavioral theories and the neurophysiology of reward. *Annu Rev Psychol* 57:87–115.

Sesack, S. R., C. Aoki, and V. M. Pickel. 1994. Ultrastructural localization of D2 receptor-like immunoreactivity in midbrain dopamine neurons and their striatal targets. *J Neurosci* 14 (1):88–106.

Sesack, S. R., A. Y. Deutch, R. H. Roth, and B. S. Bunney. 1989. Topographical organization of the efferent projections of the medial prefrontal cortex in the rat: An anterograde tract-tracing study with *Phaseolus vulgaris* leucoagglutinin. *J Comp Neurol* 290 (2):213–242.

Shen, W., S. E. Hamilton, N. M. Nathanson, and D. J. Surmeier. 2005. Cholinergic suppression of KCNQ channel currents enhances excitability of striatal medium spiny neurons. *J Neurosci* 25 (32):7449–7458.

Smith, A. D. and J. P. Bolam. 1990. The neural network of the basal ganglia as revealed by the study of synaptic connections of identified neurones. *Trends Neurosci* 13 (7):259–265.

Stuber, G. D., T. S. Hnasko, J. P. Britt, R. H. Edwards, and A. Bonci. 2010. Dopaminergic terminals in the nucleus accumbens but not the dorsal striatum corelease glutamate. *J Neurosci* 30 (24):8229–8233.

Sugita, S., N. Uchimura, Z. G. Jiang, and R. A. North. 1991. Distinct muscarinic receptors inhibit release of gamma-aminobutyric acid and excitatory amino acids in mammalian brain. *Proc Natl Acad Sci USA* 88 (6):2608–2611.

Surmeier, D. J., J. Ding, M. Day, Z. Wang, and W. Shen. 2007. D1 and D2 dopamine-receptor modulation of striatal glutamatergic signaling in striatal medium spiny neurons. *Trends Neurosci* 30 (5):228–235.

Surmeier, D. J., W. J. Song, and Z. Yan. 1996. Coordinated expression of dopamine receptors in neostriatal medium spiny neurons. *J Neurosci* 16 (20):6579–6591.

Tayebati, S. K., M. A. Di Tullio, and F. Amenta. 2004. Age-related changes of muscarinic cholinergic receptor subtypes in the striatum of Fisher 344 rats. *Exp Gerontol* 39 (2):217–223.

Tecuapetla, F., J. C. Patel, H. Xenias, D. English, I. Tadros, F. Shah, J. Berlin, K. Deisseroth, M. E. Rice, J. M. Tepper, and T. Koos. 2010. Glutamatergic signaling by mesolimbic dopamine neurons in the nucleus accumbens. *J Neurosci* 30 (20):7105–7110.

Thomson, A. M. 2000a. Facilitation, augmentation and potentiation at central synapses. *Trends Neurosci* 23 (7):305–312.

Thomson, A. M. 2000b. Molecular frequency filters at central synapses. *Prog Neurobiol* 62 (2):159–196.

Threlfell, S., M. A. Clements, T. Khodai, I. S. Pienaar, R. Exley, J. Wess, and S. J. Cragg. 2010. Striatal muscarinic receptors promote activity dependence of dopamine transmission via distinct receptor subtypes on cholinergic interneurons in ventral versus dorsal striatum. *J Neurosci* 30 (9):3398–3408.

Tobler, P. N., A. Dickinson, and W. Schultz. 2003. Coding of predicted reward omission by dopamine neurons in a conditioned inhibition paradigm. *J Neurosci* 23 (32):10402–10410.

Tzavara, E. T., F. P. Bymaster, R. J. Davis, M. R. Wade, K. W. Perry, J. Wess, D. L. McKinzie, C. Felder, and G. G. Nomikos. 2004. M4 muscarinic receptors regulate the dynamics of cholinergic and dopaminergic neurotransmission: Relevance to the pathophysiology and treatment of related CNS pathologies. *Faseb J* 18 (12):1410–1412.

Voorn, P., L. J. Vanderschuren, H. J. Groenewegen, T. W. Robbins, and C. M. Pennartz. 2004. Putting a spin on the dorsal-ventral divide of the striatum. *Trends Neurosci* 27 (8):468–474.

Wang, Z., L. Kai, M. Day, J. Ronesi, H. H. Yin, J. Ding, T. Tkatch, D. M. Lovinger, and D. J. Surmeier. 2006. Dopaminergic control of corticostriatal long-term synaptic depression in medium spiny neurons is mediated by cholinergic interneurons. *Neuron* 50 (3):443–452.

Weiner, D. M., A. I. Levey, and M. R. Brann. 1990. Expression of muscarinic acetylcholine and dopamine receptor mRNAs in rat basal ganglia. *Proc Natl Acad Sci USA* 87 (18):7050–7054.

Wess, J., A. Duttaroy, W. Zhang, J. Gomeza, Y. Cui, T. Miyakawa, F. P. Bymaster, L. McKinzie, C. C. Felder, K. G. Lamping, F. M. Faraci, C. Deng, and M. Yamada. 2003. M1–M5 muscarinic receptor knockout mice as novel tools to study the physiological roles of the muscarinic cholinergic system. *Receptors Channels* 9 (4):279–290.

Whiteaker, P., J. M. McIntosh, S. Luo, A. C. Collins, and M. J. Marks. 2000. 125I-alpha-conotoxin MII identifies a novel nicotinic acetylcholine receptor population in mouse brain. *Mol Pharmacol* 57 (5):913–925.

Wickens, J. R., J. N. Reynolds, and B. I. Hyland. 2003. Neural mechanisms of reward-related motor learning. *Curr Opin Neurobiol* 13 (6):685–690.

Wilson, C. J. 2004. Basal ganglia. In ed. G. M. Shepherd *The Synaptic Organization of the Brain*, Oxford, U.K.: Oxford University Press.

Wise, R. A. 2004. Dopamine, learning and motivation. *Nat Rev Neurosci* 5 (6):483–494.

Wonnacott, S., S. Kaiser, A. Mogg, L. Soliakov, and I. W. Jones. 2000. Presynaptic nicotinic receptors modulating dopamine release in the rat striatum. *Eur J Pharmacol* 393 (1–3):51–58.

Woolf, N. J. 1991. Cholinergic systems in mammalian brain and spinal cord. *Prog Neurobiol* 37 (6):475–524.

Xu, M., F. Mizobe, T. Yamamoto, and T. Kato. 1989. Differential effects of M1- and M2-muscarinic drugs on striatal dopamine release and metabolism in freely moving rats. *Brain Res* 495 (2):232–242.

Yan, Z., J. Flores-Hernandez, and D. J. Surmeier. 2001. Coordinated expression of muscarinic receptor messenger RNAs in striatal medium spiny neurons. *Neuroscience* 103 (4):1017–1024.

Yan, Z. and D. J. Surmeier. 1996. Muscarinic (m2/m4) receptors reduce N- and P-type Ca^{2+} currents in rat neostriatal cholinergic interneurons through a fast, membrane-delimited, G-protein pathway. *J Neurosci* 16 (8):2592–2604.

Zhang, W., A. S. Basile, J. Gomeza, L. A. Volpicelli, A. I. Levey, and J. Wess. 2002a. Characterization of central inhibitory muscarinic autoreceptors by the use of muscarinic acetylcholine receptor knock-out mice. *J Neurosci* 22 (5):1709–1717.

Zhang, H. and D. Sulzer. 2004. Frequency-dependent modulation of dopamine release by nicotine. *Nat Neurosci* 7 (6):581–582.

Zhang, W., M. Yamada, J. Gomeza, A. S. Basile, and J. Wess. 2002b. Multiple muscarinic acetylcholine receptor subtypes modulate striatal dopamine release, as studied with M1–M5 muscarinic receptor knock-out mice. *J Neurosci* 22 (15):6347–6352.

Zhou, F. M., Y. Liang, and J. A. Dani. 2001. Endogenous nicotinic cholinergic activity regulates dopamine release in the striatum. *Nat Neurosci* 4 (12):1224–1229.

Zhou, F. M., C. J. Wilson, and J. A. Dani. 2002. Cholinergic interneuron characteristics and nicotinic properties in the striatum. *J Neurobiol* 53 (4):590–605.

Zhou, F. M., C. Wilson, and J. A. Dani. 2003. Muscarinic and nicotinic cholinergic mechanisms in the mesostriatal dopamine systems. *Neuroscientist* 9 (1):23–36.

Zoli, M., M. Moretti, A. Zanardi, J. M. McIntosh, F. Clementi, and C. Gotti. 2002. Identification of the nicotinic receptor subtypes expressed on dopaminergic terminals in the rat striatum. *J Neurosci* 22 (20):8785–8789.

Zoli, M., C. Torri, R. Ferrari, A. Jansson, I. Zini, K. Fuxe, and L. F. Agnati. 1998. The emergence of the volume transmission concept. *Brain Res Rev* 26 (2–3):136–147.

7 Electrophysiology of the Corticostriatal Network *in Vivo*

Morgane Pidoux, Séverine Mahon,
and Stéphane Charpier

CONTENTS

7.1 INTRODUCTION AND OVERVIEW OF THE CHAPTER

The basal ganglia (BG) are a recipient for cortical information. Cortical projections to the BG are composed of two distinct pathways, the corticostriatal (CS) and corticosubthalamic systems, which connect monosynaptically the main input nuclei of the BG by glutamatergic synapses (Charpier et al. 2010; Wilson 2004). The striatum receives inputs from nearly all functional subdivisions of the cortical mantle, including neocortical and allocortical areas. These cortical inputs are topographically ordered and processed by the GABAergic striatal projection neurons, the main (~90%) neuronal population in the striatum morphologically defined as medium-sized spiny neurons (MSNs) (Chang et al. 1982) (see Figure 7.2A), and by most of the striatal interneurons which, in turn, modulate the responsiveness of MSNs (Kreitzer 2009). Two main CS macrocircuits can be outlined according to their different global functions. Whereas the "limbic" hippocampal (and prefrontal cortex)-ventral striatal subsystem has been delineated to mediate motivational and reward processes, the CS subsystem implicating the dorsolateral part of the striatum is involved in the correct completion of sensorimotor behaviors and is essential for learned motor sequences to become habitual (Graybiel 1995, 2008; Chapter 9). The cortical information processing in these two striatal functional subsystems is under an extrinsic modulatory control mediated by afferent dopaminergic fibers originating from the ventral tegmental area and substantia nigra pars compacta (Wilson 2004). Consistent with their normal behavioral functions, abnormal activities in CS systems are implicated in various neurological disorders related to action compulsion (such as obsessive compulsive disorders and drug addiction) and action disability (such as Parkinson's disease [PD] and Huntington's disease) (Graybiel 2008; Chapter 10). Thus, understanding how cortical inputs are processed by MSNs and transformed into behaviorally relevant outputs is a decisive challenge for the elucidation of the cellular basis of normal and pathological functions related to the BG.

In this chapter, we review the main morphofunctional features of the CS system linking the sensorimotor neocortex to the dorsal striatum, as unveiled from *in vivo* investigations that maintain intact the functional connectivity within the CS networks and preserve the structural and electrophysiological properties of their individual neuronal elements. A special attention will be paid to the dynamic interactions between the excitatory synaptic activity in MSNs, mainly generated by the CS inputs, and the voltage-gated membrane properties of striatal neurons, which provide a determinant "intrinsic device" for sculpting their membrane potential fluctuations and patterning their brain state-dependent firing.

7.2 MORPHOFUNCTIONAL DIVERSITY OF CORTICOSTRIATAL NEURONS

7.2.1 CORTICAL LAYER OF ORIGIN, PROJECTION RULES, AND MORPHOLOGY OF CORTICOSTRIATAL NEURONS

CS projections, originating from virtually all cortical areas (McGeorge and Faull 1989; Deniau et al. 1996), transmit to the striatum functional information that is encoded through distribution patterns of glutamatergic synapses within the target striatal sectors and by the density of connectivity onto individual MSNs (Kincaid et al. 1998; Zheng and Wilson 2002). Early investigations on the spatial organization of CS projections suggested a highly segregated topographical arrangement in which adjacent cortical regions connect nearby but distinct regions in the striatum (Webster 1961). More recent and detailed anatomofunctional studies have further elaborated these "CS channels" by revealing that small cortical regions can innervate relatively large striatal sectors with discontinuous and overlapping innervations that could establish, within the striatum, associative links for converging cortical areas that are functionally related (Flaherty and Graybiel 1991, 1993, 1994; Parthasarathy et al. 1992).

Anatomical and electrophysiological *in vivo* investigations, including retrograde labeling from the striatum and intracellular staining of single CS neurons identified by their antidromic activation (see Figure 7.1C), indicate that CS neurons can have ipsilateral, contralateral, or bilateral striatal projections with possible axonal bifurcations toward other cortical and subcortical regions. As summarized by the simplistic diagram in Figure 7.1A, the projection sites of CS neurons depend on their laminar localization. Single-neuron reconstructions from prelimbic (Levesque and Parent 1998), medial agranular, cingulate, and lateral agranular cortices (Wilson 1987; Cowan and Wilson 1994; Kincaid et al. 1998; Zheng and Wilson 2002) demonstrated that neurons located in the deep part of layer III and the superficial part of layer V (Va) project to ipsilateral and/or contralateral striatum, the crossed CS neurons having either ipsilateral and/or contralateral cortical projections (Wilson 1987; Cowan and Wilson 1994) (Figure 7.1A). This layer-specific location of crossed CS neurons has also been found in the motor system using antidromic activation, from the contralateral striatum, of intracellularly recorded primary motor cortex neurons (Mahon et al. 2001; Slaght et al. 2004). However, layer III/Va CS neurons do not have equivalent bilateral projections but rather exhibit prominent ipsilateral projections (Wilson 1987), as evidenced by the elevated number of cells labeled in the ipsilateral cortex following an intrastriatal injection of a retrograde tracer (see Figure 2 in Wilson 1987). In contrast, CS neurons located at the boundary of layers V (Vb) and VI exhibit extensive intracortical axonal ramifications and emit large-diameter axons that travel in the ipsilateral internal capsule, toward the thalamus, brainstem, and spinal cord, and give off collaterals within the ipsilateral striatum (Cowan and Wilson 1994) (Figure 7.1A).

The diverse groups of CS neurons, according to their location in the different layers and to their subcortical projections, can also be distinguished by their somatodendritic morphological characteristics. The crossed CS neurons from layers III/Va display relatively homogenous morphological features. They are all medium-sized pyramidal neurons with a thin apical dendrite having sparse oblique branches (Figure 7.1B) and reaching layer I with short ramifications extending horizontally (Wilson 1987; Cowan and Wilson 1994). Their basilar dendritic field is relatively small, extending about 400–500 µm horizontally, but with profuse branches that mainly spread within the layer of origin (Wilson 1987) (Figure 7.1B). Deeper CS neurons (layer Vb), with exclusive ipsilateral striatal projections and responding antidromically to brainstem stimulation, have a thicker and more densely branched dendritic tree compared to the crossed CS cells.

They exhibit a high density of spines on the main apical dendritic shaft as well as along basilar and oblique dendrites (Cowan and Wilson 1994).

7.2.2 Corticostriatal Neuron as an Electrophysiological Unit

In vivo intracellular recordings from the rat have provided a detailed description of the electrophysiological properties of CS neurons, including axonal conduction velocities, somatic electrical membrane properties, and synaptic responses to electrical stimulation of various brain regions.

Antidromic activation of CS neurons (Figure 7.1C), which allows the determination of their sites of axonal projections, reveals a high cell-to-cell, as well as a cortical area-dependent, variability in the propagation velocity of CS influx. The conduction velocity of ipsilaterally projecting CS axons, measured from the medial agranular cortex on the basis of the antidromic latencies in response to

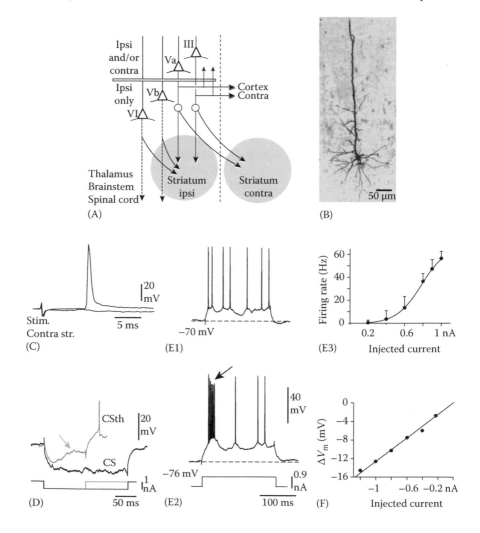

brainstem stimulation, ranges from 0.9 to 8.3 m s^{-1} with a mean value of about 4 m s^{-1} (Cowan and Wilson 1994). The axonal conduction velocity of crossed CS cells, intrasomatically recorded from motor cortex (Charpier et al. 1999a; Mahon et al. 2001; Slaght et al. 2004), is also variable (from 0.4 to 3 m s^{-1}) with an average value of 1.5 m s^{-1} (Mahon et al. 2001). The functional relevance of this discrepancy between the propagation speeds of CS information arising from different cortical regions remains unclear. Interestingly, the axonal conduction velocity of motor cortex corticosubthalamic neurons (7 m s^{-1}, Kitai and Deniau 1981; Paz et al. 2005, 2006) is considerably higher than that of CS neurons from the same cortical region (1.5 m s^{-1}). This suggests a timing-dependent unbalance between the two main corticobasal projections, which could relatively amplify the functional impact of cortical synchronized oscillations, arising from a given cortical area, onto the subthalamic neurons (Paz et al. 2005, 2006, 2007).

Intrinsic membrane properties of CS neurons, as examined *in vivo* from intrasomatic recordings, can also differ according to the cortical region and possibly to the anesthesia procedures. Electrophysiological analysis of crossed and ipsilaterally projecting CS neurons, from medial or lateral agranular cortex of urethane anesthetized rats (Cowan and Wilson 1994; Stern et al. 1997), indicate a baseline membrane potential that ranges between −53 and −72 mV. These values, which correspond to the down state of the two-state electrical behavior of CS cells under such anesthesia (see Section 7.5.1), are consistent with those measured for motor cortex crossed CS

FIGURE 7.1 Anatomical and electrophysiological features of corticostriatal neurons *in vivo*. (A) Basic rules of CS projections as a function of their cortical layer location. Whereas the more superficial CS neurons, located in the layer III or Va, can have ipsilateral and/or contralateral striatal projections, the deepest CS cells (from layers Vb and VI) strictly target the ipsilateral striatum with possible axon collaterals projecting to other subcortical structures, including thalamus, brainstem, and spinal cord. (B) Microphotograph of a neurobiotin-filled layer V motor cortex pyramidal neuron identified *in vivo* as a CS cell by its antidromic activation from the contralateral striatum. (C) All-or-none responses of the CS neuron shown in (B) to threshold antidromic electrical stimulations from the contralateral striatum (*Stim. Contra Str.*). In this cell, the antidromic latency was of 8.6 ms, corresponding to an axonal conduction velocity of ~1.5 m s^{-1}. (Modified from Slaght, S.J. et al., *J. Neurosci.*, 24, 6816, 2004. With permission.) (D) Negative current-induced responses of a somatosensory CS neuron (black trace) and of a corticosubthalamic (CSth) neuron (gray trace). Whereas the CSth neuron exhibits a sag potential (arrow) and a post-anodal suprathreshold excitation, the CS cell does not show any detectable intrinsic rebound properties. (E–F) Intrinsic membrane properties of CS neurons. A positive current pulse (bottom trace), delivered from a membrane potential of −70 mV, evokes in the CS cell a "regular spiking"-like pattern (E1) whereas the same stimulus applied during spontaneous hyperpolarization of −6 mV (E2) generates a short-latency "intrinsic bursting" (arrow). In the same cell, the firing rate-injected current relationship (E3) constructed from a wide range of current intensities reveals a sigmoïdal-like function ($r^2 = 0.99$) with an extrapolated maximal discharge of 67 Hz. The membrane potential changes as a function of the intensities of injected current pulses (200 ms duration, see (D)) does not exhibit any membrane rectification (F), and the apparent input resistance (15 MΩ) can be thus extrapolated from the regression line of data points ($r^2 = 0.99$). Here and in the following figures, the baseline membrane potential (dashed line) values are indicated below traces.

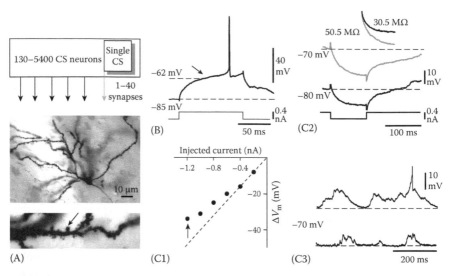

FIGURE 7.2 Corticostriatal inputs are controlled by postsynaptic intrinsic properties. (A) Statistical approximations of CS synaptic connectivity onto single MSN. Microphotograph of a striatal MSN stained *in vivo* by intracellular injection of neurobiotin (middle panel). This cell displays the distinctive morphological features of MSNs, that is, those of the type I medium spiny cells (Chang et al. 1982): a soma of ~15 μm of diameter and five primary dendrites densely covered with spines apart from their most proximal regions. The expansion of a distal dendritic region (bottom panel) clearly shows numerous dendritic spines (arrow), which are connected by the CS axons. The range of CS neurons connecting one MSN is indicated, together with the statistical variability of synaptic connectivity arising from individual presynaptic CS cell (see text for details). (B) Voltage response of an MSN (upper trace) recorded *in vivo* to injection of a positive current pulse (lower trace). Note the slow ramp depolarization that leads to a long latency of spike discharge (arrow). This delayed excitation is due to the slow kinetics of a voltage-gated potassium current (I_{As}) available at around −60 mV. (C) Inward rectification and its impact on the integration of cortical inputs. (Modified from Mahon, S. et al., *Trends Neurosci.*, 27, 460, 2004. With permission.) (C1) Plot of voltage changes (ΔV_m) in an MSN as a function of the injected current, showing, as indicated by the vertical arrow, a pronounced inward rectification (due to the striatal I_{Kir}) of the membrane potential in response to hyperpolarizing currents more negative than −0.4 nA. (Modified from Slaght, S.J. et al., *J. Neurosci.*, 24, 6816, 2004. With permission.) (C2) Voltage responses in the same MSN to a hyperpolarizing current pulse applied from resting potential (−80 mV) and during membrane depolarization (−70 mV). The increase of input resistance (the values are indicated) and apparent membrane time constant induced by the depolarization is clearly evident on the superposition of the initial voltage drops (inset). (From Mahon, S. et al., *J. Physiol. Paris*, 97, 557, 2003b. With permission.) (C3) Epochs of spontaneous synaptic activity recorded in an MSN from a baseline potential of −70 mV and during injection of a negative DC current (−100 mV; I_{DC} = −1 nA). The increase of membrane polarization, amplifying the shunting effect produced by I_{Kir}, leads to a considerable voltage attenuation of synaptic depolarizations, which mainly originate from CS inputs.

cells in the rat under barbiturate (−60 to −77 mV), ketamine-xylazine (−65 to −71 mV), and fentanyl analgesia (−61 to −74 mV) (Charpier et al. 1999a; Mahon et al. 2001). Despite these relatively homogenous values of baseline membrane potential across various cortical areas and anesthesia, membrane excitability of CS cells seems to differ according to their cortical origin. Crossed CS neurons recorded from the medial agranular cortex display prominent membrane nonlinearity, including an anomalous rectification that prevents membrane hyperpolarization below −110 mV and an increased input resistance and membrane time constant in the depolarizing direction (see Figure 8A in Cowan and Wilson 1994). This contrasts with the relative voltage insensitivity of membrane input resistance in motor (Mahon et al. 2001) and somatosensory (Pidoux et al. 2011) CS neurons, as attested by the linearity of their current–voltage relationship (Figure 7.1F), making reliable the measurement of their input resistance (from 20 to 30 MΩ) over a large range of membrane potential (Mahon et al. 2001; Slaght et al. 2004). Moreover, motor cortex CS neurons often exhibit a postinhibitory rebound of excitation, reminiscent of a low-threshold calcium potential (see Figure 3B in Mahon et al. 2001), whereas this has not been described in CS neurons from the frontal agranular (Cowan and Wilson 1994) or somatosensory (unpublished data) cortex (Figure 7.1D, black trace).

The firing rate of CS neurons in response to current pulses of increasing intensity follows a sigmoid-like curve (Figure 7.1E3). The most frequently observed current-evoked firing pattern, whatever the cortical localization of recorded CS neurons and the type of anesthesia used, is characteristic of the regular spiking neurons (Connors and Gutnick 1990; Steriade 2004) with rapid or slow spike frequency adaptation (Figure 7.1E1; Cowan and Wilson 1994; Mahon et al. 2001; Slaght et al. 2004). However, a slight spontaneous membrane hyperpolarization can promote the generation of typical intrinsic bursting, i.e., a cluster of action potentials with spike inactivation followed by a transient neuronal silence (Figure 7.1E2). This change in the firing pattern of cortical neurons, from regular spiking to intrinsic bursting, can naturally occur during transition from waking or paradoxical sleep to slow-wave sleep (SWS) and could result from the modulation of voltage-gated conductances (Steriade 2004).

Various types of evoked synaptic potentials have been described in CS neurons *in vivo*. Orthodromic stimulation of crossed and ipsilaterally projecting CS cells, *via* axon collaterals in the contralateral striatum or cortex, results in a complex synaptic response composed of a short-latency excitatory synaptic depolarization, partly overlapped by a fast inhibitory potential, followed by a prolonged (50–200 ms) electrical silence (or hyperpolarization) and subsequent long-lasting (100–300 ms duration) synaptic depolarization, which can exhibit an oscillatory-like shape (Wilson 1987; Cowan and Wilson 1994; Charpier et al. 1999a,b). These findings suggest that CS neurons can be activated by various ipsi- and contralateral cortical pathways, including local inhibitory and excitatory synaptic networks and reciprocal mono- or polysynaptic activation between nearby CS neurons. The coordinated activity, resulting from intracortical synaptic connectivity, of CS neurons converging into the same striatal sector may be determinant in the patterning of synaptic and firing activity of MSNs. This crucial issue of the CS physiology will be examined in Section 7.5.

7.3 INTEGRATION OF CORTICAL INFORMATION IN THE STRIATUM: BASIC MECHANISMS

7.3.1 CORTICAL TARGETING OF INDIVIDUAL MSNs

The CS system is a monosypatic and unidirectional excitatory network characterized by a considerable reduction in the number of neurons, from the presynaptic pool to the postsynaptic targets. In the rat, the total number of CS neurons was estimated to be 17 million, about 10 times the number of MSNs (Oorschot 1996; Zheng and Wilson 2002), suggesting *per se* a convergent profile in the CS connectivity. Refined anatomic studies performed by C. Wilson and colleagues, based on a reliable tridimensional reconstruction of single CS axons and collaterals together with an estimation of the striatal volume occupied by individual MSNs, provided a fine spatial scale model of the CS synaptic connectivity (Kincaid et al. 1998; Zheng and Wilson 2002). The striatal innervation volume of individual CS axons varies widely, occupying on average 4% of the striatum with a number of synaptic boutons formed by individual axons ranging from 25 to 2900 and being sparsely and nonhomogeneously distributed (Zheng and Wilson 2002). It is estimated that ~380,000 cortical axons innervate the volume occupied by the somatodendritic field of an MSN, each CS axon contacting at maximum 1.4% of cells located within its arborization (Oorschot 1996; Kincaid et al. 1998). A further quantitative estimation of the CS connectivity rules indicates that each MSN is connected, on its dendritic spines (Kemp and Powell 1971; Kincaid et al. 1998) (Figure 7.2A, bottom, oblique arrow), by ~130–5400 cortical neurons with a number of synaptic contacts per afferent CS cell ranging from 1 to 40 (Kincaid et al. 1998) (Figure 7.2A).

7.3.2 MEMBRANE PROPERTIES OF MSNs AS AN ELECTROPHYSIOLOGICAL SHIELD

As suggested by the CS combinatorics described earlier, each MSN is continuously exposed to excitatory synaptic bombardment arising from an elevated number of converging cortical inputs, incremented by a nearly equal number of glutamatergic synapses from thalamostriatal inputs (Smith et al. 2004). In theory, such a glutamatergic synaptic avalanche may produce a dramatic increase of intracellular calcium concentration and subsequent neuronal degeneration (Rodríguez et al. 2009). MSNs possess a bunch of active intrinsic membrane properties, operating at distinct levels of membrane potential, which resist this pathologically potent depolarization and sculpt their membrane potential fluctuations and firing patterns (see Section 7.5). Here, we focus on the two main intrinsic potassium conductances described in MSNs, which sequentially, and efficiently, control the excitability of MSNs over membrane depolarization, from resting to threshold potential.

For an electrophysiologist exploring the striatum, MSNs can be easily and unambiguously recognized, immediately after cell impalement, by their characteristic electrical response to the injection of a positive current pulse. When the membrane potential of MSNs approaches the spike threshold, the activation of a slowly inactivating voltage-dependent potassium current (I_{As}) induces a slowing of the rate of depolarization, from a membrane potential near $-60\,$mV, leading to a long latency of action potential discharge (Nisenbaum et al. 1994; Nisenbaum and Wilson 1995; Mahon et al. 2000a,b, 2003a,b, 2004) (Figure 7.2B). Although this neuronal behavior is also observed in hippocampal (Storm 1988), thalamic relay (McCormick 1991), and prefrontal cortical neurons (Hammond and Crepel 1992), which exhibit slowly inactivating potassium currents similar to I_{As}, it is distinctive to the MSNs among the striatal neuron population (Kreitzer 2009). We will see that the striatal I_{As}, in addition to acting as an electrical hindrance for suprathreshold depolarization and its participation in the maintenance of a relatively stable membrane potential during the up state (see Section 7.5.1), also provides, due to its slow kinetics, a potent intrinsic mechanism favoring the discharge of MSNs during successive cortical inputs.

Another electrophysiological distinctive feature of MSNs is the presence of a powerful inwardly rectifying potassium selective current (I_{Kir}), which accounts for most of their membrane conductance at rest. It is thus responsible for the highly polarized baseline membrane potential of striatal neurons (around $-80\,$mV) (Figure 7.2B and C2) and for a pronounced inward rectification in the hyperpolarizing direction (Nisenbaum and Wilson 1995) (Figure 7.2C1). This puissant potassium conductance significantly contributes to the relatively low input resistance and short apparent membrane time constant of MSNs when measured from the resting potential (Wilson 1992; Nisenbaum and Wilson 1995) (Figure 7.2C2). The little inactivation of the striatal I_{Kir} (Nisenbaum and Wilson 1995) together with its fast deactivation following significant depolarization, theoretically implies that this current should dynamically shape the membrane potential fluctuations of MSNs when processing cortical synaptic inputs. Indeed, as illustrated in Figure 7.2C2, a DC depolarization of MSNs *in vivo*, from the rest ($-80\,$mV) to a membrane potential of $\sim -70\,$mV, leads to a substantial increase in input resistance and membrane time constant (Nisenbaum and Wilson 1995; Mahon et al. 2003b) as well as to a reduction of the dendritic electrotonic length, due to the deactivation of the inwardly rectifying current (Wilson 1995a).

What is the functional consequence of such a voltage-dependent modulation of membrane excitability on the integration of cortical information? According to the dendritic cable theory (Wilson 1992, 1995a), which is supported by *in vivo* intracellular records (Wilson and Kawaguchi 1996; Mahon et al. 2001, 2003b, 2004), the following scenario is proposed. In the membrane potential range where the excitability of MSNs is governed by the inward rectifier, the membrane is highly polarized, the input resistance is low, and the time and dendritic space constants are short. The synergistic effect of these membrane parameters produces, despite large-amplitude synaptic currents due to an elevated ion driving force, a powerful shunt on synaptic potentials (Figure 7.2C3, bottom trace) and thus reduces the efficacy of temporally and/or spatially isolated CS inputs (Wilson 1995a,b; Wilson and Kawaguchi 1996; Mahon et al. 2001, 2003b, 2004). When the amount of synchronized cortical inputs is sufficient to significantly depolarize the MSN to around $-70\,$mV and deactivate I_{Kir}, both membrane input resistance and time constant increase, together with a collapse of the dendritic electrotonic length. These explosive activity-dependent changes in membrane excitability produce a positive feedback in the input–output relationship of striatal cells, allowing an efficient temporal integration of subsequent synaptic potentials (Wilson 1995a,b; Wilson and Kawaguchi 1996, Mahon et al. 2001, 2003b, 2004) (Figure 7.2C3, top trace).

The pattern of CS synaptic contacts onto single MSNs, characterized by the convergence of a large number of cortical cells with a relatively weak individual connectivity, together with postsynaptic membrane properties acting as a "discarder" of isolated weak inputs, led to the hypothesis that MSNs operate as a "detector" of spatially and/or temporally correlated cortical inputs (Houk 1995; Charpier et al. 1999a). We will develop the parametric dimension of this conceptual model by adding supplemental spatiotemporal features in light of recent *in vivo* experiments combining intracellular recordings of CS neurons and MSNs during periods of cortical synchronization.

7.4 INTEGRATION OF CORTICAL INFORMATION IN THE STRIATUM: SPATIOTEMPORAL DYNAMICS AND INTRINSIC PLASTICITY

7.4.1 Corticostriatal Pathway as a "Spatiotemporal Funnel"

The CS pathway is characterized by a reduction in the number of neurons from cortex to striatum, an extraordinary convergence of presynaptic cells onto postsynaptic neurons and a considerable imbalance between the excitability and firing rate of both cell populations (Stern et al. 1997; Kincaid et al. 1998; Mahon et al. 2001, 2003b, 2004; Zheng and Wilson 2002). The remarkable electrophysiological discrepancy between CS neurons and MSNs is highlighted by the experiment illustrated in Figure 7.3A (unpublished data) in which both cell types were intracellularly recorded *in vivo* in the same rat, from two connected regions and in standardized conditions of cerebral activity (barbiturate anesthesia). The current-evoked firing, in response to the same amount of injected current, is more than double in the CS cell compared to the MSN (Figure 7.3A1). This is correlated with a more depolarized membrane potential in the CS cell and an elevated spontaneous discharge whereas the striatal neuron exhibits large-amplitude fluctuating synaptic depolarizations, which however remain subthreshold for action potential discharge (Figure 7.3A2). According to these anatomical and electrophysiological features, the CS pathway can be conceptually outlined as an "anatomic funnel" in which the flow of information is spatially condensed (reduction in the number of cells) and functionally subsided (reduction in cell activity). In theory, the compactness of the striatal representation of cortical patterns would be responsible for either a relatively reduced encoding power in the striatum or a refinement of cortical information by removing cortical redundancy and/or rejecting functionally irrelevant combinations of cortical inputs arising from distinct cortical areas.

The cogency of the "spatiotemporal filtering" theory has been demonstrated by the dynamic properties of the CS funnel unveiled *in vivo* by the examination of the intracellular activity of CS neurons and MSNs simultaneously recorded with a surface local field potential (LFP) of the related cortical region (Figure 7.3B, top traces), which is indicative of the degree of spatiotemporal synchrony in the CS inputs (Mahon et al. 2001, 2004; Tseng et al. 2001). The measurement of the relative firing

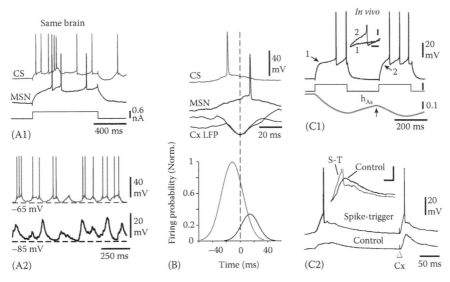

FIGURE 7.3 Information processing in the corticostriatal pathway: dynamic properties and intrinsic plasticity. (A) Comparative electrophysiology of pre- and postsynaptic elements in the CS pathway. *In vivo* intracellular recordings from CS and MSN neurons, in the same brain (barbiturate anesthesia), demonstrate (A1) a more elevated firing rate in the CS cell compared to the MSN, in response to the same current pulse (bottom trace), which is accompanied (A2) by a sustained spontaneous discharge in the cortical neuron whereas the striatal cell displays only subthreshold synaptic fluctuations. (B) Superimposition of the firing probability densities of CS neurons (gray line) and MSNs (black line). The relative timing of spontaneous action potentials in both neuronal populations (top records) is measured with respect to the peak negativity of the corresponding cortical surface LFP (Cx LFP), which is used as the zero time reference. (C) Short-term intrinsic plasticity in MSN and its effect on the input–output function of the CS network. (C1) *In vivo* intracellular injection of a suprathreshold depolarizing pre-pulse in an MSN results in an increase in the number of evoked spikes and a decrease in the first spike latency in response to a subsequent test pulse of the same intensity. (Inset) The expansion of the records (as indicated by the corresponding numbers) clearly shows that the increase in excitability is associated with an increase in slope membrane depolarization preceding the first spike. Calibration bars: 15 mV, 20 ms. The lower trace depicts the corresponding relative change in the inactivation of the striatal I_{As} (hAs) during the pairing protocol, as determined from a realistic biophysical modeling of MSN. (From Mahon, S. et al., *Trends Neurosci.*, 27, 460, 2004. With permission.) The vertical arrow indicates a substantial level of I_{As} inactivation at the onset of the second pulse, which is responsible for the increased excitability. (C2) Synaptic responses of an MSN *in vivo* to electrical stimulation of the corresponding cortical region (Cx) in control condition and 200 ms after the occurrence of a spontaneous action potential in the recorded striatal cell [Spike-Trigger (S-T)]. As shown by the superimposed expanded records (inset), the increase in the striatal cell firing probability, induced by the S-T protocol, is correlated with a steeper slope in the synaptic depolarization. Calibration bars: 10 mV, 10 ms.

probability and spike timing in CS cells and MSNs during recurrent epochs of tight cortical synchronization, sequentially isolated by the occurrence of sharp and stereo-typed negative waves in the cortical LFP (Steriade et al. 1990), reveals that the firing probability of MSNs is considerably lower than that of CS cells, likely due to the shunting effect produced by the active membrane properties of MSNs (see Section 7.3.2), with a lower temporal dispersion of action potentials in the striatum (Mahon et al. 2001, 2004) (Figure 7.3B, bottom). This finding introduces a supplemental tem-poral dimension in the CS funnel model by demonstrating that the decrease of firing probability in striatal cells, which might also reflect a diminution in the number of activated neurons in the related striatal sector, is associated with a relative narrowing of the spike timing variability, making the striatal output within a given striatal cells assembly statistically more reliable than the corresponding CS input (Mahon et al. 2001). These dynamic properties of the CS pathway during transient cortical syn-chrony, which could now be considered as a "spatiotemporal funnel" (Figure 7.4A),

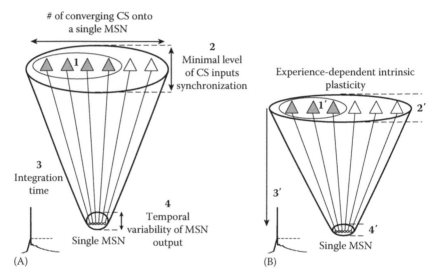

FIGURE 7.4 Dynamic contraction of the "CS spatiotemporal funnel" during intrinsic plas-ticity. (A) Diagrammatic representation of the CS funnel efficient to generate action potential in the related MSN. Among the total number of cortical cells converging onto a single MSN, only a fraction of CS neurons (numbered in hundreds, see Wilson 1992, 1995b) need to fire (gray triangles, **1**), and with a minimal level of synchronization (**2**), to bring the MSN to its firing threshold (bottom trace, dashed line). The integration time (**3**) in the funnel cor-responds to the temporal delay between the discharge of CS neurons and of the related MSN (see Figure 7.3B). The corresponding temporal variability in MSN firing (**4**) is significantly tightened compared to that of CS cells (see Figure 7.3B). (B) Alteration of CS funnel param-eters during activity-dependent intrinsic plasticity. When the MSN has been previously acti-vated by its CS inputs, it is predicted that the subsequent time-dependent increase in MSN membrane excitability will reshape the dynamic properties of the CS funnel as follows: first, the pool of active CS neurons (**1′**) and/or their minimal level of synchrony (**2′**) required to fire the MSN will be reduced; second, the firing delay between CS and the corresponding MSN will be reduced (**3′**) and, third, the temporal spreading of MSN discharge will be com-pressed (**4′**). See text for more detailed explanations.

are in accordance with the combinatorial neural code of central networks having anatomofunctional connectivity similar to that of the CS system, that is, a large number of synchronized neurons converging onto a smaller set of target cells (Diesmann et al. 1999).

The functional pertinence of the proposed CS model, expected to operate as a "spatiotemporal funnel," is further attested by our recent investigations (Pidoux et al. 2011) specifically designed to elucidate the mechanisms of propagation and processing of natural sensory information within the somatosensory CS system linking the barrel cortex and the corresponding sector of the dorsal striatum (Alloway et al. 1999, 2006; Wright et al. 2001) (Figure 7.5A through C). The application of air-puffs on contralateral vibrissae produces in the barrel cortex and related striatal sector highly correlated evoked field potentials (Figure 7.5D), demonstrating that sensory inputs are transformed into synchronized synaptic potentials in both structures. These "global" cortical and striatal activities reflect in fact distinct intracellular events in CS cells and MSNs that are in accordance with the functional funnel designed earlier. Indeed, all recorded cortical cells located in the layer V, where somatosensory CS cells are concentrated (Wright et al. 2001; Alloway et al. 2006) (Figure 7.5A), display short-latency synaptic depolarizations mostly suprathreshold for action potential firing (Figure 7.5E, Layer V cortical neuron). In contrast, only a fraction of MSNs exhibit sensory-evoked excitatory synaptic responses, which could generate action potentials (Figure 7.5E, Activated MSN), with however a lower probability compared to CS cells, or remain subthreshold (Figure 7.5E, Excited MSN). The remaining MSNs display either no detectable responses (Figure 7.5E, Nonresponding MSN) or a slight hyperpolarization. Altogether, these novel findings indicate that the integration of external information in the somatosensory CS pathway results in a "refined" processing in the striatum, consisting in a decrease in the number of activated cells and a reduced probability of neuronal responsiveness. The behavioral significance of this peculiar sensory integration in the striatum remains to be elucidated.

7.4.2 Intrinsic Plasticity in MSNs and Contraction of the "Spatiotemporal Funnel"

As detailed earlier, the integration of cortical information in individual MSNs is governed by subtle interactions between the numerous excitatory synaptic connections arising from converging CS neurons and a set of intrinsic active membrane properties that dynamically hinder the cell depolarization, from the rest to the firing threshold. Recent studies indicate that the striatal slowly inactivating potassium current I_{As}, which counteracts membrane depolarization just below spike threshold, could also provide, as a function of the prior activity, a potent time-dependent boost for action potential generation in MSNs (Mahon et al. 2000a,b, 2003a, 2004). This has been initially evidenced by intracellular injection of a suprathreshold current pre-pulse that causes, within a time window of ~1 s and with an exponential decay of ~350 ms, an increase in intrinsic excitability expressed as an augmented number of evoked spikes and a decrease in the first spike latency in response to a subsequent test pulse of same intensity (Mahon et al. 2000a,b, 2004) (Figure 7.3C1, top). This activity-dependent facilitation in membrane responsiveness, which is correlated with an increase in the slope depolarization preceding firing (Figure 7.3C1, inset), results from the slow kinetics of recovery from inactivation of I_{As} (Mahon et al. 2000b, 2003a) (Figure 7.3C1, lowest trace). The functional relevance of such a short-term intrinsic plasticity is attested by the time-dependent facilitation observed *in vivo* in MSNs during temporally ordered cortical inputs. Indeed, a spontaneous action potential in a striatal cell, resulting from a sudden synchronization in CS neurons, increases, over a time scale of 1 s, the probability that subsequent cortically-evoked synaptic depolarization induces firing and reduces the latency and the temporal dispersion of the evoked action potentials (Mahon et al. 2003a) (Figure 7.3C2). This activity-dependent intrinsic facilitation of striatal neurons discharge in response to a given cortical input is

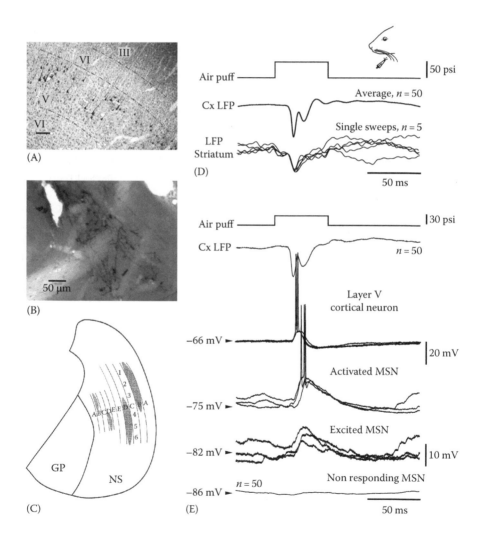

causally correlated with a lowering of the voltage firing threshold due to a steeper pre-spike synaptic depolarization (Figure 7.3C2, inset) (Mahon et al. 2003a).

The intrinsic plasticity of MSNs has a profound effect on the CS "spatiotemporal funnel" by affecting its dynamic and parametric features as a function of the past activity. First, the time-dependent inactivation of I_{As}, which reduces the voltage spike threshold of the striatal cell, might also diminish the number of activated CS neurons required to fire the targeted MSN. Moreover, the firing level of previously active MSNs will be enhanced together with a reduction in the corresponding temporal variability of discharge in individual cells, a process that would favor the synchronized discharge of striatal cells that have concomitantly experienced the intrinsic plasticity. Finally, the decrease of spike latency on cortically-evoked synaptic potentials in MSNs will reduce the effective integration time of cortical information in the striatum. A diagrammatic representation of the multiparametric changes occurring in the "spatiotemporal CS funnel" during intrinsic plasticity is given in the Figure 7.4B.

The conceptual model developed earlier, designed from the point of view of a single striatal cell considered as the "neck" of the funnel, and regardless of synaptic inputs other than cortical ones, provides a "snapshot" of the CS interactions controlling the discharge of an MSN. In the next part of the chapter, we examine the mechanisms underlying the spontaneous, and temporally maintained, activities in MSNs, which depend upon the brain state and the various levels of vigilance.

FIGURE 7.5 Sensory integration in the corticostriatal network *in vivo*. (A) Coronal section of the barrel cortex showing that CS cells, retrogradely labeled via an injection a cholera toxin in the somatosensory part of the striatum, are distributed in layer V with a predominance in the upper part. Scale bar = 100 µm. (Modified from Wright, A.K. et al., *Neuroscience*, 103, 87, 2001. With permission.) (B) Microphotograph of CS fibers endings in the striatum anterogradely labeled by an injection of *Phaseolus vulgaris* into the ipsilateral barrel cortex. CS projections typically form dense clusters of synaptic terminals in the posterior part of the dorsolateral striatum. (C) Schematic diagram illustrating the topographical organization in the striatum of cortical projections from the barrel cortex. Whisker rows (A to E) are represented in the striatum by curved parallel lamellae. The striatal projection fields of the barrel columns B2 and D5, indicated by the gray areas, illustrate the multiple and discontinuous representations of individual whiskers within each row. A reduced mirror-image of rows representation is observed more medially. GP, Globus pallidus; NS, neostriatum. (From Alloway, K.D., et al., *J. Neurosci.*, 19, 10908, 1999. With permission.) (D) Sensory-evoked field potentials in the barrel cortex (Cx LFP, middle trace) and in the related striatal region (LFP striatum, bottom trace) in response to sensory stimulation consisting in the application of air puffs to the controlateral whiskers (Air Puff, top trace).(E) Air puff-evoked (top) responses in the controlateral cortical field potential (Cx LFP), in a corresponding layer V cortical cell and in three different MSNs, which exhibit suprathreshold (activated MSN) or subthreshold (excited MSN) synaptic potentials, or a lack of evoked potential (nonresponding MSN) (Pidoux et al. 2011). In (E), recorded neurons are from the same experiment except the activated MSN.

7.5 GENERATION OF SPONTANEOUS MEMBRANE POTENTIAL FLUCTUATIONS AND FIRING PATTERNS IN STRIATAL MEDIUM SPINY NEURONS

The first studies attempting to describe the electrical activity of striatal neurons, using single-unit extracellular recordings, reported relatively low firing rates and a diversity of firing patterns, including tonic, phasic, or rhythmic bursting activities (Wilson 1993 for review). However, most of these recordings were performed from nonidentified striatal cells, making the attribution of these different firing patterns to the MSNs likely mistaken. The development of *in vivo* intracellular recordings and cellular staining techniques allowed a more rigorous characterization of MSN spontaneous firing profiles and the uncovering of the underlying synaptic and intrinsic electrical mechanisms.

Owing to the technical difficulties inherent in intracellular recordings in awake animals, most of our knowledge on the intracellular activity of MSNs comes from anesthetized preparations. Initial intracellular recordings in the rat anesthetized with urethane and ketamine/xylazine revealed that the membrane potential of MSNs exhibited *in vivo* recurrent spontaneous fluctuations between a highly hyperpolarized quiescent state, later called the "down state," and a depolarized state generating spike discharges, the "up state" (Wilson 1992, 1993). Since then, transitions between up and down states have been considered as the characteristic pattern of spontaneous activity of MSNs, and a large number of studies were devoted to the elucidation of the cellular and synaptic origin of this peculiar intracellular activity. After a description of the synaptic and intrinsic mechanisms giving rise to up and down states in MSNs, we highlight how this electrical behavior has influenced our conceptual view of striatum-related functions and dysfunctions. Thereafter, we will summarize recent findings obtained from unanesthetized animals demonstrating that the electrical activity of MSNs is not stereotyped, as generally assumed, but is rather variable with a tight dependence on the state of vigilance.

7.5.1 TWO-STATE ELECTRICAL BEHAVIOR IN THE CORTICOSTRIATAL PATHWAY

7.5.1.1 Cortical and Striatal Up and Down States

While it was initially proposed that up and down states in MSNs arise from the sequential combination of tonic excitatory synaptic drive and recurrent inhibition between the GABAergic striatal neurons (Wilson 1993 for review), it is now recognized that the transitions between the two states and the maintenance of each state result from the interplay between excitatory synaptic inputs (mainly cortical) and intrinsic membrane properties.

CS neurons and MSNs recorded *in vivo* from rats anesthetized with urethane or ketamine/xylazine display qualitatively comparable patterns of spontaneous membrane potential fluctuations and firing. The corresponding intracellular activity of both CS and striatal neurons is characterized by an alternation of depolarizing plateaus and hyperpolarizing periods (Figure 7.6A, left), resulting in a bimodal distribution of their membrane potential (Figure 7.6A, right) (Wilson 1993; Cowan and Wilson 1994; Wilson and Kawaguchi 1996; Stern et al. 1997; Mahon et al. 2001, 2003b). Up and down states described in CS neurons are similar to those recorded from other cortical pyramidal neurons in rats and cats under urethane or ketamine/xylazine anesthesia (Steriade et al. 1993; Contreras and Steriade 1995; Petersen et al. 2003;

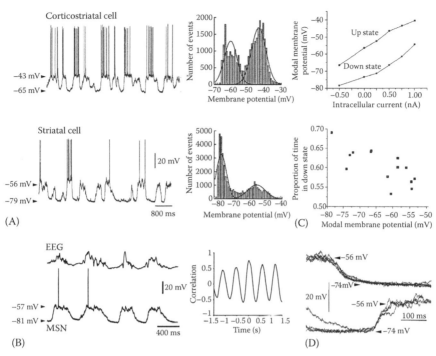

FIGURE 7.6 Up and down state transitions in MSNs *in vivo*. (A) In the rat anesthetized with urethane and ketamine/xylazine, the spontaneous intracellular activities of CS neurons (top trace) and MSNs (bottom trace) are characterized by recurrent membrane potential transitions between a hyperpolarized downstate and a depolarized upstate (left), leading to a bimodal membrane potential distribution (right). Mean membrane potential values of the up and down states are indicated at the left of the traces. Striatal neurons exhibit a highly polarized resting membrane potential and a lower firing frequency on up state episodes than CS cells. (Modified from Stern, E.A., et al., *J. Neurophysiol.*, 77, 1697, 1997. With permission.) (B) State transitions in an MSN under ketamine–xylazine anesthesia (bottom trace) and their relation to EEG waves (top trace). The cross-correlation between both signals indicates that striatal membrane potential fluctuations are temporally correlated with the slow cortical field potentials (right). (Modified from Mahon, S., et al., *J. Physiol. Paris*, 97, 557, 2003b. With permission.) (C) The intracellular injection of constant depolarizing or hyperpolarizing current into a striatal neuron affects the mean membrane potential value and the time spent in each state. Top, current–voltage relationships of the neuron at each state. Note the inward rectification associated with the down state at hyperpolarized potentials. The relationship between the amount of injected current and the mean membrane potential reached during the up state is more linear even if a slight outward rectification can be observed at potentials close to spike threshold. Bottom, plot of the time spent in down state versus the modal down state membrane potential for different levels of injected current. The proportion of time spent in the down state is increased by hyperpolarization. (Modified from Wilson, C.J. and Kawaguchi, Y., *J. Neurosci.*, 16, 2397, 1996. With permission.) (D) Superimposed traces showing that transitions between the two states have specific and stereotyped shapes. Transitions to the upstate are more abrupt than transitions to the down state, which exhibit an exponential like decay. Note the small-amplitude membrane potential fluctuations associated with the up state (see also panel B, bottom trace).

Léger et al. 2005). A quantitative comparison of the properties of up and down states in CS neurons and MSNs indicates that the average potential of the down state is significantly more negative in striatal neurons (~-80 mV) than in cortical neurons (~-65 mV) but that the average value of the up state (~-50 mV) is similar in both cell types (Figure 7.6A) (Stern et al. 1997; Mahon et al. 2001). Membrane potential fluctuations are of larger amplitude in the up state compared to the down state in the two neuronal populations. Finally, striatal neurons display a higher proportion of subthreshold up states and a lower frequency of spike discharge during suprathreshold depolarized episodes (Figure 7.6A and B) (Stern et al. 1997; Mahon et al. 2001, 2003b).

7.5.1.2 Construction of the Two-State Striatal Activity

Since CS neurons provide the major source of excitatory synaptic inputs to the striatum (see Sections 7.2.1 and 7.3.1), the two-state behavior concomitantly observed in cortical and striatal cells under the same condition of anesthesia suggests that membrane potential shifts in MSNs will be imposed by the firing rate and pattern of cortical neurons. Ketamine/xylazine and/or urethane anesthesia induces in the electroencephalogram (EEG) slowly recurring (around 1 Hz) cortical waves of large amplitude, with small- and high-frequency deflections superimposed on each slow-wave cycle (Figure 7.6B, top trace). This EEG pattern is reminiscent of the slow cortical oscillations present throughout the resting sleep in naturally sleeping mammals (Steriade 2000; Steriade et al. 2001; Timofeev et al. 2001). Recent *in vivo* investigations, conducted in ketamine/xylazine anesthetized rats, report that the onset of up states in CS neurons are closely correlated with the high-amplitude positivity of the corresponding EEG waves (Mahon et al. 2001, 2003b), suggesting that transitions to the up state occur in a highly coherent manner in a large population of CS cells (Steriade et al. 1993; Amzica and Steriade 1995; Mahon et al. 2001, 2003b). Moreover, the firing probability of CS neurons is elevated in the first part (200 ms) of the up state while it progressively decreases during the depolarized episode (Mahon et al. 2001, 2003b). Thus, the synchronized entry in up state of CS neurons likely generates a highly correlated firing among CS neurons, at least during the initial phase of the depolarization, resulting in temporally correlated excitatory inputs in the striatum and significant depolarizations in MSNs. Consistent with this assumption, the simultaneous recordings of cortical EEG and intracellular activity of MSNs show that striatal up and down states are precisely timed with the slow cortical activity (Figure 7.6B) (Mahon et al. 2001, 2003b; Tseng et al. 2001). In addition, simultaneous paired recordings of MSNs *in vivo* demonstrate that up state transitions occur mostly simultaneously in neighboring striatal neurons, presumably receiving the same combination of cortical inputs (Stern et al. 1998; see Section 7.3.1). It is therefore very likely that coherent firing within a large number of converging CS cells is responsible for the sharp depolarization that causes transition to the up state and action potential discharge in MSNs (Stern et al. 1997, 1998; Mahon et al. 2001, 2003b; Tseng et al. 2001).

Membrane potential shifts in MSNs are not strictly determined by the pattern of excitatory synaptic inputs but also by their nonlinear membrane properties. This was first demonstrated by the injection of positive or negative constant current in MSNs, which alters the time spent in both states and affects differently the modal value of membrane potential in each state according to the voltage-dependent membrane conductances activated (Figure 7.6C) (Wilson and Kawaguchi 1996). Accordingly, a functional scenario, combining synaptic and membrane properties, has been proposed to account for up and down state transitions in striatal neurons. During the down state, in the absence of excitatory inputs, the membrane polarization is maintained close to the potassium equilibrium potential by the inwardly rectifying current I_{Kir} (Wilson 1993, 1995b; Nisenbaum and Wilson 1995; Wilson and Kawaguchi 1996). Given the weak intrinsic excitability of striatal cells at this level of polarization (see Section 7.3.2), only temporally coincident CS inputs are able to produce significant depolarization causing the deactivation of I_{Kir}, and subsequent increase in striatal cell intrinsic responsiveness (see Figure 7.2C2). These synaptic and membrane interactions are likely responsible for the rapid transition from the down state to the up state (Figure 7.6D, lower traces). The relatively stable level of depolarization during the up state is controlled by the interplay between the depolarizing effect of incoming cortical inputs and the hyperpolarizing influence of outwardly rectifying potassium conductances (Nisenbaum et al. 1994; Wilson 1995a,b; Wilson and Kawaguchi 1996). The slowly inactivating potassium current I_{As}, acting in concert with the persistent non-inactivating potassium current I_{Krp} (Nisenbaum et al. 1996), will limit the level of depolarization through an active hyperpolarization and by decreasing the input resistance and time constant (Nisenbaum and Wilson 1995; Wilson 1995a,b; Wilson and Kawaguchi 1996). Approaching the termination of the up state, the hyperpolarizing effect of potassium currents together with the lowering and temporal spreading of firing probability in CS neurons (Mahon et al. 2001, 2003b) will initiate the repolarization of striatal neurons. The subsequent activation of I_{Kir} will drive the cell toward the potassium equilibrium potential, re-establishing the down state (Figure 7.6D, upper traces).

7.5.1.3 Action Potential Generation and Firing Patterns on MSN Up State

The spontaneous depolarized plateaus in MSNs provide an enabling state for spike generation. Action potentials in the up state are triggered by additional noise-like membrane potential fluctuations superimposed on the sustained depolarization (Figure 7.6A and B, bottom traces). This presumed synaptic noise probably reflects slight perturbations in the degree of synchrony among the CS cells responsible for the transition to the striatal up state. Alternatively, it could be caused by other groups of synaptic inputs arising from relatively uncorrelated CS and/or thalamostriatal cells. Consistently, the fine temporal pattern (on the time scale of a few milliseconds) of action potentials within the up state is generally not synchronous between nearby striatal neurons (Stern et al. 1998), further supporting the participation of noisy-like synaptic processes. Nevertheless, another study indicates that the firing of MSNs *in vivo* is statistically confined in the first 200 ms of the up state (Figure 7.6B), which corresponds to the highest firing probability in CS neurons (Mahon et al. 2001, 2003b).

Fast-spiking GABAergic interneurons are also likely to play a significant role in the regulation of up state–related striatal output. Indeed, paired recordings from GABAergic striatal interneuron and MSNs in brain slices revealed that spiking in a single interneuron can significantly delay or even abolish the discharge in the MSN (Koos and Tepper 1999). Consistently, blockade of GABA$_A$ receptors by local injection of picrotoxin enhances MSN firing *in vivo* (Mallet et al. 2005). Moreover, in organotypic cocultures (Plenz and Kitai 1998), as well as in urethane-anesthetized rats (Mallet et al. 2005), the firing of striatal interneurons is correlated with up states in MSNs, strongly suggesting the participation of GABAergic synaptic events in the membrane potential fluctuations on the up state. Given that the equilibrium potential of chloride ions in MSNs ($\sim-60\,mV$; Plenz 2003) is close to the average membrane potential value during the up state, the repeated firing in GABAergic striatal interneurons, generated by their CS inputs (Ramanathan et al. 2002; Mallet et al. 2005), might dynamically modulate the firing of MSNs firing through an active inhibition (hyperpolarization) or an increase in membrane conductance. It is also plausible that the generation of action potentials in the up state involves additional membrane ionic conductances operating just below spike threshold, such as calcium and sodium voltage-gated conductances (Galarraga et al. 1989; Bargas et al. 1994; see also Cantrell et al. 1995 for review).

The findings described earlier indicate that the firing pattern of MSN during the up state is continuously adjusted by complex time- and voltage-dependent interactions between excitatory and inhibitory synaptic events and active membrane properties. As an additional control mechanism, many of the ionic conductances that shape striatal spontaneous activity are sensitive to extrinsic or intrastriatal neuromodulation arising from dopaminergic and cholinergic systems. Although a definitive picture of the postsynaptic effects of dopamine on MSN excitability is not yet available (Nicola et al. 2000), it is proposed that the activation of dopamine D1 receptor could promote firing on the up state by enhancing L-type calcium currents (Surmeier et al. 1995) and/or by a blockade of I_{As} (Nisenbaum et al. 1998). In contrast, the activation of dopamine D2 receptors has a tendency to decrease firing rate by reducing L-type calcium current (Hernandez-Lopez et al. 2000). Moreover, the activation of D1 and D2 receptors could also affect the transitions from the down to up state in an opposite manner by enhancing and reducing I_{Kir}, respectively (Pacheco-Cano et al. 1996; Uchimura and North 1990). Cholinergic modulation, through M1 receptor activation, would enhance the likelihood of state transitions via an inhibition of I_{Kir} (Shen et al. 2007).

7.5.1.4 Putative Implications in Striatal Functions and Dysfunctions

The two-state behavior observed under urethane and ketamine/xylazine anesthesia provides the most described spontaneous activity of striatal MSNs *in vivo*. It is widely considered as the "functionally relevant" activity pattern of these neurons and, as a consequence, a number of experimental and theoretical models for striatum-related physiological and pathological processes are based on this assumption.

1. *State transitions in MSNs and long-term CS synaptic plasticity.* The striatum is involved in various forms of learning and memory processes (Graybiel 1995; Schultz et al. 2003) that could be supported, at the cellular level, by use-dependent short- and long-term changes in the efficacy of CS synaptic connections (Reynolds et al. 2001; see also Section 7.6 and Chapters 4 and 5). *In vivo* and *in vitro* studies indicate that long-term potentiation (LTP) and long-term depression (LTD) at CS synapses would be differentially induced depending on the type of glutamate receptors activated, the source of postsynaptic calcium rise, and the level of activity in dopaminergic inputs (Calabresi et al. 1996; Lovinger and Tyler 1996; Mahon

et al. 2004). Briefly, activation of α-amino-3-hydroxy-5-methyl-4-isolaxone propionate (AMPA) and metabotropic glutamate receptors together with an influx of calcium through L-type channels, associated with a relatively weak level of dopamine, would preferentially result in the induction of CS LTD. Conversely, induction of LTP would rather require the activation of *N*-methyl-D-aspartate (NMDA) and metabotropic glutamate receptors as well as high levels of dopamine.

Recently, it has been shown that state transitions in MSNs affect the calcium dynamics in MSNs dendrites, a process that might control the induction of synaptic plasticity at CS synapses. In organotypic cocultures, up state transitions in MSNs can generate calcium transients in soma and dendrites that are enhanced by back propagating action potentials (Kerr and Plenz 2002, 2004). In brain slices, a somatic membrane depolarization to −50 mV in MSNs (through injection of a constant positive current), expected to mimic the up state, can activate, in dendrites and spines, L-type voltage-sensitive calcium channels and produce a shift in the source of synaptic-dependent calcium influx, from calcium-permeable AMPA to NMDA receptor channels (Carter and Sabatini 2004). The activation of NMDA receptors during the depolarized state is further supported by *in vivo* experiments showing that the amplitude of state transitions and the firing frequency of MSNs are decreased following local application of a competitive NMDA receptor antagonist (Pomata et al. 2008). In accordance with these findings that demonstrate the implication of up states in synaptic and ion channel-induced calcium influx in MSNs, a recent *in vitro* study shows that the pairing of moderate frequency afferent stimulations with subthreshold membrane depolarization from −70 to −50 mV, a protocol designed to simulate down to up state transitions observed *in vivo*, elicits a form of endocannabinoid-mediated LTD at CS synapses, which depends on the activation of L-type calcium channels during the artificial upstate (Kreitzer and Malenka 2005).

2. *Alteration of state transitions in MSNs as a cellular model for PD?* The severe motor deficits that characterize PD, including akinesia, bradykinesia, muscle rigidity, and tremor at rest, are associated with the degeneration of midbrain dopaminergic neurons projecting to the striatum. Recent studies from dopamine-depleted animal and human patients with PD have reported abnormally synchronized oscillatory activities in various structures of the BG, as well as in the cerebral cortex (see Bergman et al. 1998; Bevan et al. 2002; Hammond et al. 2007; Charpier et al. 2010 for reviews). Specifically, an excessive bursting activity, associated or not with an increase in the mean firing rate, has been observed in BG output nuclei in two well-characterized animal models of PD, the 1-methyl-4-phenyl-1,2,3,6-tetrahydropyridine (MPTP)-treated monkey (Bergman et al. 1998; Boraud et al. 1998; Wichmann and Soares 2006; Hammond et al. 2007) and the 6-hydroxydopamine (6-OHDA)-lesioned rats (Murer et al. 2002; Belluscio et al. 2003; Hammond et al. 2007). This pathological activity in BG output nuclei, which is transferred to premotor structures, could initially result from an altered

processing of cortical information by the input structures of the BG. In MSNs recorded from 6OHDA-lesioned rats anesthetized with urethane, the lack of dopaminergic striatal innervation alters both the two-state behavior in striatal cells and their firing rate. Indeed, MSNs deprived of dopamine exhibit more depolarized up and down states leading to an augmented firing frequency correlated with the cortical slow waves (Tseng et al. 2001, 2004; Murer et al. 2002). Hence, it was suggested that this increased sensitivity of MSNs to cortical inputs (see also Mallet et al. 2006), which could be associated with an increased coupling between cortical and subthalamic activities (Magill et al. 2001), would facilitate and amplify the transfer of cortical rhythmicity to BG output nuclei. Consistently, an increased intrinsic excitability of MSNs, expressed as an increased firing response to depolarizing current pulses and a decrease in the threshold current, has been reported *in vitro* following dopamine depletion (Fino et al. 2007; Azdad et al. 2009). This increased striatal excitability, which probably results from a decrease in voltage-gated potassium currents including I_{As} (Azdad et al. 2009), could thus explain, at least in part, the enhanced striatal firing rate during the up state in parkinsonian rats. However, the analogy between the anesthetized rat model and the human disease is questionable since the frequency of cortical synchronization induced by urethane in these *in vivo* studies is substantially lower than that observed in patients with PD or from awake animal models (Sharott et al. 2005; Hammond et al. 2007; Degos et al. 2009).

7.5.2 MEMBRANE POTENTIAL FLUCTUATIONS AND FIRING PATTERNS IN MSNs CRITICALLY DEPEND UPON VIGILANCE STATES

Current models on the "reading" of cortical inputs by the striatum are based on the assumption that MSNs continuously express *in vivo* a two-state electrical behavior. However, this strong assumption originates from intracellular recordings obtained under ketamine and/or urethane anesthesia, which induce a relatively stereotyped slow pattern of cortical waves. This could have introduced a dramatic bias in our conceptual view on CS information processing and conceal higher computational capabilities of striatal neurons. A first questioning of the "up and down states hypothesis" arose from a comparison of the intracellular activity of MSNs under different anesthetics that revealed various shapes of synaptic depolarizations depending upon the different patterns of CS activity (Charpier et al. 1999a; Mahon et al. 2001, 2003b). Moreover, spontaneous or electrically evoked cortical desynchronization in urethane-anesthetized rats suppress up and down state transitions in MSNs and lead to the appearance of small- and high-frequency synaptic events (Kasanetz et al. 2002, 2006). Altogether, these findings open up the possibility that the profile of intracellular activity in MSNs could change as a function of brain states and, in particular, as a function of the various types of activity that can be naturally expressed by the cerebral cortex. This is now confirmed.

Using an anesthetic free rat preparation (Souliere et al. 2000), we recently unmasked and characterized distinct patterns of activity in MSNs associated with the different states of vigilance (Mahon et al. 2006). During active waking, identified by a low-amplitude desynchronized EEG correlated with a sustained or phasic EMG activity (Figure 7.7A1, top traces) (Timo-Iaria et al. 1970; Gottesmann 1992), the intracellular activity of MSNs is characterized by depolarizing envelopes of variable duration and amplitude, on which are superimposed high-frequency small-amplitude noise-like fluctuations (Figure 7.7A1, bottom trace), leading to a unimodal distribution

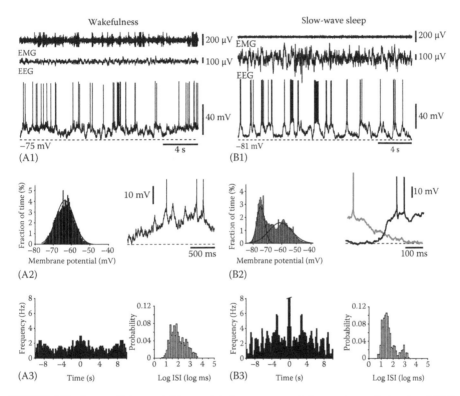

FIGURE 7.7 Intracellular activities and firing patterns of the awake and sleeping MSN. (A) Membrane potential fluctuations and irregular firing pattern in MSNs during wakefulness. (A1) Striatal neuron intracellularly recorded (bottom trace) together with the corresponding EEG and EMG activities during the waking state. (A2) The membrane potential distribution (bin size, 0.5 mV) is unimodal with a mean value of −63.5 ± 5.2 mV. Inset, expanded view of an intracellular epoch showing high-frequency small-amplitude membrane potential fluctuations superimposed on a depolarizing envelope. Spikes are truncated. (A3) Left, autocorrelogram of spike discharge for the recording period shown in (A1) (bin size, 0.2 s). Right, averaged histogram (*n* = 5 cells) of the log ISI distribution associated with wakefulness (bin width, 0.1 log ms). (B) Up and down states and rhythmic firing in MSNs during slow-wave sleep. (B1) Spontaneous intracellular activity of a striatal neuron during a period of SWS recorded simultaneously with the EEG and EMG. (B2) Corresponding histogram of membrane potential distribution (bin size, 0.5 mV). The histogram could be fitted by a double Gaussian. Inset, representative examples of spontaneous transitions to the up state (black trace) and the down state (gray trace). Spikes are truncated. (B3) Left, autocorrelogram (bin size, 0.2 s) of spike discharge for the recording shown in (B1) illustrating the rhythmic firing pattern of MSN during SWS. Right, averaged histogram of the log ISIs (bin width, 0.1 log ms, *n* = 4 cells). (Modified from Mahon, S., et al., *J. Neurosci.*, 26, 12587, 2006. With permission.)

of membrane potential (Figure 7.7A2, left) (Mahon et al. 2006). Most recorded MSNs were spontaneously active during these epochs of waking, with however a quite variable mean firing rate (range = 0.1–10.8 Hz). Action potentials are triggered by the temporal summation of the small-amplitude membrane potential fluctuations on the depolarizing envelopes (Figure 7.7A2, right). As shown by the relatively flat autocorrelogram of spike discharge (Figure 7.7A3, left), the waking state is reflected in MSNs by an irregular firing pattern with a positively skewed unimodal distribution of inter-spike intervals (ISIs) (Figure 7.7A3, right). No significant correlation between the low-amplitude desynchronized EEG and the intracellular striatal activity could be detected (Mahon et al. 2006), a finding in accordance with the lack of large-amplitude oscillations in both signals. Consistent with this uncorrelated activity in the CS pathway, the desynchronized EEG activity during waking (Steriade 2000) is associated in cortical pyramidal neurons with a sustained membrane potential depolarization generating a tonic spike discharge (Steriade et al. 2001; Timofeev et al. 2001).

Because MSNs are innervated by a large number of CS neurons with few synaptic contacts per afferent axon (see Section 7.3.1), the high-frequency noise-like membrane potential fluctuations likely originate in the uncorrelated firing of the CS neurons. The slowly depolarizing envelopes, sculpted by the temporal summation of synaptic potentials, might reflect on their part periods of relatively coherent activity within ensembles of related cortical inputs.

In the rat, the waking state is occasionally followed by a period of drowsiness preceding the appearance of SWS. The electrophysiological correlates of drowsiness are the attenuation of the EMG activity and the appearance of slow, though slightly disorganized, waves in the EEG. This intermediate state of vigilance is accompanied in MSNs with larger amplitude and less variable membrane potential fluctuations and by a broad multipeak distribution of membrane potential (Mahon et al. 2006).

Transitions to SWS are evidenced by the appearance of slowly recurring large-amplitude EEG waves and a mild muscle activity without phasic contractions (Figure 7.7B1, top traces) (Timo-Iaria et al. 1970; Gottesmann 1992). Most MSNs recorded during SWS exhibit periodic transitions between a hyperpolarized and a depolarized state (Figure 7.7B1, bottom trace), which are reflected in the bimodal distribution of membrane potential (Figure 7.7B2, left) (Mahon et al. 2006). As observed in urethane-anesthetized animals (see Section 7.5.1), transitions to the up state have fast kinetics and sigmoïdal shape whereas the return to the down state is drawn by an exponential like waveform (compare Figures 7.6D and 7.7B2, right). In most recorded MSNs, small depolarizing perturbations in the up state can generate action potential firing, with a high cell-to-cell variability in the mean firing rate (range = 0.6–16.1 Hz) (Mahon et al. 2006). Although the average rate of discharge during SWS does not significantly differ from that measured during wakefulness, the sleep-related firing activity in MSNs shows a distinctive pattern (Figure 7.7B1) characterized by rhythmic brisk discharges (Figure 7.7B3, left) as reflected in the corresponding bimodal distribution of ISIs (Figure 7.7B3, right).

The high-voltage slow-wave cortical activity, characteristic of the natural SWS (Steriade 2000), results from the synchronized alternation of suprathreshold up states and quiescent down states in cortical neurons (Steriade et al. 2001; Timofeev et al.

2001). Given the strong temporal correlation between the rhythmic cortical field potentials and the intracellular activity of MSNs during SWS, it is very likely, as described for the striatal up and down states during anesthesia, that the rhythmic ongoing CS activity interacts with the active membrane properties of MSNs to determine the membrane potential shifts in striatal cells (see Section 7.5.1). Although the two-state intracellular activity observed in MSNs during SWS is relatively less stereotyped than during anesthesia, with membrane potential fluctuations showing a higher temporal and voltage variability (Mahon et al. 2006), the findings described earlier strongly suggest that "natural" up and down states in MSNs reflect the basic pattern of CS activity during SWS.

Contrasting with the relatively stereotyped electrical activity generated by anesthesia, membrane fluctuations and firing patterns in MSNs are naturally versatile due to their strong dependence upon the states of vigilance. This suggests that neuronal computation achieved by striatal neurons would vary in a brain state-dependent manner according to the physiological or pathological processes in which the CS systems are engaged.

7.6　CONCLUSIONS AND FUTURE CHALLENGES

This chapter highlights a few of the key advances in research on the sensorymotor CS circuits. We made the choice to focus on their fundamental anatomical and electrophysiological properties that may provide a comprehensive "cellular basis" for the mechanisms of cortical information processing in the striatum and explain the transformation of CS inputs into functionally relevant striatal outputs. As stated earlier, recent *in vivo* investigations performed at single-cell level unveiled the complex spatial organization of the cortical projections within the striatum and allowed a refined quantification of CS synaptic connectivity: thousands of CS synapses on a single MSN arising from numerous converging cortical afferents, each afferent giving off a small number of synapses. This peculiar profile of synaptic organization is associated with an extraordinary electrical "nonlinearity" of MSNs membrane, mainly due to different sets of voltage-gated potassium channels, which makes isolated CS inputs functionally irrelevant. This led to a mechanistic, and predictive, cellular model for a "dynamic shaping" of membrane potential fluctuations and firing patterns in MSNs, which continuously operates as a function of the spatiotemporal pattern of CS inputs and the concomitant level of membrane polarization of the MSN.

The ultimate challenge of the research on CS physiology will be to establish causal and specific links between the various behaviors that engage neural circuits interconnecting the neocortex with the striatum and the intimate synaptic and electrical membrane phenomena that concomitantly occur in the different neuronal elements of the network. This would require to further develop an emerging experimental technology (Chorev et al. 2009), allowing multiple *in vivo* intracellular recordings from freely moving animals and during appropriate behavioral procedures and conditioning paradigms. However, in a recent study, our group could for the first time correlate the various states of vigilance with distinctive intracellular electrical behaviors in MSNs, notably demonstrating that wakefulness is reflected in MSNs by temporally disorganized depolarizing synaptic events of variable amplitude (Mahon et al. 2006) (Figure 7.7), a finding that refutes the ubiquitous two-state behavior of MSNs hypothesized from anesthetized preparations and extends the "computational field" available for MSNs (see Section 7.5.2). Because these experiments were performed from unanesthetized but head-restrained rats, the synaptic and intrinsic interactions that sculpt the intracellular activity of MSNs during a given behavioral task remain unknown.

As an alternative approach to resolve the critical issue of the cellular basis of CS-related functions, we designed here a new theoretical, though realistic, representation of the CS system, the so-called "spatiotemporal CS funnel," which takes into account the excitability properties of MSNs and the basic morphofunctional features of the network (see Figure 7.4A). Based on this model, we proposed that the experience-dependent increase in intrinsic excitability that occurs in MSNs shortly after their activation by convergent and temporally coherent cortical inputs will produce a time-dependent "contraction" of the CS funnel, making more efficient and reliable the integration of cortical information by MSNs (see Figure 7.4B). As a behavioral corollary, this could amplify and optimize the temporal link between striatal cells following temporally ordered activity of converging cortical inputs, normally associated during the completion of a given task. This may provide a basic and elementary cellular mechanism for the dorsal striatum-related sensorymotor learning and habits formation (Graybiel 2008). Indeed, when the sensory cue, during learning of a sensorymotor association, becomes effective to discharge the related striatal neurons, the subsequent intrinsic plasticity could make more reliable the integration in the striatum of sequential sensory and motor information required for the correct execution of the behavioral task. Moreover, the repeated association of sensory and motor CS inputs during training would lead to repetitive and synchronized discharges in the corresponding CS funnel, a pattern of activity that is required for the induction of LTP at CS synapses (Charpier and Deniau 1997; Charpier et al. 1999a; Reynolds et al. 2001). Thus, the short-term increase in intrinsic excitability and the subsequent induction of long-term synaptic potentiation would durably reinforce the efficiency of associated CS inputs, leading to stabilization of striatal sensorimotor engrams and consequently to acquisition of behavioral routines.

ACKNOWLEDGMENTS

Our work was supported by the Collège de France, the Institut National de la Santé et de la Recherche Médicale, the Ministère Français de la Recherche et de l'Enseignement Supérieur, the Université Pierre et Marie Curie-Paris VI, the Agence Nationale de la Recherche (ANR R06274DS, 2006), and the Programme Interdisciplinaire-CNRS Neuro IC 2010.

REFERENCES

Alloway, K. D., Crist, J., Mutic, J. J. and Roy, S. A. 1999. Corticostriatal projections from rat barrel cortex have an anisotropic organization that correlates with vibrissal whisking behavior. *J Neurosci* 19: 10908–10922.
Alloway, K. D., Lou, L., Nwabueze-Ogbo, F. and Chakrabarti, S. 2006. Topography of cortical projections to the dorsolateral neostriatum in rats: Multiple overlapping sensorimotor pathways. *J Comp Neurol* 499: 33–48.

Amzica, F. and Steriade, M. 1995. Short- and long-range neuronal synchronization of the slow (<1 Hz) cortical oscillation. *J Neurophysiol* 73: 20–38.

Azdad, K., Chàvez, M., Don Bischop, P. et al. 2009. Homeostatic plasticity of striatal neurons intrinsic excitability following dopamine depletion. *PLoS One* 4: e6908.

Bargas, J., Howe, A., Eberwine, J., Cao, Y. and Surmeier D. J. 1994. Cellular and molecular characterization of Ca²⁺ currents in acutely isolated, adult rat neostriatal neurons. *J Neurosci* 14: 6667–6686.

Belluscio, M. A., Kasanetz, F., Riquelme, L. A. and Murer, M. G. 2003. Spreading of slow cortical rhythms to the basal ganglia output nuclei in rats with nigrostriatal lesions. *Eur J Neurosci* 17: 1046–1052.

Bergman, H., Feingold, A., Nini, A. et al. 1998. Physiological aspects of information processing in the basal ganglia of normal and parkinsonian primates. *Trends Neurosci* 21: 32–38.

Bevan, M. D., Magill, P. J., Terman, D., Bolam, J. P. and Wilson, C. J. 2002. Move to the rhythm: Oscillations in the subthalamic nucleus-external globus pallidus network. *Trends Neurosci* 25: 525–531.

Boraud, T., Bezard, E., Guehl, D., Bioulac, B. and Gross, C. 1998. Effects of L-DOPA on neuronal activity of the globus pallidus externalis (GPe) and globus pallidus internalis (GPi) in the MPTP-treated monkey. *Brain Res* 787: 157–160.

Calabresi, P., Pisani, A., Mercuri, N. B. and Bernardi, G. 1996. The corticostriatal projection: From synaptic plasticity to basal ganglia disorders. *Trends Neurosci* 19:19–24.

Cantrell, A. R., Carter-Russel, H., Mermelstein, P. and Surmeier, D. J. 1995. Ca²⁺ and Na²⁺ currents in acutely-isolated neostriatal neurons from the rat. In *Molecular and Cellular Mechanisms of Neostriatal Function*, eds. M. A. Ariano and D. J. Surmeier, pp. 151–163. Austin, TX: Landes Company.

Carter, A. G. and Sabatini, B. L. 2004. State-dependent calcium signaling in dendritic spines of striatal medium spiny neurons. *Neuron* 44: 483–493.

Chang, H. T., Wilson, C. J. and Kitai, S. T. 1982. A Golgi study of rat neostriatal neurons: Light microscopic analysis. *J Comp Neurol* 208: 107–126.

Charpier, S., Beurrier, C. and Paz, J. T. 2010. The subthalamic nucleus: From *in vitro* to in vivo mechanisms. In *Handbook of Basal Ganglia Structure and Function*, eds. H. Steiner and K. Tseng, pp. 259–273. Amsterdam, the Netherlands: Elsevier.

Charpier, S. and Deniau, J.-M. 1997. In vivo activity-dependent plasticity at cortico-striatal connections: Evidence for physiological long-term potentiation. *Proc Natl Acad Sci USA* 94: 7036–7040.

Charpier, S., Leresche, N., Deniau, J.-M., Mahon, S., Hughes, S.W. and Crunelli, V. 1999b. On the putative contribution of GABA(B) receptors to the electrical events occurring during spontaneous spike and wave discharges. *Neuropharmacology* 38: 1699–1706.

Charpier, S., Mahon, S. and Deniau, J.-M. 1999a. In vivo induction of striatal long-term potentiation by low-frequency stimulation of the cerebral cortex. *Neuroscience* 91: 1209–1222.

Chorev, E., Epsztein, J., Houweling, A. R., Lee, A. K. and Brecht, M. 2009. Electrophysiological recordings from behaving animals—Going beyond spikes. *Curr Opin Neurobiol* 19: 513–519.

Connors, B. W. and Gutnick, M. J. 1990. Intrinsic firing patterns of diverse neocortical neurons. *Trends Neurosci* 13: 99–104.

Contreras, D. and Steriade, M. 1995. Cellular basis of EEG slow rhythms: A study of dynamic corticothalamic relationships. *J Neurosci* 15: 604–622.

Cowan, R. L. and Wilson, C. J. 1994. Spontaneous firing patterns and axonal projections of single corticostriatal neurons in the rat medial agranular cortex. *J Neurophysiol* 71: 17–32.

Degos, B., Deniau, J.-M., Chavez, M. and Maurice, N. 2009. Chronic but not acute dopaminergic transmission interruption promotes a progressive increase in cortical beta frequency synchronization: Relationships to vigilance state and akinesia. *Cereb Cortex* 19: 1616–1630.

Deniau, J.-M., Menetrey, A. and Charpier, S. 1996. The lamellar organization of the rat substantia nigra pars reticulata: Segregated patterns of striatal afferents and relationship to the topography of corticostriatal projections. *Neuroscience* 73: 761–781.

Diesmann, M., Gewaltig, M. O. and Aertsen, A. 1999. Stable propagation of synchronous spiking in cortical neural networks. *Nature* 402: 529–533.

Fino, E., Glowinski, J. and Venance L. 2007. Effects of acute dopamine depletion on the electrophysiological properties of striatal neurons. *Neurosci Res* 58: 305–316.

Flaherty, A. W. and Graybiel, A. M. 1991. Corticostriatal transformations in the primate somatosensory system. Projections from physiologically mapped body-part representations. *J Neurophysiol* 66: 1249–1263.

Flaherty, A. W. and Graybiel, A. M. 1993. Two input systems for body representations in the primate striatal matrix: Experimental evidence in the squirrel monkey. *J Neurosci* 13: 1120–1137.

Flaherty, A. W. and Graybiel, A. M. 1994. Input-output organization of the sensorimotor striatum in the squirrel monkey. *J Neurosci* 14: 599–610.

Galarraga, E., Bargas, J., Sierra, A. and Aceves, J. 1989. The role of calcium in the repetitive firing of neostriatal neurons. *Exp Brain Res* 75: 157–168.

Gottesmann, C. 1992. Detection of seven sleep-waking stages in the rat. *Neurosci Biobehav Rev* 16: 31–38.

Graybiel, A. M. 1995. Building action repertoires: Memory and learning functions of the basal ganglia. *Curr Opin Neurobiol* 5: 733–741.

Graybiel, A. M. 2008. Habits, rituals, and the evaluative brain. *Annu Rev Neurosci* 31: 359–387.

Hammond, C., Bergman, H. and Brown, P. 2007. Pathological synchronization in Parkinson's disease: Networks, models and treatments. *Trends Neurosci* 30: 357–364.

Hammond, C. and Crepel, F. 1992. Evidence of a slowly inactivating K+ current in prefrontal cortical cells. *Eur J Neurosci* 4: 1087–1092.

Hernandez-Lopez, S., Tkatch, T. and Perez-Garci, E. et al. 2000. D2 dopamine receptors in striatal medium spiny neurons reduce L-type Ca^{2+} currents and excitability via a novel $PLC\beta1$-IP3-calcineurin-signaling cascade. *J Neurosci* 20: 8987–8995.

Houk, J. C. 1995. Information processing in modular circuits linking basal ganglia and cerebral cortex. In *Models of Information Processing in the Basal Ganglia*, eds. J. C. Houk, J. L. Davies and D. G. Beiser, pp. 3–9. Cambridge, MA: MIT Press.

Kasanetz, F., Riquelme, L. A. and Murer, M. G. 2002. Disruption of the two-state membrane potential of striatal neurones during cortical desynchronisation in anaesthetised rats. *J Physiol* 543: 577–589.

Kasanetz, F., Riquelme, L. A., O'donnell, P. and Murer, M. G. 2006. Turning off cortical ensembles stops striatal UP states and elicits phase perturbations in cortical and striatal slow oscillations in vivo. *J Physiol* 577: 97–113.

Kemp, J. M. and Powell, T. P. 1971. The termination of fibres from the cerebral cortex and thalamus upon dendritic spines in the caudate nucleus: A study with the Golgi method. *Philos Trans R Soc Lond B Biol Sci* 262: 429–439.

Kerr, J. N. and Plenz, D. 2002. Dendritic calcium encodes striatal neuron output during up-states. *J Neurosci* 22: 1499–1512.

Kerr, J. N. and Plenz, D. 2004. Action potential timing determines dendritic calcium during striatal up-states. *J Neurosci* 24: 877–885.

Kincaid, A.E., Zheng, T. and Wilson, C.J. 1998. Connectivity and convergence of single corticostriatal axons. *J Neurosci* 18: 4722–4731.

Kitai, S. T. and Deniau, J. M. 1981. Cortical inputs to the subthalamus: Intracellular analysis. *Brain Res* 214: 411–415.

Koós, T. and Tepper, J. M. 1999. Inhibitory control of neostriatal projection neurons by GABAergic interneurons. *Nat Neurosci* 2: 467–472.

Kreitzer, A. C. 2009. Physiology and pharmacology of striatal neurons. *Annu Rev Neurosci* 32: 127–147.

Kreitzer, A. C. and Malenka, R. C. 2005. Dopamine modulation of state-dependent endocannabinoid release and long-term depression in the striatum. *J Neurosci* 25: 10537–10545.

Léger, J. F., Stern, E. A., Aertsen, A. and Heck, D. 2005. Synaptic integration in rat frontal cortex shaped by network activity. *J Neurophysiol* 93: 281–293.

Levesque, M. and Parent, A. 1998. Axonal arborization of corticostriatal and corticothalamic fibers arising from prelimbic cortex in the rat. *Cereb Cortex* 8: 602–613.

Lovinger, D. M. and Tyler, E. 1996. Synaptic transmission and modulation in the neostriatum. *Int Rev Neurobiol* 39:77–111.

Magill, P. J., Bolam, J. P. and Bevan, M. D. 2001. Dopamine regulates the impact of the cerebral cortex on the subthalamic nucleus-globus pallidus network. *Neuroscience* 106: 313–330.

Mahon, S., Casassus, G., Mulle, C. and Charpier, S. 2003a. Spike-dependent intrinsic plasticity increases firing probability in rat striatal neurons in vivo. *J Physiol* 550: 947–959.

Mahon, S., Delord, B., Deniau, J.-M., Charpier, S. 2000a. Intrinsic properties of rat striatal output neurones and time-dependent facilitation of cortical inputs in vivo. *J Physiol* 527: 345–354.

Mahon, S., Deniau, J.-M. and Charpier, S. 2001. Relationship between EEG potentials and intracellular activity of striatal and cortico-striatal neurons: An in vivo study under different anesthetics. *Cereb Cortex* 11: 360–373.

Mahon, S., Deniau, J.-M. and Charpier, S. 2003b. Various synaptic activities and firing patterns in cortico-striatal and striatal neurons in vivo. *J Physiol Paris* 97: 557–566.

Mahon, S., Deniau J.-M. and Charpier, S. 2004. Corticostriatal plasticity: Life after the depression. *Trends Neurosci* 27: 460–467.

Mahon, S., Deniau, J.-M., Charpier, S. and Delord, B. 2000b. Role of a striatal slowly inactivating potassium current in short-term facilitation of corticostriatal inputs: A computer simulation study. *Learn Mem* 7: 357–362.

Mahon, S., Vautrelle, N., Pezard, L. et al. 2006. Distinct patterns of striatal medium spiny neuron activity during the natural sleep-wake cycle. *J Neurosci* 26: 12587–12595.

Mallet, N., Ballion, B., Le Moine, C. and Gonon, F. 2006. Cortical inputs and GABA interneurons imbalance projection neurons in the striatum of parkinsonian rats. *J Neurosci* 26: 3875–3884.

Mallet, N., Le Moine, C., Charpier, S. and Gonon, F. 2005. Feedforward inhibition of projection neurons by fast-spiking GABA interneurons in the rat striatum in vivo. *J Neurosci* 25: 3857–3869.

McCormick, D. A. 1991. Functional properties of a slowly inactivating potassium current in guinea pig dorsal lateral geniculate relay neurons. *J Neurophysiol* 66: 1176–1189.

McGeorge, A. J. and Faull, R. L. M. 1989. The organization from the projection from the cerebral cortex to the striatum in the rat. *Neuroscience* 29: 503–537.

Murer, M. G., Tseng, K. Y., Kasanetz, F., Belluscio, M. and Riquelme, L. A. 2002. Brain oscillations, medium spiny neurons, and dopamine. *Cell Mol Neurobiol* 22: 611–632.

Nicola, S. M., Surmeier, D. J. and Malenka, R. C. 2000. Dopaminergic modulation of neuronal excitability in the striatum and nucleus accumbens. *Annu Rev Neurosci* 23: 185–215.

Nisenbaum, E. S., Mermelstein, P. G., Wilson, C. J. and Surmeier D. J. 1998. Selective blockade of a slowly inactivating potassium current in striatal neurons by (+/−) 6-chloro-APB hydrobromide (SKF82958). *Synapse* 29: 213–224.

Nisenbaum, E. S. and Wilson, C. J. 1995. Potassium current responsible for inward and outward rectification in rat neostriatal spiny projection neurons. *J Neurosci* 15: 4449–4463.

Nisenbaum, E. S., Wilson, C. J., Foehring, R. C. and Surmeier, D. J. 1996. Isolation and characterization of a persistent potassium current in neostriatal neurons. *J Neurophysiol* 76: 1180–1194.

Nisenbaum, E. S., Xu, Z. C. and Wilson, C. J. 1994. Contribution of a slowly inactivating potassium current to the transition to firing of neostriatal spiny projections neurons. *J Neurophysiol* 71: 1174–1189.

Oorschot, D. E. 1996. Total number of neurons in the neostriatal, pallidal, subthalamic, and substantia nigral nuclei of the rat basal ganglia: A stereological study using the cavalieri and optical disector methods. *J Comp Neurol* 366: 580–599.

Pacheco-Cano, M. T., Bargas, J., Hernandez-Lopez, S., Tapia, D. and Galarraga, E. 1996. Inhibitory action of dopamine involves a subthreshold Cs(+)-sensitive conductance in neostriatal neurons. *Exp Brain Res* 110: 205–211.

Parthasarathy, H. B., Schall, J. D. and Graybiel, A. M. 1992. Distributed but convergent ordering of corticostriatal projections: Analysis of the frontal eye field and the supplementary eye field in the macaque monkey. *J Neurosci* 12: 4468–4488.

Paz, J. T., Chavez, M., Saillet, S., Deniau, J.-M. and Charpier, S. 2007. Activity of ventral medial thalamic neurons during absence seizures and modulation of cortical paroxysms by the nigrothalamic pathway. *J Neurosci* 27: 929–941.

Paz, J. T., Deniau, J.-M. and Charpier, S. 2005. Rhythmic bursting in the cortico-subthalamo-pallidal network during spontaneous genetically determined spike and wave discharges. *J Neurosci* 25: 2092–2101.

Paz, J. T., Polack, P.-O., Slaght, S. J., Deniau, J.-M., Mahon, S. and Charpier, S. 2006. Propagation and dynamic processing of cortical paroxysms in the basal ganglia networks during absence seizures. In *Generalized Seizures: From Clinical Phenomenology to Underlying Systems and Networks, Progress in Epileptic Disorders*, eds. E. Hirsch, F. Andermann, P. Chauvel, J. Engel, F. Lopes da Silva and H. Luders, pp. 75–91. Montrouge, France: John Libbey Eurotext.

Petersen, C. C., Hahn, T. T., Mehta, M., Grinvald, A. and Sakmann, B. 2003. Interaction of sensory responses with spontaneous depolarization in layer 2/3 barrel cortex. *Proc Natl Acad Sci USA* 100: 13638–13643.

Pidoux, M., Mahon, S., Deniau, J.-M. and Charpier S. 2011. Integration and propagation of somatosensory responses in the corticostriatal pathway: An intracellular study in vivo. *J Physiol* 589: 263–281.

Plenz, D. 2003. When inhibition goes incognito: Feedback interaction between spiny projection neurons in striatal function. *Trends Neurosci* 26: 436–443.

Plenz, D. and Kitai, S. T. 1998. Up and down states in striatal medium spiny neurons simultaneously recorded with spontaneous activity in fast-spiking interneurons studied in cortex-striatum-substantia nigra organotypic cultures. *J Neurosci* 18: 266–283.

Pomata, P. E., Belluscio, M. A., Riquelme, L. A. and Murer, M. G. 2008. NMDA receptor gating of information flow through the striatum in vivo. *J Neurosci* 28: 13384–13389.

Ramanathan, S., Hanley, J. J, Deniau, J.-M. and Bolam, J. P. 2002. Synaptic convergence of motor and somatosensory cortical afferents onto GABAergic interneurons in the rat striatum. *J Neurosci* 22: 8158–8169.

Reynolds, J. N., Hyland, B. I. and Wickens, J. R. 2001. A cellular mechanism of reward-related learning. *Nature* 413: 67–70.

Rodríguez, M. J., Pugliese, M. and Mahy, N. 2009. Drug abuse, brain calcification and glutamate-induced neurodegeneration. *Curr Drug Abuse Rev* 2: 99–112.

Schultz, W., Tremblay, L. and Hollerman, J. R. 2003. Changes in behavior-related neuronal activity in the striatum during learning. *Trends Neurosci* 26: 321–328.

Sharott, A., Magill, P. J., Harnack, D., Kupsch, A., Meissner, W. and Brown, P. 2005. Dopamine depletion increases the power and coherence of beta-oscillations in the cerebral cortex and subthalamic nucleus of the awake rat. *Eur J Neurosci* 21: 1413–1422.

Shen, W., Tian, X., Day, M. et al. 2007. Cholinergic modulation of Kir2 channels selectively elevates dendritic excitability in striatopallidal neurons. *Nat Neurosci* 10: 1458–1466.

Slaght, S. J., Paz, J. T., Chavez, M., Deniau, J.-M., Mahon, S. and Charpier, S. 2004. On the activity of the corticostriatal networks during spike-and-wave discharges in a genetic model of absence epilepsy. *J Neurosci* 24: 6816–6825.

Smith, Y., Raju, D. V., Pare, J. F. and Sidibe, M. 2004. The thalamostriatal system: A highly specific network of the basal ganglia circuitry. *Trends Neurosci* 27: 520–527.

Souliere, F., Urbain, N., Gervasoni, D., Schmitt, P. et al. 2000. Single-unit and polygraphic recordings associated with systemic or local pharmacology: A multi-purpose stereotaxic approach for the awake, anaesthetic-free, and head-restrained rat. *J Neurosci Res* 61: 88–100.

Steriade, M. 2000. Corticothalamic resonance, states of vigilance and mentation. *Neuroscience* 101: 243–276.

Steriade, M. 2004. Neocortical cell classes are flexible entities. *Nat Rev Neurosci* 5: 121–134.

Steriade, M., Gloor, P., Llinás, R. R., Lopes da Silva, F. H. and Mesulam, M. M. 1990. Basic mechanisms of cerebral rhythmic activities. *Electroencephalogr Clin Neurophysiol* 76: 481–508.

Steriade, M., Nunez, A. and Amzica, F. 1993. A novel slow (<1 Hz) oscillation of neocortical neurons in vivo: Depolarizing and hyperpolarizing components. *J Neurosci* 13: 3252–3265.

Steriade, M., Timofeev, I. and Grenier, F. 2001. Natural waking and sleep states: A view from inside neocortical neurons. *J Neurophysiol* 85: 1969–1985.

Stern, E. A., Jaeger, D. and Wilson, C. J. 1998. Membrane potential synchrony of simultaneously recorded striatal spiny neurons in vivo. *Nature* 394: 475–478.

Stern, E. A., Kincaid, A. E. and Wilson, C. J. 1997. Spontaneous subthreshold membrane potential fluctuations and action potential variability of rat corticostriatal and striatal neurons in vivo. *J Neurophysiol* 77: 1697–1715.

Storm, J. F. 1988. Temporal integration by a slowly inactivating K+ current in hippocampal neurons. *Nature* 336: 379–381.

Surmeier, D. J., Bargas, J., Hemmings, H. C. Jr., Nairn, A. C. and Greengard P. 1995. Modulation of calcium currents by a D1 dopaminergic protein kinase/phosphatase cascade in rat neostriatal neurons. *Neuron* 14: 385–397.

Timofeev, I., Grenier, F. and Steriade, M. 2001. Disfacilitation and active inhibition in the neocortex during the natural sleep-wake cycle: An intracellular study. *Proc Natl Acad Sci USA* 98: 1924–1929.

Timo-Iaria, C., Negrao, N., Schmidek, W. R., Hoshino K., Lobato de Menezes, C. E. and Leme da Rocha, T. 1970. Phases and states of sleep in the rat. *Physiol Behav* 5: 1057–1062.

Tseng, K. Y., Kasanetz, F., Kargieman, L., Riquelme, L. A. and Murer, M. G. 2001. Cortical slow oscillatory activity is reflected in the membrane potential and spike trains of striatal neurons in rats with chronic nigrostriatal lesions. *J Neurosci* 21: 6430–6439.

Tseng, K. Y., Riquelme, L. A. and Murer, M. G. 2004. Impact of D1-class dopamine receptor on striatal processing of cortical input in experimental parkinsonism in vivo. *Neuroscience* 123: 293–298.

Uchimura, N. and North, R. A. 1990. Actions of cocaine on rat nucleus accumbens neurones in vitro. *Br J Pharmacol* 99: 736–740.

Webster, K. E. 1961. Cortico-striate interrelations in the albino rat. *J Anat* 95: 532–545.

Wichmann, T. and Soares, J. 2006. Neuronal firing before and after burst discharges in the monkey basal ganglia is predictably patterned in the normal state and altered in parkinsonism. *J Neurophysiol* 95: 2120–2133.

Wilson, C. J. 1987. Morphology and synaptic connections of crossed corticostriatal neurons in the rat. *J Comp Neurol* 263: 567–580.

Wilson, C. J. 1992. Dendritic morphology, inward rectification, and the functional properties of neostriatal neurons. In *Single Neuron Computation*, eds. T. McKenna, J. Davis and S. F. Zornetzer, pp. 141–171. San Diego, CA: Academic Press.

Wilson, C. J. 1993. The generation of natural firing patterns in neostriatal neurons. *Prog Brain Res* 99: 277–297.

Wilson, C. J. 1995a. Dynamic modification of dendritic cable properties and synaptic transmission by voltage-gated potassium channels. *J Comput Neurosci* 2: 91–115.

Wilson, C. J. 1995b. The contribution of cortical neurons to the firing pattern of striatal spiny neurons. In *Models of Information Processing in the Basal Ganglia*, eds. J. C. Houk, J. L. Davies and D. G. Beiser, pp. 29–50. Cambridge, MA: MIT Press.

Wilson, C. J. 2004. Basal ganglia. In *The Synaptic Organization of the Brain*, ed. G. M. Shepherd, pp. 361–415. New York: Oxford University Press.

Wilson, C. J. and Kawaguchi, Y. 1996. The origins of two-state spontaneous membrane potential fluctuations of neostriatal spiny neurons. *J Neurosci* 16: 2397–2410.

Wright, A. K., Ramanathan, S. and Arbuthnott, G. W. 2001. Identification of the source of the bilateral projection system from cortex to somatosensory neostriatum and an exploration of its physiological actions. *Neuroscience* 103: 87–96.

Zheng, T. and Wilson, C. J. 2002. Corticostriatal combinatorics: The implications of corticostriatal axonal arborizations. *J Neurophysiol* 87: 1007–1017.

8 Functional Organization of the Midbrain Substantia Nigra

Jennifer Brown

CONTENTS

8.1 INTRODUCTION

The substantia nigra (SN) is one of the subcortical nuclei that make up the basal ganglia (BG) network, which has a role in the control of voluntary movement. Classically, the SN is divided into two functional subregions, the substantia nigra pars reticulata (SNr) and the substantia nigra pars compacta (SNc). In rodents, the SNr is the main output nucleus of the BG, receiving input from both the "direct" striatonigral and "indirect" striato-pallidal-subthalamo-nigral pathway through the BG, and sending γ-amino-butyric acid (GABA)ergic inhibitory projections to the thalamus, superior colliculus (SC), and brain-stem (BS) tegmentum (Beckstead et al. 1979; Alexander and Crutcher 1990; Parent 1990; Deniau and Chevalier 1992; Deniau et al. 1996; Mailly et al. 2001; Middleton and Strick 2002; Cebrián et al. 2005). In contrast, the SNc predominantly consists of dopa-minergic neurons that receive input from nuclei both within and outside the BG and in turn send extensive dendritic and axonal projections throughout the BG as well as a number of external targets (Grofová 1975; Bunney and Aghajanian 1976; Grofová et al. 1982; Bolam and Smith 1990; Tepper and Lee 2007; Lee and Tepper 2009). In addition to the output targets listed earlier, there is also extensive dopaminergic and GABAergic connectivity within and between subregions of the SN. In this regard, although afferent glutamatergic input does play an important role in this nucleus, a chapter on the SN is primarily concerned with GABA–dopamine interactions (Figure 8.1).

The focus of this chapter is to describe the function and topography of inputs to and outputs from the SN, the properties and heterogeneity of cells within the SN, and

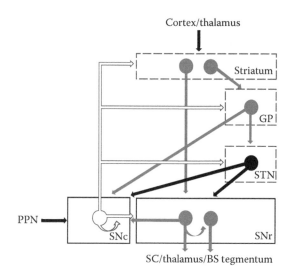

FIGURE 8.1 Wiring diagram of BG circuitry focusing on the SN network. SNc dopamine neurons receive input from BG nuclei (GP, STN, and SNr) and nuclei outside the BG includ-ing the PPN. SNc neurons send axonal and dendritic projections throughout the BG. SNr GABA neurons receive input from BG nuclei (striatum, GP, STN, and SNc) and form the pri-mary output nucleus of the BG, sending axonal projections to target nuclei including the SC, thalamus, BS tegmentum, and local axonal collaterals within the SN. Open circles/arrows (dopaminergic), light gray circles/arrows (GABAergic), and dark gray circles/arrows (gluta-matergic). Boxes indicate nuclei within BG.

the intranigral connectivity between SN cells. It provides a companion to Chapter 7, where the topographical organization of corticostriatal processing is considered. Here, the functional organization of the SN is reviewed.

8.2 SUBSTANTIA NIGRA PARS RETICULATA CONNECTIVITY

8.2.1 SUBSTANTIA NIGRA PARS RETICULATA: INPUTS

A principal property of SNr GABAergic neurons is their spontaneous high-frequency activity (Figures 8.2 and 8.3), which provides tonic inhibition of their target nuclei, including the thalamus, SC, and BS tegmentum. Tonic inhibition from the SNr appears to play a critical role in suppressing the generation of unwanted movements (Alexander and Crutcher 1990; Chevalier and Deniau 1990). Conversely, relief of tonic inhibition from the SNr is important for permitting movement generation (Deniau et al. 1978; Chevalier and Deniau 1990). Synaptic input to SNr neurons from other BG nuclei modulates SNr activity (Figure 8.1) and may either enhance or inhibit

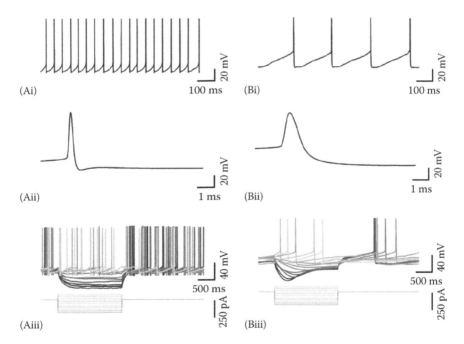

FIGURE 8.2 Electrophysiological properties of SNr GABAergic and SNc dopamine neurons recorded *in vitro*. Whole-cell recording of spontaneously firing SNr GABAergic (Ai) and SNc dopamine neuron (Bi) *in vitro*. SNr GABAergic neuron displays a much higher spontaneous firing rate than the dopamine neuron. The average action potential waveform of SNr GABAergic neuron (Aii) has a short AP half width and rapid rate of repolarization compared to SNc dopamine neuron (Bii). Response to hyperpolarizing and depolarizing current pulses (from −160 to +40 pA in steps of 20 pA) of GABAergic (Aiii) and dopamine neuron (Biii). GABAergic neuron shows little or no voltage sag response to hyperpolarizing current steps whereas SNc dopamine neuron shows a prominent voltage sag due to the activation of hyperpolarization-activated current (Ih). (From Jennifer Brown, unpublished data.)

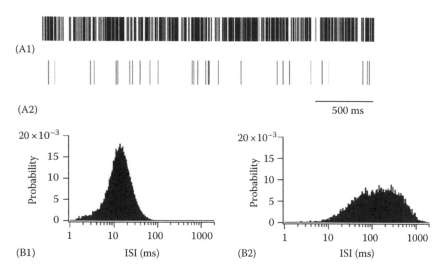

(A1)

(A2)

500 ms

FIGURE 8.3 Firing properties of SN GABAergic and dopamine neurons *in vivo*. Raster plot of a putative SN GABAergic (A1) and dopamine neuron (A2) recorded *in vivo* in a freely moving mouse. GABAergic neuron has a higher tonic firing rate compared to the low-frequency irregular firing of the dopamine neurons. Interspike interval (ISI) histogram of putative GABAergic (B1) and dopamine (B2) neuron. (From Wei-Xing Pan and Joshua Dudman, unpublished data.)

firing depending on the type of input received. Thus, the SNr can be thought of as an analogue control device for information flow through the BG, with the relative weight of inhibitory and excitatory input to SNr neurons determining the extent of BG inhibition of output targets and ultimately influencing the control of voluntary movements.

8.2.1.1 GABAergic Input

A primary source of inhibitory input to the SNr comes from a subpopulation of medium spiny neurons (MSNs) in the striatum (predominantly the D1 receptor and substance P-expressing MSNs) that send axonal projections to the SNr (the "direct" pathway) and release the inhibitory neurotransmitter GABA (Alexander and Crutcher 1990; Smith and Bolam 1990). The cortex innervates the striatum in a topographic manner (Veening et al. 1980; Alexander et al. 1986; McGeorge and Faull 1989; Alexander and Crutcher 1990), and such topography is in part maintained through to the striatonigral pathway (Mailly et al. 2001), providing a mechanism for action-specific selection of movement via SNr outputs. Cortical input to the striatum generates burst activity in the normally silent MSNs to inhibit SNr firing and thus the disinhibition of target nuclei (Chevalier et al. 1985; Chevalier and Deniau 1990; Smith and Bolam 1991; Hikosaka et al. 2000). This direct striatonigral inhibitory pathway, which is facilitated by dopamine release (for review, see Gerfen and Surmeier 2010), appears to be critical for the suppression of tonic firing in SNr neurons. By inhibiting the GABAergic projection neurons of the SNr, activation of the direct striatonigral pathway is thought to relieve inhibition of target outputs and thus facilitate the generation of voluntary movement.

In awake rodents, cats, and primates, injection of the GABA$_A$ receptor agonist muscimol into the SNr has been demonstrated to inhibit firing and generate behavioural effects such as fast saccadic eye movements to the contralateral side (Boussaoud and Joseph 1985; Hikosaka and Wurtz 1985; Sakamoto and Hikosaka 1989) and contralateral turning behavior (Martin et al. 1978; Arnt and Scheel-Kruger 1979; Ikeda et al. 2010). These observations are consistent with a model in which disinhibition of target neurons in the SC and thalamus is sufficient for the generation of movement (Figure 8.1). Direct measurement of activity in SNr target regions has provided further support for this model. For example, it has been demonstrated that intranigral application of GABA or stimulation of the striatonigral pathway enhanced neuron activity in the ventromedial thalamus, which projects to the motor cortex (Deniau and Chevalier 1984). Similarly, enhanced neuronal activity in the SC is detected following SNr inhibition (Hikosaka 1989).

An additional GABAergic input to the SNr comes from tonically active neurons of the globus pallidus (GP) (Smith and Bolam 1989), which fire at high frequencies (50–70 Hz) under basal conditions *in vivo* (DeLong 1971; Kita and Kitai 1991). Due to their tonic activity and location within the "indirect" BG circuitry, the functional consequence of GP input to SNr neurons is predicted to differ from that of the "direct" striatonigral projecting MSNs (Smith and Bolam 1989; Alexander and Crutcher 1990). Anatomical and electrophysiological experiments have shown that GP neurons receive inhibitory input from the D2 receptor, enkephalin-expressing MSNs of the striatum associated with the "indirect" pathway (Hazrati and Parent 1992; Wu et al. 2000). Activity in the striatum will thus inhibit GP neurons, promoting disinhibition of the SNr. Furthermore, GP neurons send inputs to the tonically active glutamatergic neurons of the subthalamic nucleus (STN), which also projects to the SNr (Kita and Kitai 1987; Smith et al. 1998). Through this disynaptic pathway, inhibition of the GP following striatal input will disinhibit the STN and so enhance glutamatergic excitatory input to the SNr. Therefore, the net effect of both the monosynaptic pallidonigral and the polysynaptic pallido-subthalamo-nigral pathway is to enhance the activity of SNr neurons through disinhibition and excitation respectively. Enhanced activity of SNr neurons will increase the inhibition onto output neurons and so inhibit unwanted movement (Alexander and Crutcher 1990).

Differences between the striatonigral and pallidonigral GABAergic projections to SNr neurons have been highlighted in electrophysiological and anatomical data. While both striatonigral and pallidonigral projections can converge onto the same SNr neurons (Smith and Bolam 1991), each innervates different somatodendritic domains. Striatonigral synapses are predominantly found at distal dendritic sites on SNr neurons while pallidonigral terminals are more commonly located at the soma or proximal dendrites (Smith and Bolam 1989, 1991; von Krosigk et al. 1992). As synapse location along the dendrite can greatly influence synaptic integration in neurons (for review, see Magee 2000), such differences may have functional implications for the processing of converging inputs to the SNr (Smith and Bolam 1991). Furthermore, differences between the properties of synaptic transmission at the two synapses have been identified; striatonigral synapses show facilitation of inhibitory postsynaptic currents (IPSCs) during a train of stimulation, whereas a depression of IPSCs is detected during a train of stimulation to the GP (Connelly et al. 2010).

Recent evidence suggests that neuromodulation of synaptic transmission via presynaptic dopamine receptors may also differ between the striatonigral and pallidonigral synapse. Presynaptic D1-like receptor activation facilitates striatonigral transmission, thus enhancing the inhibition of SNr neurons via the direct pathway (Porceddu et al. 1979; Floran et al. 1990; Radnikow and Misgeld 1998). In contrast, D2-like receptors are expressed at terminals of the pallidonigral afferents. Activation of presynaptic D2 receptors reduces the probability of GABA release and thereby reduces the input from the indirect pallidonigral pathway (de Jesús Aceves et al. 2011).

A third and largely ignored inhibitory input to the SNr comes from the lateral inhibitory connections between neighboring SNr neurons (Juraska et al. 1977; Karabelas and Purpura 1980; Deniau et al. 1982; Grofova et al. 1982; Mailly et al. 2003). Anatomical data show that each SNr neuron sends one to six axonal collaterals back into the SNr (Mailly et al. 2003). While there is little electrophysiological data on the properties of these synapses, it has been proposed that lateral inhibition within the SNr may work to enhance or refine the disinhibition of target structures by SNr projection neurons (Chevalier and Deniau 1990; Mailly et al. 2003). Several mechanisms could be engaged by lateral inhibition in the SNr, such as enhancing the contrast of activity between nigral neurons receiving different degrees of inhibitory input or desynchronizing the discharge of nigral neurons that project to similar target nuclei (Mailly et al. 2003).

8.2.1.2   Inhibition through GABA$_A$ and GABA$_B$ Receptors

GABAergic transmission may mediate its effects via one of two classes of receptors. Ionotropic GABA$_A$ receptors are ligand-gated chloride channels that generally produce membrane hyperpolarization following activation, while metabotropic GABA$_B$ receptors are coupled to G-proteins, mediating their effect via intracellular signaling pathways (for review, see Nutt 2006; Pinard et al. 2010). Predominately GABA$_A$ receptors are expressed postsynaptically within the SNr (Fujiyama et al. 2002; Boyes and Bolam 2007) while GABA$_B$ receptors are expressed both presynaptically and postsynaptically (Boyes and Bolam 2003, 2007).

Immunogold labeling for GABA$_A$ receptor subunits reveals that $\alpha 1$, $\beta 2/3$, and $\gamma 2$ receptor subunits are abundant in the SNr, located preferentially along the synaptic specialization of symmetrical (presumed inhibitory) synapses (Fujiyama et al. 2002). Postsynaptic GABA$_A$ receptor activation underlies fast synaptic transmission to generate both spontaneous and evoked IPSCs, as these may be blocked by the GABA$_A$ receptor blocker antagonist bicuculline (McNair et al. 1972; Deniau et al. 1976; Collingridge and Davies 1981; Hikosaka et al. 1993; Chan and Yung 1999). Furthermore, *in vitro* recordings demonstrate a tonic inhibitory GABA tone on SNr neurons mediated via GABA$_A$ receptor activation (Rick and Lacey 1994). Application of bicuculline significantly increased the basal firing rate of the neurons, while GABA$_B$ receptor antagonists had no effect (Rick and Lacey 1994). Additionally, spontaneous IPSCs recorded from SNr neurons had a significant tetrodotoxin (TTX)-insensitive component that is hypothesized to be due to a tonic inhibitory GABAergic tone on the SNr (Chan and Yung 1999). While the source of such tonic GABA$_A$ receptor–mediated inhibitory tone is unknown, local release of

GABA via axonal collateral projections within the SNr is likely to play an important role (Rick and Lacey 1994; Chan and Yung 1999). However, it should be noted that the functional significance of a GABA tone in the SNr has been questioned by a recent *in vivo* behavioral study (Ikeda et al. 2010). While contralateral tuning behavior is observed in the rat following unilateral injection of $GABA_A$ or $GABA_B$ receptor agonists, no behavioral effect was observed following the injection of $GABA_A$ or $GABA_B$ receptor antagonists alone, leading the authors to argue for low GABA tone in the SNr *in vivo* (Ikeda et al. 2010).

In contrast to the fast synaptic inhibition generated by $GABA_A$ receptor activation, $GABA_B$ receptors appear to play more of a modulatory role within the SNr. Pre-embedding immunogold labeling identified both $GABA_{B1}$ and $GABA_{B2}$ receptor subunits at extrasynaptic and putative GABAergic synapses both presynaptically and postsynaptically, in addition to high labeling identified in intracellular sites (Boyes and Bolam 2003). Furthermore, anterograde labeling from the striatum and GP confirmed both afferent inputs to the SNr-expressed presynaptic $GABA_B$ receptors at the axonal terminals in addition to sites along the axon (Boyes and Bolam 2003). Such anatomical data support electrophysiological studies showing presynaptic $GABA_B$ receptor–mediated modulation of GABA release (Floran et al. 1988; Giralt et al. 1990; Shen and Johnson 1997; Chan and Yung 1999; Boyes and Bolam 2003). *In vitro*, the $GABA_B$ receptor agonist baclofen reduces the amplitude of evoked IPSCs and increases the degree of paired-pulse facilitation in SNr neurons, suggesting that presynaptic $GABA_B$ receptors reduce GABA release from presynaptic terminals (Stanford and Lacey 1996). On the other hand, despite evidence demonstrating postsynaptic $GABA_B$ receptor expression in SNr neurons (Boyes and Bolam 2003), its functional role is still unclear. A weak postsynaptic effect of $GABA_B$ receptor activation in reducing cell firing rate has been observed (Rick and Lacey 1994), although other studies have found no evidence for $GABA_B$ receptor–mediated IPSC in SNr neurons *in vitro* nor has any effect of the $GABA_B$ receptor antagonist CGP 55845A on spontaneous IPSCs been observed (Rick and Lacey 1994; Chan et al. 1998). Thus, the function of postsynaptic $GABA_B$ receptors is still under investigation.

8.2.1.3 Glutamatergic Input

The main excitatory input to SNr neurons comes from the glutamatergic fibers of the STN (Kita and Kitai 1987; Smith et al. 1998). The STN receives input either indirectly from the cortex via the cortical-striato-pallido-thalamo pathway or directly via a "hyperdirect" monosynaptic pathway from the cortex to the STN (for review, see Nambu et al. 2002; Deniau et al. 2007). STN neurons are tonically active under basal conditions firing at 10–30 Hz (Bevan and Wilson 1999). Tonic excitatory input from the STN to the SNr contributes to the sustained output of the SNr at rest (DeLong et al. 1985; DeLong 1990; Smith et al. 1998). However, as the GP is also tonically active and projects to the STN (Groenewegen et al. 1993; Bell et al. 1995), the relative balance between tonically active neurons in these two nuclei will determine the level of excitatory input to the SNr. Burst activity of striatal neurons projecting through the "indirect" pathway will inhibit GP activity, thus disinhibiting the STN and enhancing the excitatory input to the SNr.

STN fibers arborize extensively throughout the SNr and form asymmetrical (presumed excitatory) synapses onto the proximal and distal dendrites of SNr neurons (Rinvik and Ottersen 1993; Bevan et al. 1994). Electrical stimulation to drive activity in STN afferents *in vitro* generates monosynaptic excitatory postsynaptic potentials (EPSPs) in SNr neurons and enhances SNr discharge rate (Nakanishi et al. 1987a). Similarly, chemical stimulation of the STN, through injection of bicuculline to block GP inhibition, enhances the activity in the STN and subsequently increases the activity of SNr neurons (Robledo and Féger 1990). *In vivo*, a reduction in SNr activity is seen following the inhibition of STN activity (Féger and Robledo 1991). Pathophysiologically, excessive burst firing of SNr neurons is associated with symptoms of parkinsonism in both patients and animal models of Parkinson's disease (PD) (DeLong 1990; Levy et al. 2002). High-frequency stimulation (HFS) of the STN is one of the techniques used to treat PD symptoms and while the mechanisms underlying the effects of HFS are still controversial, some evidence points to a reduction of excitatory input to the SNr, thus relieving the targets of SNr neurons from inhibition (Burbaud et al. 1994; Benazzouz et al. 2000; Beurrier et al. 2001; Magariños-Ascone et al. 2002; Maurice et al. 2003; Gradinaru et al. 2009).

8.2.1.4   Excitation through Glutamate Receptors

Glutamatergic input to the SNr mediates its effect principally via three types of glutamate receptors: N-methyl-D-aspartate (NMDA), α-amino-3-hydroxy-5-methyl-4-isoxaline propionic acid/kainite (AMPA), and metabotropic receptors (Nakanishi 1992; Hollmann and Heinemann 1994; Schmitt et al. 1999). Both NMDA and AMPA receptors are ionotropic glutamate receptors and are associated with fast synaptic transmission (Chapter 2). Metabotropic glutamate receptors on the other hand are linked to G-proteins and have a role in modulating synaptic transmission and synaptic plasticity (Chapter 1).

AMPA and NMDA receptors are widely expressed throughout the SNr (Yung 1998; Chatha et al. 2000). Immunolabelling for these receptors is associated with the postsynaptic membrane of asymmetrical synapses in addition to intracellular sites (Chatha et al. 2000). Interestingly, within the SNr, there is evidence for different compositions of glutamate receptor subunit expression at different synapses, suggesting specificity or compartmentalization of receptor expression within the SNr, though the functional significance of this needs to be further investigated (Schmitt et al. 1999; Tse and Yung 2000).

Microionotophoretic application of glutamate receptor agonists kainic acid, AMPA, or NMDA each enhanced the firing rate of SNr neurons *in vivo* in anesthetized rats (Schmitt et al. 1999). Furthermore, microinjection of the ionotropic glutamate receptor antagonists kynurenic acid, AP-5, and, to a lesser extent, CNQX reduced the spontaneous firing rate of SNr neurons, suggesting tonic glutamatergic excitation of the SNr (Schmitt et al. 1999). However, in the freely moving rat, glutamate has much less of an effect on the spontaneous firing rate of SNr neurons compared to that seen in the anesthetized rat, with only a small proportion of neurons showing an increase in firing rate and others showing either no change or variable responses (Windels and Kiyatkin 2004). The authors proposed that such differences may be a result of enhanced GABAergic inhibition of the SNr during anesthesia

(Windels and Kiyatkin 2004). However, behavioral changes following application of kainate, NMDA, or AMPA were observed in earlier studies in rodents, with such changes being inhibited through infusion of glutamate receptor antagonists (Porceddu et al. 1979; Arnt 1981; Albin et al. 1992). Thus, it seems that the glutamatergic effect on SNr neurons may be complex and dependent on the experimental conditions.

In contrast to the role of NMDA and AMPA receptors in fast synaptic transmission, metabotropic glutamate receptors have more of a modulatory influence. Metabotropic glutamate receptors are divided into three groups: group I mGluR (mGluR1 and mGluR5) are coupled to G_q and phosphoinositide hydrolysis, while groups II (mGluR2 and mGluR3) and III (mGluR4, mGluR6, mGluR7, and mGluR8) are coupled to G_i/G_o (for review, see Conn and Pin 1997).

Immunocytochemistry shows labeling of group I mGluRs on postsynaptic dendrites opposite symmetrical and asymmetrical synapses within the SNr while only sparse expression is found presynaptically on the axon terminals of symmetrical synapses (Testa et al. 1994; Marino et al. 2001). The functional roles of these receptors have been investigated pharmacologically; postsynaptic group I mGluR activation triggers membrane depolarization and enhanced firing of SNr neurons, while presynaptic group I mGluR activation generates a reduction in inhibitory transmission within the SNr (Marino et al. 2001). This dual effect of group I mGluRs within the SNr may thus generate both a direct excitation of SNr neurons in addition to disinhibition through a reduction in afferent GABAergic transmission (Marino et al. 2001). Group II mGluRs are expressed on presynaptic STN-SNr terminals (Bradley et al. 2000). Activation of these receptors inhibits excitatory transmission, a mechanism that the authors suggest may provide negative feedback to prevent over-excitation of SNr neurons following enhanced STN activity (Bradley et al. 2000). Similarly, immunocytochemistry has shown group III mGluR expression within the SNr (Bradley et al. 1999; Kosinski et al. 1999). Activation of group III mGluR via the selective agonist L-AP4 inhibited both excitatory and inhibition synaptic transmission within the SNr via a presumed presynaptic mechanism (Wittmann et al. 2001). Thus, it seems that mGluRs have complex and varied roles within the SNr to modulate neurotransmission within this nucleus.

8.2.1.5 Dopaminergic Input

An additional input to SNr neurons, often not included in basic wiring diagrams of the BG circuitry, is the dopaminergic input from the SNc. Extensive dendritic projections from SNc dopamine neurons extend dorsoventrally into the SNr where dopamine is released somatodendritically (Björklund and Lindvall 1975; Geffen et al. 1976; Cheramy et al. 1981). Immunocytochemical studies show dopamine receptor expression within the SNr (Richfield et al. 1987; Beckstead 1988), and while electrophysiological and pharmacological studies have been somewhat conflicting about the effect of dopamine within the SNr it seems that it may have both presynaptic and postsynaptic influences.

While some studies have found that iontophoretically applied dopamine promotes an increase in SNr GABA cell firing rate and attenuates the inhibitory response to applied GABA *in vitro* and *in vivo* (Ruffieux and Schultz 1980; Waszcak and Walters 1983; Miyazaki and Lacey 1998), others show that dopamine enhances GABA release within the SNr to inhibit SNr activity (Floran et al. 1990). An explanation for such conflicting results may be found by further consideration of dopamine receptor expression within the SNr.

Dopamine receptors are metabotropic G-protein-coupled receptors of which at least five subtypes have been identified: D1, D2, D3, D4, and D5 receptors. These can be further subdivided into two families: the D1-like family which includes D1 and D5 receptors and are coupled to $G_{\alpha s}$ and the D2-like family which includes D2, D3, and D4 receptors and are coupled to $G_{\alpha i}$ (Neves et al. 2002). Anatomical and biochemical studies show primarily D1-like receptor expression along the axon and axon terminals of putative striatonigral afferents in rodents (Yung et al. 1995) and primates (Caillé et al. 1996; Kliem et al. 2007; Kliem et al. 2009). Microinjection of the D1-like receptor agonist SKF-82958 into the SNr decreased the discharge rate of SNr neurons and increased the level of GABA neurotransmitter in the SNr, indicating that dopamine is acting presynaptically on striatonigral terminals to enhance GABAergic transmission (Kliem et al. 2007). Supporting this, *in vivo* studies using microdialysis to measure extracellular GABA concentrations have shown that local infusion of D1-like receptor agonists enhances extracellular GABA concentrations within the SNr (You et al. 1994; Timmerman and Westerink 1995).

Dopamine receptors are also expressed on the GABAergic pallidonigral terminals and the glutamatergic terminals from the STN (Flores et al. 1999; Ibañez-Sandoval et al. 2006; de Jesús Aceves et al. 2011). In contrast to the striatonigral synapses, D2-like receptors inhibit GABAergic neurotransmission at pallidonigral synapses and thus enhance SNr activity (de Jesús Aceves et al. 2011). Furthermore, application of a D2-like receptor antagonist reveals a tonic inhibitory action of dopamine on pallidonigral transmission (de Jesús Aceves et al. 2011). It is interesting to note the contrasting influence dopamine has on striatonigral verses pallidonigral GABAergic transmission, a factor that may be critical in determining the weight of influence each input has upon SNr neurons following local release of dopamine from SNc neurons. In contrast, subthalamic neurons express both D1-like and D2-like dopamine receptors (Flores et al. 1999; Ibañez-Sandoval et al. 2006). In brain slices, EPSCs in SNr cells evoked by stimulation of the STN are enhanced by the D1-like receptor agonist SKF-38393 and reduced by the D2-like receptor agonist quinpirole (Ibañez-Sandoval et al. 2006). Overall, it seems that dopamine has complex modulatory effects on both excitatory and inhibitory presynaptic neurotransmitter release within the SNr.

D1-like receptors are also expressed by SNr GABAergic projection neurons (Zhou et al. 2009). D1-like receptor agonists cause excitation of SNr GABA neurons via activation of a TRPC3 channel (Zhou et al. 2009). This direct effect of dendritically released dopamine on SNr activity was proposed to act as an "ultra-short" pathway for dopamine to modulate the BG output (Zhou et al. 2009). The consequence of this *in vivo* remains to be fully explored.

Thus, it seems that local somatodendritic dopamine release in the SNr has significant neuromodulatory effects via both presynaptic and postsynaptic actions, which directly influence BG output. This may be significant when considering neurological deficits resulting from the degeneration of dopamine neurons seen in PD, as the direct effects of dopamine on SNr neuronal activity may contribute to the neuropathology and the neurological symptoms.

8.2.2 SUBSTANTIA NIGRA PARS RETICULATA: OUTPUTS

SNr neurons project to various subcortical nuclei including the thalamus, SC, and BS tegmentum (Parent 1990; Deniau and Chevalier 1992; Middleton and Strick 2002; Cebrián et al. 2005) where under basal conditions they have a tonic inhibitory input on these nuclei. Relief of such SNr inhibition onto target output neurons is thought to be important in allowing voluntary movement programs to be relayed to higher motor areas (for review, see Hikosaka 2007).

Studies in the 1970s using the retrograde transport of horseradish peroxidase from the SC and thalamus in the rat described the organization of nigro-thalamic and nigro-collicular cells into two distinct subregions, with nigro-collicular cells residing in more ventral SNr regions than nigro-thalamic cells (Faull and Mehler 1978; Beckstead et al. 1981; Beckstead and Frankfurter 1982; Francois et al. 1984). However, single-cell tracing studies instead suggest that cells from the same area of SNr, or indeed an individual cell, can send axonal projections to several target output nuclei by means of their axonal collaterals (Deniau et al. 1978; Beckstead and Frankfurter 1982; Parent et al. 1983; Cebrián et al. 2005). A recent study tracing SNr axonal projections, using injection of biotinylated dextran amine, described four types of SNr neurons based on their axonal projection targets (Cebrián et al. 2005). Type I SNr neurons project solely to the thalamus; type II send axonal projections to the thalamus, SC, and pedunculopontine tegmental nucleus; type III project to the periaqueductal gray matter and thalamus; type IV project to the deep mesencephalic nucleus and to SC (Cebrián et al. 2005). Such a diversity and spread of projection targets suggest that target location does not predict topographical organization within the SNr. Interestingly, though, the degree of axonal arborization within different target areas varied for each type of neuron; as the authors note, this may determine the relative strength of these projections (Cebrián et al. 2005).

8.3 SUBSTANTIA NIGRA PARS COMPACTA CONNECTIVITY

SNc dopamine neurons are known to play an important neuromodulatory role throughout the BG nuclei, the significance of which can perhaps best be appreciated when considering the neuropathology of PD where loss of SNc dopamine neurons is the hallmark (for review, see Dawson and Dawson 2003). *In vivo*, SNc dopamine neurons typically fire at 1–5 Hz and may display a range of firing patterns from irregular or regular tonic firing to phasic burst firing (Grace and Bunney 1983a, 1984a,b; Freeman et al. 1985; Clark and Chiodo 1988; Tepper and Lee 2007) (Figures 8.2 and 8.3). The transition from tonic to phasic burst firing in response to a conditioned stimulus or reward is believed to be important for modulating axonal dopamine release (Ljungberg et al. 1992; Nissbrandt et al. 1994; Schultz 2007). The trigger for this transition is still under debate, although synaptic input to SNc cells is likely to make a significant contribution.

8.3.1 SUBSTANTIA NIGRA PARS COMPACTA: INPUTS

SNc dopamine cells receive a range of inputs, including glutamatergic, GABAergic, cholinergic, serotonergic, and noradrenergic (Grillner and Mercuri 2002). This chapter primarily focuses on GABAergic and glutamatergic inputs in addition to local interactions between neighboring dopamine neurons.

8.3.1.1 GABAergic Input

A significant majority of inputs to SNc dopamine neurons are GABAergic, with the primary afferents originating from the striatum, GP, and from local axonal collaterals of the SNr (Grofová 1975; Hattori et al. 1975; Fonnum et al. 1978; Bolam and Smith 1990; Smith and Bolam 1991; Tepper et al. 1995; Mailly et al. 2001). Both $GABA_A$ and $GABA_B$ receptors are expressed in the SNc (Bowery et al. 1987; Nicholson et al. 1992; Charara et al. 2000), and GABA receptor agonists inhibit SNc firing (Engberg et al. 1993; Tepper et al. 1995), though the precise role of GABAergic input to SNc is complex and is still being explored (for review, see Tepper and Lee 2007).

In vivo stimulation of the striatum, GP, or SNr inhibits SNc dopamine neurons and enhances the occurrence of burst firing; this effect can be blocked by local application of $GABA_A$ receptor antagonists (Grace and Bunney 1979; Erhardt and Engberg 2002). In contrast, *in vivo* application of the $GABA_B$ receptor antagonist 2-OH-saclofen promotes a reduction in firing rate and a shift in firing mode from burst to pacemaker like (Grace and Bunney 1979; Tepper et al. 1995) (for review, see Tepper and Lee 2007). Until recently, only a short-latency $GABA_A$ receptor–mediated inhibitory response to afferent stimulation was identified in SNc neurons *in vivo* (Tepper et al. 1995; Paladini and Tepper 1999), suggesting that $GABA_B$ receptors must modulate synaptic transmission mainly presynaptically (Tepper and Lee 2007). However, in a recent *in vivo* study in the mouse, both $GABA_A$- and $GABA_B$-mediated inhibitions were recorded in SNc neurons; striatal stimulation elicited an early inhibitory response in SNc neurons that was sensitive to picrotoxin and a late inhibitory response which was not affected by picrotoxin but was blocked by CGP-55845A, a $GABA_B$ receptor antagonist (Brazhnik et al. 2008). This supports *in vitro* results where both $GABA_A$- and $GABA_B$-mediated IPSP has been recorded in SNc dopamine neurons following striatal stimulation (Häusser and Yung 1994).

In addition to direct, monosynaptic GABAergic inputs, a strong disynaptic pathway (from the GP through the SNr to the SNc) significantly influences SNc activity (Celada et al. 1999). Single-unit recordings in anesthetized rats showed that, following pharmacological inhibition of the GP, SNc dopamine neurons responded with a decrease in firing rate and a shift to a regular pacemaker-like mode. This effect is thought to be due to GP disinhibition of the SNr and consequent increased inhibition of the SNc (Celada et al. 1999). Interestingly, SNr GABA neurons are more sensitive to GABA receptor–mediated inhibition than SNc dopamine neurons, due to the expression of different $GABA_A$ receptor subunits (Nicholson et al. 1992; Waldvogel et al. 2008) and due to SNr GABA neurons having a more hyperpolarized $GABA_A$ reversal potential (Gulácsi et al. 2003). Therefore, under certain conditions, the indirect disynaptic disinhibitory effect of GP stimulation will predominate over the direct inhibitory GP-SNc effect (Iribe et al. 1999; Tepper and Lee 2007).

8.3.1.2 Glutamatergic Input

Glutamatergic input to SNc dopamine neurons appears to play a pivotal role in regulating SNc activity. *In vitro*, SNc cells mainly fire in a pacemaker-like mode rather than bursting mode, suggesting that afferent input is important in generating a burst-like firing mode (Grace and Onn 1989). Evidence suggests direct excitatory input to SNc dopamine neurons may in part contribute to burst-like firing mode in these neurons (Grace and Bunney 1984b; Overton and Clark 1992; Smith and Grace 1992). Anatomical studies show glutamatergic afferents originating from the STN, frontal cortex, and pedunculopontine nucleus (PPN) synapse on SNc dopamine neurons (Kanazawa et al. 1976; Van Der Kooy and Hattori 1980; Jackson and Crossman 1983; Kita and Kitai 1987; Smith et al. 1990; Naito and Kita 1994; Parent et al. 1999). Immunohistochemistry and *in situ* hybridization studies show that dopamine neurons express ionotropic and metabotropic glutamate receptors (see Chapters 1 and 2).

The STN in particular has a powerful input to SNc, although as with GP input to SNc, conflicting results have been presented regarding the functional consequences of STN stimulation. Stimulation of the STN or removal of inhibition to the STN (using bicuculline) promotes both excitation and inhibition of SNc neurons (Hammond et al. 1978; Robledo and Féger 1990; Smith and Grace 1992). Such a bidirectional effect of STN stimulation is again thought to be due to both a direct and an indirect polysynaptic pathway from the STN through the SNr to the SNc (Iribe et al. 1999).

Another source of glutamatergic input to the SNc comes from the PPN (Parent et al. 1999). The PPN has strong reciprocal connections with BG nuclei (Garcia-Rill 1991; Inglis and Winn 1995) and consists of both cholinergic and glutamatergic neurons, both of which send inputs to the SNc (Inglis and Winn 1995). Electrical stimulation of the PPN in rats generates a glutamate-mediated excitation of SNc dopamine neurons (Scarnati et al. 1984; Lokwan et al. 1999) which may be blocked following the application of excitatory amino acid receptor blockers (Futami et al. 1995). Behaviorally, it has been proposed that the PPN sends sensory signals to the SNc to evoke phasic activity in these neurons during reward-mediated associative learning (Schultz 1998; Pan and Hyland 2005).

8.3.1.3 Dopaminergic Input

Local dopamine release within the SNc can modulate SNc activity, providing a mechanism for local interactions between SNc dopamine neurons. Morphological studies using Golgi staining in addition to intracellular labeling show that dendritic projections of SNc dopamine neurons traverse in mediolateral, dorsoventral, and rostrocaudal directions within the SNc (Juraska et al. 1977; Preston et al. 1981; Kita et al. 1986; Tepper et al. 1987). This wide dendritic spread, combined with the capacity of SNc neurons to release dopamine somatodendritically, provides a mechanism for local dopaminergic modulation (Rice et al. 1997). Furthermore, it seems the neuronal mechanisms associated with somatodendritic dopamine release differ from those of axonal dopamine release (Cragg and Greenfield 1997; Cobb and Abercrombie 2003), suggesting dendritic dopamine release may have a specific and unique role compared to that of terminal dopamine release (Geffen et al. 1976; Robertson and Robertson 1989; Rice et al. 1997; Bergquist et al. 2003).

SNc dopamine neurons express D2-like receptors, largely extrasynaptically (Bouthenet et al. 1987; Sesack et al. 1994; Yung et al. 1995). Somatodendritic dopamine release acts both on D2 autoreceptors and on receptors expressed on neighboring cells to inhibit cell firing (Chiodo 1992; Cragg et al. 2001; Falkenburger et al. 2001). The limited number of synaptic contacts between dopamine neurons (Bayer and Pickel 1990), in addition to the high number of extrasynaptic release and uptake sites (Sesack et al. 1994; Jaffe et al. 1998; Nirenberg et al. 1998; Cragg and Rice 2004), suggests that the majority of chemical communication between SNc neurons occurs via volume transmission (Cragg et al. 2001); however, there is some evidence for chemical synaptic connectivity between dopamine neurons (Vandecasteele et al. 2008).

In addition, the presence of electrical synapses between dopamine neurons has been proposed (Grace and Bunney 1983b; Leung et al. 2002). Lucifer yellow injection into a single SN dopamine neuron revealed labeling of between two and five neighboring dopamine neurons, indicating dye transfer across gap junctions between dopamine cells (Grace and Bunney 1983b). Furthermore, connexin36 (a gap junction protein) expression has been identified in the SNc (Leung et al. 2002); however, it should be noted that another study found no evidence for dye coupling or connexin expression (Lin et al. 2003). Simultaneous recordings from pairs of dopamine neurons support the concept of electrical connectivity (Vandecasteele et al. 2005).

If electrical synapses are present in SNc dopamine neurons, what is their role? There is some evidence for synchronized firing between SNc dopamine cells (Grace and Bunney 1983b; Berretta et al. 2010). Simultaneously recording from multiple SNc neurons in slices using multielectrode arrays found that 12% of pairs of SNc cells had synchronized firing (Berretta et al. 2010). This low incidence of paired cells would make them difficult to detect using paired recordings (possibly explaining why other studies have failed to find them). Furthermore, the application of dopamine to the perfusion solution was found to enhance synchronicity between dopamine neurons (Berretta et al. 2010). Electrical synapses could play a significant role in coordinating dopaminergic activity, and Berretta et al. suggest that synchronized firing may operate to enhance the overall dopamine signal within the target neurons following a rewarding task, a theory which is supported by an *in vivo* study in monkeys, which found that exposure of a rewarding stimulus increased dopamine neuronal synchronicity (Joshua et al. 2009).

8.3.2 Substantia Nigra Pars Compacta: Outputs

SNc dopamine neurons send extensive axonal projections to the striatum, as well as extrastriatal nuclei including the GP and STN, where terminal dopamine release plays an important neuromodulatory role in information processing. Furthermore, as already discussed, somatodendritic dopamine release provides a mechanism for local dopaminergic neuromodulation within the SN.

Striatal dopamine release is believed to modulate cortical and thalamic glutamatergic synaptic integration in MSNs via effects on D1-like and D2-like receptors (Chapters 3 through 5). Activation of D1-like receptors expressed on MSNs of the "direct" pathway enhances dendritic glutamatergic signaling, while D2-like receptor activation expressed on MSNs of the "indirect" pathway inhibits glutamate signaling (Surmeier et al. 2007). A change in dopaminergic activity from tonic to burst firing enhances dopamine release within the striatum (Gonon and Buda 1985; Ljungberg et al. 1992;

Manley et al. 1992; Nissbrandt et al. 1994; Schultz 2007), which is believed to be important for signaling the presence of salient stimuli (for review, see Schultz 2010).

The mosaic organization of the striatum has been described as comprising patch and matrix compartments distinguished from one another by their expression of neurochemical markers and by the topography of inputs and outputs (Graybiel and Ragsdale 1978; Herkenham and Pert 1981; Gerfen 1984, 1985). Early studies distinguished dorsal and ventrally located SN dopamine neurons based on their prominent projections to the matrix and patch regions of the striatum respectively (Gerfen et al. 1987; Jimenez-Castellanos and Graybiel 1987; Langer and Graybiel 1989). More recently, single neuron labeling of SNc dopamine neurons has suggested that axonal arbors innervating the striatum are much wider and denser than had been previously reported (Matsuda et al. 2009). Furthermore, while single dorsal and ventral tier dopamine neurons were found to show a preference for matrix and patch compartments respectively, all SNc neurons traced sent axonal projection to both compartments, suggesting that axonal dopamine release within the striatum has much broader effects than previously thought (Matsuda et al. 2009).

A further distinction between SNc dopamine cells has been proposed based on their projection targets throughout the rest of the BG. Tracing studies show that SNc dopamine neurons also send axonal projections to GP and STN, either directly (Smith et al. 1989) or via the striatum (Lindvall and Björklund 1979). Injecting anterograde tracers into single SNc neurons in the rat provided evidence for two subpopulations of dopamine neurons based on axonal projections (Gauthier et al. 1999). Over half of SNc cells traced had axons branching and terminating extensively within the dorsal striatum, but also sent an additional thin, short axon collateral to GP. Other neurons were found to branch mainly extrastriatally within GP and/or entopeduncular nucleus or STN and had only minimal projections to the striatum (Gauthier et al. 1999). Additionally, SNc neurons that send projections out of the BG to the thalamus and the peduncular nucleus have also been described as distinct from the SNc neurons projecting within the BG (Cebrián and Prensa 2010). On the basis of axonal projections of ventral tier SNc dopamine neurons, three populations of cells were described: type I project to BG nuclei; type II project both to BG nuclei and to the thalamus; type III neurons project solely to the thalamus (Cebrián and Prensa 2010). While evidence is required for the functional significance of these projections, based on this anatomical data it appears that SNc dopamine neurons may promote a neuromodulatory influence beyond the BG, which has important implications for the wider impact of the loss of dopamine neurons in PD.

8.4 PROPERTIES AND HETEROGENEITY OF SUBSTANTIA NIGRA NEURONS

The SN has been considered a relatively homogenous structure comprised of two neuronal populations: the SNr GABAergic cells and the SNc dopaminergic cells. However, morphological, electrophysiological, and biochemical evidence has provided support for some important heterogeneity within these two populations of cells. After considering the basic properties of SNc and SNr neurons, this heterogeneity will be considered.

8.4.1 GABAergic Neurons in Substantia Nigra Pars Reticulata

8.4.1.1 Firing Properties

SNr GABAergic neurons tonically fire action potentials (APs) at high frequencies (5–50 Hz) both *in vivo* (Sanghera et al. 1984; Chen and Ramirez 1988) and *in vitro* (Nakanishi et al. 1987; Lacey et al. 1989; Richards et al. 1997; Lee and Tepper 2007b) (Figures 8.2 and 8.3), with this sustained firing being critical to provide tonic inhibition of target nuclei (Chevalier and Deniau 1990). While SNr cell firing can be modulated by synaptic transmission and/or neuromodulators (see Section 8.2), high-frequency firing of SNr neurons is maintained *in vitro* following the addition of synaptic blockers, suggesting that these neurons are autoactive and thus able to maintain tonic firing in the absence of synaptic input (Atherton and Bevan 2005). In some autoactive cells, subthreshold oscillatory inward currents, carried by L-type Ca^{2+} channels, play an important role in depolarizing the membrane to AP threshold (Pennartz et al. 2002; Durante et al. 2004; Puopolo et al. 2007). However, when spiking is blocked in SNr neurons *in vitro* through the application of TTX or through membrane hyperpolarization, the membrane potential remains relatively stable, with no subthreshold oscillations (Atherton and Bevan 2005). This argues against membrane oscillatory currents as a causative factor in providing the depolarizing drive to AP threshold. As yet, the conductance required to drive these neurons to fire APs is not known; however, the role of a voltage-independent, TTX-insensitive background cation current in conjunction with a subthreshold, slowly inactivating, voltage-dependent, TTX-sensitive Na^+ current has been proposed (Atherton and Bevan 2005).

The AP waveform of SNr GABA neurons is ideally suited to the generation of high-frequency firing, having a depolarized threshold (−45 to −50 mV), short duration (AP half width 0.2–0.4 ms), rapid repolarization [~1.2 ms from AP peak to afterhyperpolarization potential (AHP) peak], and a small amplitude AHP (~22 mV) (Richards et al. 1997; Atherton and Bevan 2005; Lee and Tepper 2007a,b). In addition, SNr GABA neurons do not show much spike frequency adaptation following depolarizing current injection, suggesting that these cells can maintain high-frequency firing without any attenuation (Richards et al. 1997).

8.4.1.2 Morphology

A range of SNr GABAergic neurons has been described based on differences in cell soma shape and size as well as on dendritic spread and branching patterns (Gulley and Wood 1971; Juraska et al. 1977; Grofova et al. 1982; Mailly et al. 2001; Cebrián et al. 2007). In a morphological study, single SNr neurons located in different axes of the SNr were reconstructed following the injection of biotinylated dextran amine, and their somatodendritic domain was analyzed (Cebrián et al. 2007). Five cell types were described, including the fusiform, polygonal, triangular, ovoid, and round types, though no correlation was found between cell soma shape and their distribution within the SNr (Cebrián et al. 2007). Interestingly, and in support of previous observations (Gulley and Wood 1971; Juraska et al. 1977; Grofova et al. 1982), a correlation between the cell soma size and the location within the dorsoventral axis of the SNr was found, with more ventrally located neurons having smaller cell somas than more dorsally located neurons (Cebrián et al. 2007). Furthermore, looking at dendritic properties, dorsally located SNr neurons were found to have a greater number of primary dendrites and dendritic projections, both in the SNr and in the SNc, than more ventrally located neurons (Cebrián et al. 2007). As discussed in Section 8.2.1, input to the SNr is topographically organized, with dorsal and ventral SNr being targeted by distinct striatal, pallidal, and subthalamic regions (Gerfen 1985; Deniau et al. 1996; Wu et al. 2000). Therefore, it has been proposed that the differences between cell soma size and dendritic arborizations of dorsally and ventrally located SNr neurons may have a functional role in determining the capacity of the neurons to receive and integrate information from the other BG nuclei (Cebrián et al. 2007). However, the functional significance of the range of cell soma shapes throughout the SNr is as yet unclear; no differences in the electrophysiological properties of morphologically distinct SNr neurons have been found (Lee and Tepper 2007b).

SNr GABAergic neurons send a primary axon from the soma or a primary dendrite to target neurons in the thalamus, SC, and BS tegmentum (see Section 8.2). In addition, SNr neurons send between one and six axonal collaterals back to the SN, making boutons either en passant or in clusters (Deniau et al. 1982; Grofova et al. 1982; Mailly et al. 2003). Juxtacellular labeling and 3D reconstruction analysis of the axonal collateral network from GABAergic neurons located within the lateral sensorimotor region of the SNr identified much axonal heterogeneity between cells, including differences in the projection, direction, and distance of axonal collaterals spread within the SN (Mailly et al. 2003). Such structural variability in the spatial relationship of the local axonal collateral network of SNr neurons may have functional relevance for SN processing. It has been proposed that recurrent axon collateral networks may contribute to improving the information processing power of SNr through lateral inhibitory mechanisms, with the extent of axon projections within the SN determining the inhibitory connectivity between SN neurons (Mailly et al. 2003; see Section 8.2.1).

8.4.1.3 Neurochemical Expression

In some brain areas such as the hippocampus, neocortex, and neostriatum, the differential expression of Ca^{2+}-binding proteins has become a valuable tool used to discriminate between heterogeneous subpopulations of neurons, complimenting differences in morphological and electrophysiological characteristics (Kawaguchi 1993; Hof et al. 1999; Schwaller et al. 2002; Somogyi and Klausberger 2005). In the SNr, differential expression of neurochemicals between cells has also been described: for example, high expression of the calcium-binding protein calretinin (CR) is found in rostromedial SNr, while more caudomedial and rostrolateral neurons express the calcium-binding protein parvalbumin (PV) (González-Hernández and Rodríguez 2000). However, electrophysiological recordings from rat SNr GABAergic neurons have found very similar morphological and electrophysiological properties between CR- and PV-positive neurons identified by post-hoc immunolabeling (Lee and Tepper 2007). Furthermore, no differences in the projection targets of PV and CR neurons have been identified by retrograde labeling from the thalamus and SC (Lee and Tepper 2007), leaving the functional significance still open to question. It should be noted that most studies comparing electrophysiological properties of subpopulations of SNr neurons have been performed *in vitro*, where many afferent inputs are lost; thus, it is possible that neurochemical expression is more significant *in vivo*.

8.4.2 DOPAMINERGIC NEURONS IN SUBSTANTIA NIGRA

8.4.2.1 Electrophysiological Properties

In vivo, SNc dopamine neurons display predominantly two modes of firing, irregular, low-frequency activity (1–5 Hz) and burst activity, while *in vitro* dopamine cells fire tonically in a regular pacemaker-like mode (Figures 8.1 and 8.2; see Section 8.3). Detailed electrophysiological studies have identified and characterized some of the conductances involved in the generation of spontaneous activity in SNc dopamine cells. Subthreshold intrinsic membrane potential oscillations which drive the SNc dopamine neurons to AP threshold are in part due to a low-threshold, non-inactivating Ca^{2+} conductance and a Ca^{2+}-activated K^+ conductance (Kang and Kitai 1993; Nedergaard et al. 1993; Wilson and Callaway 2000). Dopamine neuron APs have specific features, including a relatively depolarized AP threshold ($\sim$−40 mV), a long duration (AP half width 1–2 ms), and a large-amplitude and long-lasting apamin-sensitive AHP, which influences the regularity of dopamine cell firing and extends the AP refractory period (Shepard and Bunney 1991; Ping and Shepard 1996). When SNc neurons fire bursts of APs, these occur on top of depolarizing oscillatory current; however, the conductance responsible for burst firing has not yet been identified. Whole-cell current clamp recordings have shown that SNc dopamine neurons show strong frequency adaptation following depolarizing current injections, which is accompanied by changes in the AP duration and amplitude, with dopamine neurons entering inactivation block at firing frequencies of above $\sim$10 Hz (Richards et al. 1997).

8.4.2.2 Neurochemical Expression

SNc dopamine neurons also show heterogeneity in electrophysiological, morphological, and neurochemical properties. Early on, the concept of dorsal and ventral compartments or tiers within the SNc was proposed based on differential afferent projections and expression of neurochemicals in each region (Fallon and Moore 1978; Veening et al. 1980; Nemoto et al. 1999). For example, dopamine neurons located in the rostrodorsal SNc mainly express the Ca^{2+}-binding protein calbindin (CB) and send projections to the striatal matrix, while those located in the ventral SNc, in addition to SNr dopamine neurons, do not express CB and project to the patch region of the striatum (Gerfen et al. 1987; Lynd-Balta and Haber 1994).

Some differences between the electrophysiological properties of the CB positive (CB^+) and negative (CB^-) neurons have been identified. In response to hyperpolarizing current pulses. CB^- neurons have a pronounced hyperpolarization-activated current (Ih) followed by a short rebound delay and a transient increase in firing rate, while CB^+ neurons have a much smaller Ih, a prolonged rebound delay, and no transient increase in spike frequency (Neuhoff et al. 2002).

Also, Ca^{2+}-activated SK3 channels and subunits of the A-type K^+ channels, both of which have important roles in spike firing, have different expression patterns between dorsal and ventral SNc (Liss et al. 2001; Wolfart et al. 2001). These differences may influence their responses to afferent input and to neuromodulators (Neuhoff et al. 2002).

8.4.2.3 Morphology

Differences in dendritic morphology between dorsal and ventral tier dopamine neurons have also been identified, with dorsal tier dendrites projecting mainly in a mediolateral direction through SNc, while ventral tier dendrites project more ventrally to SNr and ventral borders (for review, see Joel and Weiner 1997). In addition to the obvious implications for the dendritic integration of inputs, the consequences of somatodendritic dopamine release from each subset could be markedly different.

A small number of dopamine neurons are also found within the SNr. While no differences in the electrophysiology or morphology of dopamine neurons recorded in the SNc and SNr have been identified (Richards et al. 1997), it is possible that their roles may show target specificity. For example, extensive dendritic projections from SNr-located dopamine cells are sent both throughout the SNr and to the SNc, while dopamine neurons located in the ventral tier of the SNc, but not in the dorsal tier, send dendritic projections into the SNr (Björklund and Lindvall 1975). Further investigation of the inputs and outputs of SNr dopamine neurons is needed in order to correctly classify these neurons.

8.4.3 LOCAL INTERNEURONS IN SUBSTANTIA NIGRA?

The presence of local nigral interneurons both in the SNc and in the SNr has been proposed based on some experimental observations; however, conclusive evidence for either their existence or their functional role is still to be established.

Morphological studies have described small neurons (10–12 μm diameter) with short axons residing in the SN with a greater relative abundance in the SNr (40%) verses SNc (10%) (Gulley and Wood 1971; Schwyn and Fox 1974; Francois et al. 1979; Lacey et al. 1989). However, whether these are true interneurons or whether they send axonal projections outside of the SN is difficult to confirm. A population of cells with distinct electrophysiological characteristics to SN dopamine and GABA cells have been described. For example, so-called "bursty cells" were identified in the SNc of anesthetized rats; these fired a burst of 6–12 spikes before falling silent for 1–10 s and were not immunopositive for tyrosine hydroxylase (Wilson et al. 1977; Yung et al. 1991). In addition, a recent *in vitro* study using multielectrode array recordings identified a small population of cells (~5%) that fired at regular frequencies of 5–10 Hz and were insensitive to dopamine (Berretta et al. 2010). SN projection neurons are typically sensitive to dopamine, suggesting that this small population of neurons may comprise a local, dopamine-insensitive inhibitory network (Berretta et al. 2010). However, other studies have found no evidence for local interneurons; thus, conclusive evidence supporting their existence is still required.

8.5 CONCLUSIONS

This chapter has attempted to provide an overview of the functional organization of the SN. Extending from the basic wiring diagram of the inputs and outputs of the SN within the BG circuitry, we have seen that both the SNc and SNr receive a complex range of projections that are topographically arranged. Furthermore, intranigral communication appears to be an important aspect of SN processing and can influence the subsequent output of the SN. To truly understand such processing, a collective and comprehensive understanding of the topographical, electrophysiological, morphological, and neurochemical heterogeneity of the SN needs to be established.

ACKNOWLEDGMENTS

The author thanks C.D. Brown, J.T. Dudman, and S. Jones for their comments on this chapter, and Wei-Xing Pan and Joshua T. Dudman for providing the data shown in Figure 8.3.

REFERENCES

Albin, R.L. et al., 1992. Excitatory amino acid binding sites in the basal ganglia of the rat: A quantitative autoradiographic study. *Neuroscience*, 46(1), 35–48.

Alexander, G.E. and Crutcher, M.D., 1990. Functional architecture of basal ganglia circuits: Neural substrates of parallel processing. *Trends in Neurosciences*, 13(7), 266–271.

Alexander, G.E., DeLong, M.R., and Strick, P.L., 1986. Parallel organization of functionally segregated circuits linking basal ganglia and cortex. *Annual Review of Neuroscience*, 9, 357–381.

Arnt, J., 1981. Turning behaviour and catalepsy after injection of excitatory amino acids into rat substantia nigra. *Neuroscience Letters*, 23(3), 337–342.

Arnt, J. and Scheel-Kruger, J., 1979. Behavioral differences induced by muscimol selectively injected into pars compacta and pars reticulata of substantia nigra. *Naunyn-Schmiedeberg's Archives of Pharmacology*, 310(1), 43–51.

Atherton, J.F. and Bevan, M.D., 2005. Ionic mechanisms underlying autonomous action potential generation in the somata and dendrites of GABAergic substantia nigra pars reticulata neurons in vitro. *The Journal of Neuroscience: The Official Journal of the Society for Neuroscience*, 25(36), 8272–8281.

Bayer, V.E. and Pickel, V.M., 1990. Ultrastructural localization of tyrosine hydroxylase in the rat ventral tegmental area: Relationship between immunolabeling density and neuronal associations. *The Journal of Neuroscience: The Official Journal of the Society for Neuroscience*, 10(9), 2996–3013.

Beckstead, R.M., 1988. Association of dopamine D1 and D2 receptors with specific cellular elements in the basal ganglia of the cat: The uneven topography of dopamine receptors in the striatum is determined by intrinsic striatal cells, not nigrostriatal axons. *Neuroscience*, 27(3), 851–863.

Beckstead, R.M., Domesick, V.B., and Nauta, W.J., 1979. Efferent connections of the substantia nigra and ventral tegmental area in the rat. *Brain Research*, 175(2), 191–217.

Beckstead, R., Edwards, S., and Frankfurter, A., 1981. A comparison of the intranigral distribution of nigrotectal neurons labeled with horseradish peroxidase in the monkey, cat, and rat. *Journal of Neuroscience*, 1(2), 121–125.

Beckstead, R.M. and Frankfurter, A., 1982. The distribution and some morphological features of substantia nigra neurons that project to the thalamus, superior colliculus and pedunculopontine nucleus in the monkey. *Neuroscience*, 7(10), 2377–2388.

Bell, K., Churchill, L., and Kalivas, P.W., 1995. GABAergic projection from the ventral pallidum and globus pallidus to the subthalamic nucleus. *Synapse*, 20(1), 10–18.

Benazzouz, A. et al., 2000. Effect of high-frequency stimulation of the subthalamic nucleus on the neuronal activities of the substantia nigra pars reticulata and ventrolateral nucleus of the thalamus in the rat. *Neuroscience*, 99(2), 289–295.

Bergquist, F., Shahabi, H.N., and Nissbrandt, H., 2003. Somatodendritic dopamine release in rat substantia nigra influences motor performance on the accelerating rod. *Brain Research*, 973(1), 81–91.

Berretta, N., Bernardi, G., and Mercuri, N.B., 2010. Firing properties and functional connectivity of substantia nigra pars compacta neurones recorded with a multi-electrode array in vitro. *The Journal of Physiology*, 588(Pt 10), 1719–1735.

Beurrier, C. et al., 2001. High-frequency stimulation produces a transient blockade of voltage-gated currents in subthalamic neurons. *Journal of Neurophysiology*, 85(4), 1351–1356.

Bevan, M.D., Bolam, J.P., and Crossman, A.R., 1994. Convergent synaptic input from the neostriatum and the subthalamus onto identified nigrothalamic neurons in the rat. *The European Journal of Neuroscience*, 6(3), 320–334.

Bevan, M.D. and Wilson, C.J., 1999. Mechanisms underlying spontaneous oscillation and rhythmic firing in rat subthalamic neurons. *The Journal of Neuroscience: The Official Journal of the Society for Neuroscience*, 19(17), 7617–7628.

Björklund, A. and Lindvall, O., 1975. Dopamine in dendrites of substantia nigra neurons: Suggestions for a role in dendritic terminals. *Brain Research*, 83(3), 531–537.

Bolam, J.P. and Smith, Y., 1990. The GABA and substance P input to dopaminergic neurones in the substantia nigra of the rat. *Brain Research*, 529(1–2), 57–78.

Boussaoud, D. and Joseph, J.P., 1985. Role of the cat substantia nigra pars reticulata in eye and head movements. II. Effects of local pharmacological injections. *Experimental Brain Research*, 57(2), 297–304.

Bouthenet, M.L. et al., 1987. A detailed mapping of dopamine D-2 receptors in rat central nervous system by autoradiography with [125I]iodosulpride. *Neuroscience*, 20(1), 117–155.

Bowery, N.G., Hudson, A.L., and Price, G.W., 1987. GABAA and GABAB receptor site distribution in the rat central nervous system. *Neuroscience*, 20(2), 365–383.

Boyes, J. and Bolam, J.P., 2003. The subcellular localization of GABAB receptor subunits in the rat substantia nigra. *European Journal of Neuroscience*, 18(12), 3279–3293.

Boyes, J. and Bolam, J.P., 2007. Localization of GABA receptors in the basal ganglia. *Progress in Brain Research*, 160, 229–243.

Bradley, S.R. et al., 1999. Distribution of group III mGluRs in rat basal ganglia with subtype-specific antibodies. *Annals of the New York Academy of Sciences*, 868, 531–534.

Bradley, S.R. et al., 2000. Activation of group II metabotropic glutamate receptors inhibits synaptic excitation of the substantia nigra pars reticulata. *Journal of Neuroscience*, 20(9), 3085–3094.

Brazhnik, E., Shah, F., and Tepper, J.M., 2008. GABAergic afferents activate both GABAA and GABAB receptors in mouse substantia nigra dopaminergic neurons in vivo. *The Journal of Neuroscience: The Official Journal of the Society for Neuroscience*, 28(41), 10386–10398.

Bunney, B.S. and Aghajanian, G.K., 1976. The precise localization of nigral afferents in the rat as determined by a retrograde tracing technique. *Brain Research*, 117(3), 423–435.

Burbaud, P., Gross, C., and Bioulac, B., 1994. Effect of subthalamic high frequency stimulation on substantia nigra pars reticulata and globus pallidus neurons in normal rats. *Journal of Physiology, Paris*, 88(6), 359–361.

Caillé, I., Dumartin, B., and Bloch, B., 1996. Ultrastructural localization of D1 dopamine receptor immunoreactivity in rat striatonigral neurons and its relation with dopaminergic innervation. *Brain Research*, 730(1–2), 17–31.

Cebrián, C., Parent, A., and Prensa, L., 2005. Patterns of axonal branching of neurons of the substantia nigra pars reticulata and pars lateralis in the rat. *The Journal of Comparative Neurology*, 492(3), 349–369.

Cebrián, C., Parent, A., and Prensa, L., 2007. The somatodendritic domain of substantia nigra pars reticulata projection neurons in the rat. *Neuroscience Research*, 57(1), 50–60.

Cebrián, C. and Prensa, L., 2010. Basal ganglia and thalamic input from neurons located within the ventral tier cell cluster region of the substantia nigra pars compacta in the rat. *The Journal of Comparative Neurology*, 518(8), 1283–1300.

Celada, P., Paladini, C.A., and Tepper, J.M., 1999. GABAergic control of rat substantia nigra dopaminergic neurons: Role of globus pallidus and substantia nigra pars reticulata. *Neuroscience*, 89(3), 813–825.

Chan, P.K., Leung, C.K., and Yung, W.H., 1998. Differential expression of pre- and postsynaptic GABA(B) receptors in rat substantia nigra pars reticulata neurones. *European Journal of Pharmacology*, 349(2–3), 187–197.

Chan, P.K.Y. and Yung, W., 1999. Inhibitory postsynaptic currents of rat substantia nigra pars reticulata neurons: Role of GABA receptors and GABA uptake. *Brain Research*, 838(1–2), 18–26.

Charara, A. et al., 2000. Pre- and postsynaptic localization of GABA(B) receptors in the basal ganglia in monkeys. *Neuroscience*, 95(1), 127–140.

Chatha, B.T. et al., 2000. Synaptic localization of ionotropic glutamate receptors in the rat substantia nigra. *Neuroscience*, 101(4), 1037–1051.

Chen, J. and Ramirez, V.D., 1988. In vivo dopaminergic activity from nucleus accumbens, substantia nigra and ventral tegmental area in the freely moving rat: Basal neurochemical output and prolactin effect. *Neuroendocrinology*, 48(4), 329–335.

Cheramy, A., Leviel, V., and Glowinski, J., 1981. Dendritic release of dopamine in the substantia nigra. *Nature*, 289(5798), 537–542.

Chevalier, G. and Deniau, J.M., 1990. Disinhibition as a basic process in the expression of striatal functions. *Trends in Neurosciences*, 13(7), 277–280.

Chevalier, G. et al., 1985. Disinhibition as a basic process in the expression of striatal functions. I. The striato-nigral influence on tecto-spinal/tecto-diencephalic neurons. *Brain Research*, 334(2), 215–226.

Chiodo, L.A., 1992. Dopamine autoreceptor signal transduction in the DA cell body: A "current view." *Neurochemistry International*, 20 Suppl, 81S–84S.

Clark, D. and Chiodo, L.A., 1988. Electrophysiological and pharmacological characterization of identified nigrostriatal and mesoaccumbens dopamine neurons in the rat. *Synapse*, 2(5), 474–485.

Cobb, W.S. and Abercrombie, E.D., 2003. Differential regulation of somatodendritic and nerve terminal dopamine release by serotonergic innervation of substantia nigra. *Journal of Neurochemistry*, 84(3), 576–584.

Collingridge, G.L. and Davies, J., 1981. The influence of striatal stimulation and putative neurotransmitters on identified neurones in the rat substantia nigra. *Brain Research*, 212(2), 345–359.

Conn, P.J. and Pin, J.P., 1997. Pharmacology and functions of metabotropic glutamate receptors. *Annual Review of Pharmacology and Toxicology*, 37, 205–237.

Connelly, W.M. et al., 2010. Differential short-term plasticity at convergent inhibitory synapses to the substantia nigra pars reticulata. *Journal of Neuroscience*, 30(44), 14854–14861.

Cragg, S.J. and Greenfield, S.A., 1997. Differential autoreceptor control of somatodendritic and axon terminal dopamine release in substantia nigra, ventral tegmental area, and striatum. *The Journal of Neuroscience: The Official Journal of the Society for Neuroscience*, 17(15), 5738–5746.

Cragg, S.J. and Rice, M.E., 2004. DAncing past the DAT at a DA synapse. *Trends in Neurosciences*, 27(5), 270–277.

Cragg, S.J. et al., 2001. Dopamine-mediated volume transmission in midbrain is regulated by distinct extracellular geometry and uptake. *Journal of Neurophysiology*, 85(4), 1761–1771.

Dawson, T.M. and Dawson, V.L., 2003. Molecular pathways of neurodegeneration in Parkinson's disease. *Science*, 302(5646), 819–822.

DeLong, M.R., 1971. Activity of pallidal neurons during movement. *Journal of Neurophysiology*, 34(3), 414–427.

DeLong, M.R., 1990. Primate models of movement disorders of basal ganglia origin. *Trends in Neurosciences*, 13(7), 281–285.

DeLong, M.R., Crutcher, M.D., and Georgopoulos, A.P., 1985. Primate globus pallidus and subthalamic nucleus: Functional organization. *Journal of Neurophysiology*, 53(2), 530–543.

Deniau, J.M. and Chevalier, G., 1984. Synaptic organization of the basal ganglia: An electroanatomical approach in the rat. *Ciba Foundation Symposium*, 107, 48–63.

Deniau, J. and Chevalier, G., 1992. The lamellar organization of the rat substantia nigra pars reticulata: Distribution of projection neurons. *Neuroscience*, 46(2), 361–377.

Deniau, J., Feger, J., and Le Guyader, C., 1976. Striatal evoked inhibition of identified nigrothalamic neurons. *Brain Research*, 104(1), 152–156.

Deniau, J.M., Menetrey, A., and Charpier, S., 1996. The lamellar organization of the rat substantia nigra pars reticulata: Segregated patterns of striatal afferents and relationship to the topography of corticostriatal projections. *Neuroscience*, 73(3), 761–781.

Deniau, J.M. et al., 1978. Electrophysiological properties of identified output neurons of the rat substantia nigra (pars compacta and pars reticulata): Evidences for the existence of branched neurons. *Experimental Brain Research*, 32(3), 409–422.

Deniau, J.M. et al., 1982. Neuronal interactions in the substantia nigra pars reticulata through axon collaterals of the projection neurons. An electrophysiological and morphological study. *Experimental Brain Research*, 47(1), 105–113.

Deniau, J.M. et al., 2007. The pars reticulata of the substantia nigra: A window to basal ganglia output. *Progress in Brain Research*, 160, 151–172.

Durante, P. et al., 2004. Low-threshold L-type calcium channels in rat dopamine neurons. *Journal of Neurophysiology*, 91(3), 1450–1454.

Engberg, G., Kling-Petersen, T., and Nissbrandt, H., 1993. GABAB-receptor activation alters the firing pattern of dopamine neurons in the rat substantia nigra. *Synapse*, 15(3), 229–238.

Erhardt, S. and Engberg, G., 2002. Increased phasic activity of dopaminergic neurones in the rat ventral tegmental area following pharmacologically elevated levels of endogenous kynurenic acid. *Acta Physiologica Scandinavica*, 175(1), 45–53.

Falkenburger, B.H., Barstow, K.L., and Mintz, I.M., 2001. Dendrodendritic inhibition through reversal of dopamine transport. *Science*, 293(5539), 2465–2470.

Fallon, J.H. and Moore, R.Y., 1978. Catecholamine innervation of the basal forebrain. IV. Topography of the dopamine projection to the basal forebrain and neostriatum. *The Journal of Comparative Neurology*, 180(3), 545–580.

Faull, R.L. and Mehler, W.R., 1978. The cells of origin of nigrotectal, nigrothalamic and nigrostriatal projections in the rat. *Neuroscience*, 3(11), 989–1002.

Féger, J. and Robledo, P., 1991. The effects of activation or inhibition of the subthalamic nucleus on the metabolic and electrophysiological activities within the pallidal complex and substantia nigra in the rat. *The European Journal of Neuroscience*, 3(10), 947–952.

Floran, B. et al., 1988. Presynaptic modulation of the release of GABA by GABAA receptors in pars compacta and by GABAB receptors in pars reticulata of the rat substantia nigra. *European Journal of Pharmacology*, 150(3), 277–286.

Floran, B. et al., 1990. Activation of D1 dopamine receptors stimulates the release of GABA in the basal ganglia of the rat. *Neuroscience Letters*, 116(1–2), 136–140.

Flores, G. et al., 1999. Expression of dopamine receptors in the subthalamic nucleus of the rat: Characterization using reverse transcriptase-polymerase chain reaction and autoradiography. *Neuroscience*, 91(2), 549–556.

Fonnum, F., Gottesfeld, Z., and Grofova, I., 1978. Distribution of glutamate decarboxylase, choline acetyl-transferase and aromatic amino acid decarboxylase in the basal ganglia of normal and operated rats. Evidence for striatopallidal, striatoentopeduncular and striato-nigral GABAergic fibres. *Brain Research*, 143(1), 125–138.

Francois, C., Percheron, G., and Yelnik, J., 1984. Localization of nigrostriatal, nigrothalamic and nigrotectal neurons in ventricular coordinates in macaques. *Neuroscience*, 13(1), 61–76.

Francois, C. et al., 1979. Demonstration of the existence of small local circuit neurons in the Golgi-stained primate substantia nigra. *Brain Research*, 172(1), 160–164.

Freeman, A.S., Meltzer, L.T., and Bunney, B.S., 1985. Firing properties of substantia nigra dopaminergic neurons in freely moving rats. *Life Sciences*, 36(20), 1983–1994.

Fujiyama, F., Stephenson, F.A., and Bolam, J.P., 2002. Synaptic localization of GABA(A) receptor subunits in the substantia nigra of the rat: Effects of quinolinic acid lesions of the striatum. *The European Journal of Neuroscience*, 15(12), 1961–1975.

Futami, T., Takakusaki, K., and Kitai, S.T., 1995. Glutamatergic and cholinergic inputs from the pedunculopontine tegmental nucleus to dopamine neurons in the substantia nigra pars compacta. *Neuroscience Research*, 21(4), 331–342.

Garcia-Rill, E., 1991. The pedunculopontine nucleus. *Progress in Neurobiology*, 36(5), 363–389.

Gauthier, J. et al., 1999. The axonal arborization of single nigrostriatal neurons in rats. *Brain Research*, 834(1–2), 228–232.

Gerfen, C.R., 1984. The neostriatal mosaic: Compartmentalization of corticostriatal input and striatonigral output systems. *Nature*, 311(5985), 461–464.

Gerfen, C.R., 1985. The neostriatal mosaic. I. Compartmental organization of projections from the striatum to the substantia nigra in the rat. *The Journal of Comparative Neurology*, 236(4), 454–476.

Gerfen, C.R., Baimbridge, K.G., and Thibault, J., 1987. The neostriatal mosaic: III. Biochemical and developmental dissociation of patch-matrix mesostriatal systems. *The Journal of Neuroscience: The Official Journal of the Society for Neuroscience*, 7(12), 3935–3944.

Gerfen, C., Herkenham, M., and Thibault, J., 1987. The neostriatal mosaic: II. Patch- and matrix-directed mesostriatal dopaminergic and non-dopaminergic systems. *Journal Neuroscience*, 7(12), 3915–3934.

Gerfen, C.R. and Surmeier, D.J., 2010. Dichotomous modulation of striatal direct and indirect pathway neurons by dopamine. *Annual Review of Neuroscience*, 34(1), 110301101035033.

Geffen, L.B. et al., 1976. Release of dopamine from dendrites in rat substantia nigra. *Nature*, 260(5548), 258–260.

Giralt, M.T., Bonanno, G., and Raiteri, M., 1990. GABA terminal autoreceptors in the pars compacta and in the pars reticulata of the rat substantia nigra are GABAB. *European Journal of Pharmacology*, 175(2), 137–144.

Gonon, F.G. and Buda, M.J., 1985. Regulation of dopamine release by impulse flow and by autoreceptors as studied by in vivo voltammetry in the rat striatum. *Neuroscience*, 14(3), 765–774.

González-Hernández, T. and Rodríguez, M., 2000. Compartmental organization and chemical profile of dopaminergic and GABAergic neurons in the substantia nigra of the rat. *The Journal of Comparative Neurology*, 421(1), 107–135.

Grace, A.A. and Bunney, B.S., 1979. Paradoxical GABA excitation of nigral dopaminergic cells: Indirect mediation through reticulata inhibitory neurons. *European Journal of Pharmacology*, 59(3–4), 211–218.

Grace, A.A. and Bunney, B.S., 1983a. Intracellular and extracellular electrophysiology of nigral dopaminergic neurons–1. Identification and characterization. *Neuroscience*, 10(2), 301–315.

Grace, A.A. and Bunney, B.S., 1983b. Intracellular and extracellular electrophysiology of nigral dopaminergic neurons–3. Evidence for electrotonic coupling. *Neuroscience*, 10(2), 333–348.

Grace, A. and Bunney, B., 1984a. The control of firing pattern in nigral dopamine neurons: Burst firing. *Journal of Neuroscience*, 4(11), 2877–2890.

Grace, A. and Bunney, B., 1984b. The control of firing pattern in nigral dopamine neurons: Single spike firing. *Journal of Neuroscience*, 4(11), 2866–2876.

Grace, A.A. and Onn, S.P., 1989. Morphology and electrophysiological properties of immunocytochemically identified rat dopamine neurons recorded in vitro. *The Journal of Neuroscience: The Official Journal of the Society for Neuroscience*, 9(10), 3463–3481.

Gradinaru, V. et al., 2009. Optical deconstruction of parkinsonian neural circuitry. *Science*, 324(5925), 354–359.

Graybiel, A.M. and Ragsdale, C.W., 1978. Histochemically distinct compartments in the striatum of human, monkeys, and cat demonstrated by acetylthiocholinesterase staining. *Proceedings of the National Academy of Sciences*, 75(11), 5723–5726.

Grillner, P. and Mercuri, N.B., 2002. Intrinsic membrane properties and synaptic inputs regulating the firing activity of the dopamine neurons. *Behavioural Brain Research*, 130(1–2), 149–169.

Groenewegen, H.J., Berendse, H.W., and Haber, S.N., 1993. Organization of the output of the ventral striatopallidal system in the rat: Ventral pallidal efferents. *Neuroscience*, 57(1), 113–142.

Grofová, I., 1975. The identification of striatal and pallidal neurons projecting to substantia nigra. An experimental study by means of retrograde axonal transport of horseradish peroxidase. *Brain Research*, 91(2), 286–291.

Grofová, I., Deniau, J.M., and Kitai, S.T., 1982. Morphology of the substantia nigra pars reticulata projection neurons intracellularly labeled with HRP. *The Journal of Comparative Neurology*, 208(4), 352–368.

Gulácsi, A. et al., 2003. Cell type-specific differences in chloride-regulatory mechanisms and GABA(A) receptor-mediated inhibition in rat substantia nigra. *The Journal of Neuroscience: The Official Journal of the Society for Neuroscience*, 23(23), 8237–8246.

Gulley, R.L. and Wood, R.L., 1971. The fine structure of the neurons in the rat substantia nigra. *Tissue and Cell*, 3(4), 675–690.

Hammond, C. et al., 1978. Electrophysiological demonstration of an excitatory subthalamonigral pathway in the rat. *Brain Research*, 151(2), 235–244.

Hattori, T., Fibiger, H.C., and McGeer, P.L., 1975. Demonstration of a pallido-nigral projection innervating dopaminergic neurons. *The Journal of Comparative Neurology*, 162(4), 487–504.

Häusser, M.A. and Yung, W.H., 1994. Inhibitory synaptic potentials in guinea-pig substantia nigra dopamine neurones in vitro. *The Journal of Physiology*, 479(Pt 3), 401–422.

Hazrati, L. and Parent, A., 1992. The striatopallidal projection displays a high degree of anatomical specificity in the primate. *Brain Research*, 592(1–2), 213–227.

Herkenham, M. and Pert, C.B., 1981. Mosaic distribution of opiate receptors, parafascicular projections and acetylcholinesterase in rat striatum. *Nature*, 291(5814), 415–418.

Hikosaka, O., 1989. [Eye movement is controlled by basal ganglia-induced GABAergic inhibition]. *Rinshō Shinkeigaku = Clinical Neurology*, 29(12), 1515–1518.

Hikosaka, O., 2007. GABAergic output of the basal ganglia. *Progress in Brain Research*, 160, 209–226.

Hikosaka, O., Sakamoto, M., and Miyashita, N., 1993. Effects of caudate nucleus stimulation on substantia nigra cell activity in monkey. *Experimental Brain Research. Experimentelle Hirnforschung. Expérimentation Cérébrale*, 95(3), 457–472.

Hikosaka, O., Takikawa, Y., and Kawagoe, R., 2000. Role of the basal ganglia in the control of purposive saccadic eye movements. *Physiological Reviews*, 80(3), 953–978.

Hikosaka, O. and Wurtz, R.H., 1985. Modification of saccadic eye movements by GABA-related substances. II. Effects of muscimol in monkey substantia nigra pars reticulata. *Journal of Neurophysiology*, 53(1), 292–308.

Hof, P.R. et al., 1999. Cellular distribution of the calcium-binding proteins parvalbumin, calbindin, and calretinin in the neocortex of mammals: Phylogenetic and developmental patterns. *Journal of Chemical Neuroanatomy*, 16(2), 77–116.

Hollmann, M. and Heinemann, S., 1994. Cloned glutamate receptors. *Annual Review of Neuroscience*, 17, 31–108.

Ibañez-Sandoval, O. et al., 2006. Control of the subthalamic innervation of substantia nigra pars reticulata by D1 and D2 dopamine receptors. *Journal of Neurophysiology*, 95(3), 1800–1811.

Ikeda, H. et al., 2010. Differential role of GABAA and GABAB receptors in two distinct output stations of the rat striatum: Studies on the substantia nigra pars reticulata and the globus pallidus. *Neuroscience*, 167(1), 31–39.

Inglis, W.L. and Winn, P., 1995. The pedunculopontine tegmental nucleus: Where the striatum meets the reticular formation. *Progress in Neurobiology*, 47(1), 1–29.

Iribe, Y. et al., 1999. Subthalamic stimulation-induced synaptic responses in substantia nigra pars compacta dopaminergic neurons in vitro. *Journal of Neurophysiology*, 82(2), 925–933.

Jackson, A. and Crossman, A.R., 1983. Nucleus tegmenti pedunculopontinus: Efferent connections with special reference to the basal ganglia, studied in the rat by anterograde and retrograde transport of horseradish peroxidase. *Neuroscience*, 10(3), 725–765.

Jaffe, E.H. et al., 1998. Extrasynaptic vesicular transmitter release from the somata of substantia nigra neurons in rat midbrain slices. *The Journal of Neuroscience: The Official Journal of the Society for Neuroscience*, 18(10), 3548–3553.

de Jesús Aceves, J. et al., 2011. Dopaminergic presynaptic modulation of nigral afferents: Its role in the generation of recurrent bursting in substantia nigra pars reticulata neurons. *Frontiers in Systems Neuroscience*, 5, 6.

Jimenez-Castellanos, J. and Graybiel, A., 1987. Subdivisions of the dopamine-containing A8-A9-A10 complex identified by their differential mesostriatal innervation of striosomes and extrastriosomal matrix. *Neuroscience*, 23(1), 223–242.

Joel, D. and Weiner, I., 1997. The connections of the primate subthalamic nucleus: Indirect pathways and the open-interconnected scheme of basal ganglia-thalamocortical circuitry. *Brain Research. Brain Research Reviews*, 23(1–2), 62–78.

Joshua, M. et al., 2009. Synchronization of midbrain dopaminergic neurons is enhanced by rewarding events. *Neuron*, 62(5), 695–704.

Juraska, J.M., Wilson, C.J., and Groves, P.M., 1977. The substantia nigra of the rat: A golgi study. *The Journal of Comparative Neurology*, 172(4), 585–600.

Kanazawa, I., Marshall, G.R., and Kelly, J.S., 1976. Afferents to the rat substantia nigra studied with horseradish peroxidase, with special reference to fibres from the subthalamic nucleus. *Brain Research*, 115(3), 485–491.

Kang, Y. and Kitai, S.T., 1993. A whole cell patch-clamp study on the pacemaker potential in dopaminergic neurons of rat substantia nigra compacta. *Neuroscience Research*, 18(3), 209–221.

Karabelas, A.B. and Purpura, D.P., 1980. Evidence for autapses in the substantia nigra. *Brain Research*, 200(2), 467–473.

Kawaguchi, Y., 1993. Physiological, morphological, and histochemical characterization of three classes of interneurons in rat neostriatum. *The Journal of Neuroscience: The Official Journal of the Society for Neuroscience*, 13(11), 4908–4923.

Kita, H. and Kitai, S.T., 1987. Efferent projections of the subthalamic nucleus in the rat: Light and electron microscopic analysis with the PHA-L method. *The Journal of Comparative Neurology*, 260(3), 435–452.

Kita, H. and Kitai, S.T., 1991. Intracellular study of rat globus pallidus neurons: Membrane properties and responses to neostriatal, subthalamic and nigral stimulation. *Brain Research*, 564(2), 296–305.

Kita, T., Kita, H., and Kitai, S.T., 1986. Electrical membrane properties of rat substantia nigra compacta neurons in an in vitro slice preparation. *Brain Research*, 372(1), 21–30.

Kliem, M.A. et al., 2007a. Activation of nigral and pallidal dopamine D1-like receptors modulates basal ganglia outflow in monkeys. *Journal of Neurophysiology*, 98(3), 1489–1500.

Kliem, M.A. et al., 2009. Comparative ultrastructural analysis of D1 and D5 dopamine receptor distribution in the substantia nigra and globus pallidus of monkeys. *Advances in Behavioral Biology*, 58, 239–253.

Kosinski, C.M. et al., 1999. Localization of metabotropic glutamate receptor 7 mRNA and mGluR7a protein in the rat basal ganglia. *The Journal of Comparative Neurology*, 415(2), 266–284.

von Krosigk, M. et al., 1992. Synaptic organization of gabaergic inputs from the striatum and the globus pallidus onto neurons in the substantia nigra and retrorubral field which project to the medullary reticular formation. *Neuroscience*, 50(3), 531–549.

Lacey, M.G., Mercuri, N.B., and North, R.A., 1989. Two cell types in rat substantia nigra zona compacta distinguished by membrane properties and the actions of dopamine and opioids. *The Journal of Neuroscience: The Official Journal of the Society for Neuroscience*, 9(4), 1233–1241.

Langer, L.F. and Graybiel, A.M., 1989. Distinct nigrostriatal projection systems innervate striosomes and matrix in the primate striatum. *Brain Research*, 498(2), 344–350.

Lee, C.R. and Tepper, J.M., 2007a. A calcium-activated nonselective cation conductance underlies the plateau potential in rat substantia nigra GABAergic neurons. *The Journal of Neuroscience: The Official Journal of the Society for Neuroscience*, 27(24), 6531–6541.

Lee, C.R. and Tepper, J.M., 2007b. Morphological and physiological properties of parvalbumin- and calretinin-containing gamma-aminobutyric acidergic neurons in the substantia nigra. *The Journal of Comparative Neurology*, 500(5), 958–972.

Lee, C.R. and Tepper, J.M., 2009. Basal ganglia control of substantia nigra dopaminergic neurons. *Journal of Neural Transmission. Supplementum*, 73, 71–90.

Leung, D.S.Y., Unsicker, K., and Reuss, B., 2002. Expression and developmental regulation of gap junction connexins cx26, cx32, cx43 and cx45 in the rat midbrain-floor. *International Journal of Developmental Neuroscience: The Official Journal of the International Society for Developmental Neuroscience*, 20(1), 63–75.

Levy, R. et al., 2002. Dependence of subthalamic nucleus oscillations on movement and dopamine in Parkinson's disease. *Brain*, 125(6), 1196–1209.

Lin, J.Y. et al., 2003. Dendritic projections and dye-coupling in dopaminergic neurons of the substantia nigra examined in horizontal brain slices from young rats. *Journal of Neurophysiology*, 90(4), 2531–2535.

Lindvall, O. and Björklund, A., 1979. Dopaminergic innervation of the globus pallidus by collaterals from the nigrostriatal pathway. *Brain Research*, 172(1), 169–173.

Liss, B. et al., 2001. Tuning pacemaker frequency of individual dopaminergic neurons by Kv4.3L and KChip3.1 transcription. *The EMBO Journal*, 20(20), 5715–5724.

Ljungberg, T., Apicella, P., and Schultz, W., 1992. Responses of monkey dopamine neurons during learning of behavioral reactions. *Journal of Neurophysiology*, 67(1), 145–163.

Lokwan, S.J. et al., 1999. Stimulation of the pedunculopontine tegmental nucleus in the rat produces burst firing in A9 dopaminergic neurons. *Neuroscience*, 92(1), 245–254.

Lynd-Balta, E. and Haber, S.N., 1994. Primate striatonigral projections: A comparison of the sensorimotor-related striatum and the ventral striatum. *The Journal of Comparative Neurology*, 345(4), 562–578.

Magariños-Ascone, C. et al., 2002. High-frequency stimulation of the subthalamic nucleus silences subthalamic neurons: A possible cellular mechanism in Parkinson's disease. *Neuroscience*, 115(4), 1109–1117.

Magee, J.C., 2000. Dendritic integration of excitatory synaptic input. *Nature Reviews. Neuroscience*, 1(3), 181–190.

Mailly, P. et al., 2001. Dendritic arborizations of the rat substantia nigra pars reticulata neurons: Spatial organization and relation to the lamellar compartmentation of striatonigral projections. *The Journal of Neuroscience: The Official Journal of the Society for Neuroscience*, 21(17), 6874–6888.

Mailly, P. et al., 2003. Three-dimensional organization of the recurrent axon collateral network of the substantia nigra pars reticulata neurons in the rat. *The Journal of Neuroscience: The Official Journal of the Society for Neuroscience*, 23(12), 5247–5257.

Manley, L.D. et al., 1992. Effects of frequency and pattern of medial forebrain bundle stimulation on caudate dialysate dopamine and serotonin. *Journal of Neurochemistry*, 58(4), 1491–1498.

Marino, M.J. et al., 2001. Activation of group I metabotropic glutamate receptors produces a direct excitation and disinhibition of GABAergic projection neurons in the substantia nigra pars reticulata. *The Journal of Neuroscience: The Official Journal of the Society for Neuroscience*, 21(18), 7001–7012.

Martin, G., Papp, N., and Bacino, C., 1978. Contralateral turning evoked by the intranigral microinjection of muscimol and other GABA agonists. *Brain Research*, 155(2), 297–312.

Matsuda, W. et al., 2009. Single nigrostriatal dopaminergic neurons form widely spread and highly dense axonal arborizations in the neostriatum. *The Journal of Neuroscience*, 29(2), 444–453.

Maurice, N. et al., 2003. Spontaneous and evoked activity of substantia nigra pars reticulata neurons during high-frequency stimulation of the subthalamic nucleus. *The Journal of Neuroscience*, 23(30), 9929–9936.

McGeorge, A.J. and Faull, R.L., 1989. The organization of the projection from the cerebral cortex to the striatum in the rat. *Neuroscience*, 29(3), 503–537.

McNair, J.L., Sutin, J., and Tsubokawa, T., 1972. Suppression of cell firing in the substantia nigra by caudate nucleus stimulation. *Experimental Neurology*, 37(2), 395–411.

Middleton, F.A. and Strick, P.L., 2002. Basal-ganglia 'projections' to the prefrontal cortex of the primate. *Cerebral Cortex*, 12(9), 926–935.

Miyazaki, T. and Lacey, M.G., 1998. Presynaptic inhibition by dopamine of a discrete component of GABA release in rat substantia nigra pars reticulata. *The Journal of Physiology*, 513(3), 805–817.

Naito, A. and Kita, H., 1994. The cortico-nigral projection in the rat: An anterograde tracing study with biotinylated dextran amine. *Brain Research*, 637(1–2), 317–322.

Nakanishi, S., 1992. Molecular diversity of glutamate receptors and implications for brain function. *Science*, 258(5082), 597–603.

Nakanishi, H., Kita, H., and Kitai, S.T., 1987. Intracellular study of rat substantia nigra pars reticulata neurons in an in vitro slice preparation: Electrical membrane properties and response characteristics to subthalamic stimulation. *Brain Research*, 437(1), 45–55.

Nambu, A., Tokuno, H., and Takada, M., 2002. Functional significance of the cortico-subthalamo-pallidal 'hyperdirect' pathway. *Neuroscience Research*, 43(2), 111–117.

Nedergaard, S., Flatman, J.A., and Engberg, I., 1993. Nifedipine- and omega-conotoxin-sensitive Ca²⁺ conductances in guinea-pig substantia nigra pars compacta neurones. *The Journal of Physiology*, 466, 727–747.

Nemoto, C., Hida, T., and Arai, R., 1999. Calretinin and calbindin-D28k in dopaminergic neurons of the rat midbrain: A triple-labeling immunohistochemical study. *Brain Research*, 846(1), 129–136.

Neuhoff, H. et al., 2002. I(h) channels contribute to the different functional properties of identified dopaminergic subpopulations in the midbrain. *The Journal of Neuroscience: The Official Journal of the Society for Neuroscience*, 22(4), 1290–1302.

Neves, S.R., Ram, P.T., and Iyengar, R., 2002. G protein pathways. *Science*, 296(5573), 1636–1639.

Nicholson, L.F. et al., 1992. The regional, cellular and subcellular localization of GABAA/benzodiazepine receptors in the substantia nigra of the rat. *Neuroscience*, 50(2), 355–370.

Nirenberg, M.J. et al., 1998. Ultrastructural localization of the vesicular monoamine transporter 2 in mesolimbic and nigrostriatal dopaminergic neurons. *Advances in Pharmacology*, 42, 240–243.

Nissbrandt, H., Elverfors, A., and Engberg, G., 1994. Pharmacologically induced cessation of burst activity in nigral dopamine neurons: Significance for the terminal dopamine efflux. *Synapse*, 17(4), 217–224.

Nutt, D., 2006. GABAA receptors: Subtypes, regional distribution, and function. *Journal of Clinical Sleep Medicine: JCSM: Official Publication of the American Academy of Sleep Medicine*, 2(2), S711.

Overton, P. and Clark, D., 1992. Iontophoretically administered drugs acting at the N-methyl-D-aspartate receptor modulate burst firing in A9 dopamine neurons in the rat. *Synapse*, 10(2), 131–140.

Paladini, C.A. and Tepper, J.M., 1999. GABA(A) and GABA(B) antagonists differentially affect the firing pattern of substantia nigra dopaminergic neurons in vivo. *Synapse*, 32(3), 165–176.

Pan, W. and Hyland, B.I., 2005. Pedunculopontine tegmental nucleus controls conditioned responses of midbrain dopamine neurons in behaving rats. *The Journal of Neuroscience: The Official Journal of the Society for Neuroscience*, 25(19), 4725–4732.

Parent, A., 1990. Extrinsic connections of the basal ganglia. *Trends in Neurosciences*, 13(7), 254–258.

Parent, A., Parent, M., and Charara, A., 1999. Glutamatergic inputs to midbrain dopaminergic neurons in primates. *Parkinsonism* and *Related Disorders*, 5(4), 193–201.

Parent, A. et al., 1983. The output organization of the substantia nigra in primate as revealed by a retrograde double labeling method. *Brain Research Bulletin*, 10(4), 529–537.

Pennartz, C.M.A. et al., 2002. Diurnal modulation of pacemaker potentials and calcium current in the mammalian circadian clock. *Nature*, 416(6878), 286–290.

Pinard, A., Seddik, R., and Bettler, B., 2010. GABAB receptors: Physiological functions and mechanisms of diversity. *Advances in Pharmacology*, 58, 231–255.

Ping, H.X. and Shepard, P.D., 1996. Apamin-sensitive Ca(2+)-activated K+ channels regulate pacemaker activity in nigral dopamine neurons. *Neuroreport*, 7(3), 809–814.

Porceddu, M.L. et al., 1979. Opposite turning effects of kainic and ibotenic acid injected in the rat substantia nigra. *Neuroscience Letters*, 15(2–3), 271–276.

Preston, R.J. et al., 1981. Anatomy and physiology of substantia nigra and retrorubral neurons studied by extra- and intracellular recording and by horseradish peroxidase labeling. *Neuroscience*, 6(3), 331–344.

Puopolo, M., Raviola, E., and Bean, B.P., 2007. Roles of subthreshold calcium current and sodium current in spontaneous firing of mouse midbrain dopamine neurons. *Journal of Neuroscience*, 27(3), 645–656.

Radnikow, G. and Misgeld, U., 1998. Dopamine D1 receptors facilitate GABAA synaptic currents in the rat substantia nigra pars reticulata. *The Journal of Neuroscience: The Official Journal of the Society for Neuroscience*, 18(6), 2009–2016.

Rice, M.E., Cragg, S.J., and Greenfield, S.A., 1997. Characteristics of electrically evoked somatodendritic dopamine release in substantia nigra and ventral tegmental area in vitro. *Journal of Neurophysiology*, 77(2), 853–862.

Richards, C.D., Shiroyama, T., and Kitai, S.T., 1997. Electrophysiological and immunocytochemical characterization of GABA and dopamine neurons in the substantia nigra of the rat. *Neuroscience*, 80(2), 545–557.

Richfield, E.K., Young, A.B., and Penney, J.B., 1987. Comparative distribution of dopamine D-1 and D-2 receptors in the basal ganglia of turtles, pigeons, rats, cats, and monkeys. *The Journal of Comparative Neurology*, 262(3), 446–463.

Rick, C.E. and Lacey, M.G., 1994. Rat substantia nigra pars reticulata neurones are tonically inhibited via GABAA, but not GABAB, receptors in vitro. *Brain Research*, 659(1–2), 133–137.

Rinvik, E. and Ottersen, O.P., 1993. Terminals of subthalamonigral fibres are enriched with glutamate-like immunoreactivity: An electron microscopic, immunogold analysis in the cat. *Journal of Chemical Neuroanatomy*, 6(1), 19–30.

Robertson, G.S. and Robertson, H.A., 1989. Evidence that L-dopa-induced rotational behavior is dependent on both striatal and nigral mechanisms. *The Journal of Neuroscience: The Official Journal of the Society for Neuroscience*, 9(9), 3326–3331.

Robledo, P. and Féger, J., 1990. Excitatory influence of rat subthalamic nucleus to substantia nigra pars reticulata and the pallidal complex: Electrophysiological data. *Brain Research*, 518(1–2), 47–54.

Ruffieux, A. and Schultz, W., 1980. Dopaminergic activation of reticulata neurones in the substantia nigra. *Nature*, 285(5762), 240–241.

Sakamoto, M. and Hikosaka, O., 1989. Eye movements induced by microinjection of GABA agonist in the rat substantia nigra pars reticulata. *Neuroscience Research*, 6(3), 216–233.

Sanghera, M.K., Trulson, M.E., and German, D.C., 1984. Electrophysiological properties of mouse dopamine neurons: In vivo and in vitro studies. *Neuroscience*, 12(3), 793–801.

Scarnati, E., Campana, E., and Pacitti, C., 1984. Pedunculopontine-evoked excitation of substantia nigra neurons in the rat. *Brain Research*, 304(2), 351–361.

Schmitt, P. et al., 1999. Regulation of substantia nigra pars reticulata neuronal activity by excitatory amino acids. *Naunyn-Schmiedeberg's Archives of Pharmacology*, 360(4), 402–412.

Schultz, W., 1998. Predictive reward signal of dopamine neurons. *Journal of Neurophysiology*, 80(1), 1–27.

Schultz, W., 2007. Behavioral dopamine signals. *Trends in Neurosciences*, 30(5), 203–210.

Schultz, W., 2010. Dopamine signals for reward value and risk: Basic and recent data. *Behavioral and Brain Functions: BBF*, 6, 24.

Schwaller, B., Meyer, M., and Schiffmann, S., 2002. 'New' functions for 'old' proteins: The role of the calcium-binding proteins calbindin D-28k, calretinin and parvalbumin, in cerebellar physiology. Studies with knockout mice. *Cerebellum*, 1(4), 241–258.

Schwyn, R.C. and Fox, C.A., 1974. The primate substantia nigra: A golgi and electron microscopic study. *Journal Für Hirnforschung*, 15(1), 95–126.

Sesack, S.R., Aoki, C., and Pickel, V.M., 1994. Ultrastructural localization of D2 receptor-like immunoreactivity in midbrain dopamine neurons and their striatal targets. *The Journal of Neuroscience: The Official Journal of the Society for Neuroscience*, 14(1), 88–106.

Shen, K.Z. and Johnson, S.W., 1997. Presynaptic GABAB and adenosine A1 receptors regulate synaptic transmission to rat substantia nigra reticulata neurones. *The Journal of Physiology*, 505(Pt 1), 153–163.

Shepard, P.D. and Bunney, B.S., 1991. Repetitive firing properties of putative dopamine-containing neurons in vitro: Regulation by an apamin-sensitive Ca(2+)-activated K+ conductance. *Experimental Brain Research*, 86(1), 141–150.

Smith, Y. and Bolam, J., 1989. Neurons of the substantia nigra reticulata receive a dense GABA-containing input from the globus pallidus in the rat. *Brain Research*, 493(1), 160–167.

Smith, A.D. and Bolam, J.P., 1990. The neural network of the basal ganglia as revealed by the study of synaptic connections of identified neurones. *Trends in Neurosciences*, 13(7), 259–265.

Smith, Y. and Bolam, J.P., 1991. Convergence of synaptic inputs from the striatum and the globus pallidus onto identified nigrocollicular cells in the rat: A double anterograde labelling study. *Neuroscience*, 44(1), 45–73.

Smith, I.D. and Grace, A.A., 1992. Role of the subthalamic nucleus in the regulation of nigral dopamine neuron activity. *Synapse*, 12(4), 287–303.

Smith, Y., Hazrati, L.N., and Parent, A., 1990. Efferent projections of the subthalamic nucleus in the squirrel monkey as studied by the PHA-L anterograde tracing method. *The Journal of Comparative Neurology*, 294(2), 306–323.

Smith, Y. et al., 1998. Microcircuitry of the direct and indirect pathways of the basal ganglia. *Neuroscience*, 86(2), 353–387.

Smith, Y. et al., 1989. Evidence for a distinct nigropallidal dopaminergic projection in the squirrel monkey. *Brain Research*, 482(2), 381–386.

Somogyi, P. and Klausberger, T., 2005. Defined types of cortical interneurone structure space and spike timing in the hippocampus. *The Journal of Physiology*, 562(1), 9–26.

Stanford, I.M. and Lacey, M.G., 1996. Differential actions of serotonin, mediated by 5-HT1B and 5-HT2C receptors, on GABA-mediated synaptic input to rat substantia nigra pars reticulata neurons in vitro. *The Journal of Neuroscience: The Official Journal of the Society for Neuroscience*, 16(23), 7566–7573.

Surmeier, D.J. et al., 2007. D1 and D2 dopamine-receptor modulation of striatal glutamatergic signaling in striatal medium spiny neurons. *Trends in Neurosciences*, 30(5), 228–235.

Tepper, J.M. and Lee, C.R., 2007. GABAergic control of substantia nigra dopaminergic neurons. *Progress in Brain Research*, 160, 189–208.

Tepper, J.M., Martin, L.P., and Anderson, D.R., 1995. GABAA receptor-mediated inhibition of rat substantia nigra dopaminergic neurons by pars reticulata projection neurons. *The Journal of Neuroscience: The Official Journal of the Society for Neuroscience*, 15(4), 3092–3103.

Tepper, J.M., Sawyer, S.F., and Groves, P.M., 1987. Electrophysiologically identified nigral dopaminergic neurons intracellularly labeled with HRP: Light-microscopic analysis. *The Journal of Neuroscience: The Official Journal of the Society for Neuroscience*, 7(9), 2794–2806.

Testa, C. et al., 1994. Metabotropic glutamate receptor mRNA expression in the basal ganglia of the rat. *Journal of Neuroscience*, 14(5), 3005–3018.

Timmerman, W. and Westerink, B.H., 1995. Extracellular gamma-aminobutyric acid in the substantia nigra reticulata measured by microdialysis in awake rats: Effects of various stimulants. *Neuroscience Letters*, 197(1), 21–24.

Tse, Y.C. and Yung, K.K.L., 2000. Cellular expression of ionotropic glutamate receptor subunits in subpopulations of neurons in the rat substantia nigra pars reticulata. *Brain Research*, 854(1–2), 57–69.

Van Der Kooy, D. and Hattori, T., 1980. Single subthalamic nucleus neurons project to both the globus pallidus and substantia nigra in rat. *The Journal of Comparative Neurology*, 192(4), 751–768.

Vandecasteele, M., Glowinski, J., and Venance, L., 2005. Electrical synapses between dopaminergic neurons of the substantia nigra pars compacta. *The Journal of Neuroscience: The Official Journal of the Society for Neuroscience*, 25(2), 291–298.

Vandecasteele, M. et al., 2008. Chemical transmission between dopaminergic neuron pairs. *Proceedings of the National Academy of Sciences of the United States of America*, 105(12), 4904–4909.

Veening, J.G., Cornelissen, F.M., and Lieven, P.A., 1980. The topical organization of the afferents to the caudatoputamen of the rat. A horseradish peroxidase study. *Neuroscience*, 5(7), 1253–1268.

Waldvogel, H.J. et al., 2008. Differential localization of GABAA receptor subunits within the substantia nigra of the human brain: An immunohistochemical study. *The Journal of Comparative Neurology*, 506(6), 912–929.

Waszcak, B. and Walters, J.R. 1983. Dopamine modulation of the effects of gamma-aminobutyric acid on substantia nigra pars reticulata neurons. *Science*, 220(4593), 218–221.

Wilson, C.J. and Callaway, J.C., 2000. Coupled oscillator model of the dopaminergic neuron of the substantia nigra. *Journal of Neurophysiology*, 83(5), 3084–3100.

Wilson, C.J., Young, S.J., and Groves, P.M., 1977. Statistical properties of neuronal spike trains in the substantia nigra: Cell types and their interactions. *Brain Research*, 136(2), 243–260.

Windels, F. and Kiyatkin, E.A., 2004. GABA, not glutamate, controls the activity of substantia nigra reticulata neurons in awake, unrestrained rats. *The Journal of Neuroscience: The Official Journal of the Society for Neuroscience*, 24(30), 6751–6754.

Wittmann, M. et al., 2001. Activation of group III mGluRs inhibits GABAergic and glutamatergic transmission in the substantia nigra pars reticulata. *Journal of Neurophysiology*, 85(5), 1960–1968.

Wolfart, J. et al., 2001. Differential expression of the small-conductance, calcium-activated potassium channel SK3 is critical for pacemaker control in dopaminergic midbrain neurons. *The Journal of Neuroscience: The Official Journal of the Society for Neuroscience*, 21(10), 3443–3456.

Wu, Y., Richard, S., and Parent, A., 2000. The organization of the striatal output system: A single-cell juxtacellular labeling study in the rat. *Neuroscience Research*, 38(1), 49–62.

You, Z.B. et al., 1994. The striatonigral dynorphin pathway of the rat studied with in vivo microdialysis–II. Effects of dopamine D1 and D2 receptor agonists. *Neuroscience*, 63(2), 427–434.

Yung, K.K.L., 1998. Localization of ionotropic and metabotropic glutamate receptors in distinct neuronal elements of the rat substantia nigra. *Neurochemistry International*, 33(4), 313–326.

Yung, W.H., Häusser, M.A., and Jack, J.J., 1991. Electrophysiology of dopaminergic and non-dopaminergic neurones of the guinea-pig substantia nigra pars compacta in vitro. *The Journal of Physiology*, 436, 643–667.

Yung, K.K. et al., 1995. Immunocytochemical localization of D1 and D2 dopamine receptors in the basal ganglia of the rat: Light and electron microscopy. *Neuroscience*, 65(3), 709–730.

Zhou, F. et al., 2009. An ultra-short dopamine pathway regulates basal ganglia output. *The Journal of Neuroscience: The Official Journal of the Society for Neuroscience*, 29(33), 10424–10435.

9 Striatal Dopamine and Glutamate in Action

The Generation and Modification of Adaptive Behavior

Henry H. Yin and Rui M. Costa

CONTENTS

9.1   INTRODUCTION

Of the group of cerebral nuclei collectively known as the basal ganglia, the striatum is the major input nucleus and receives glutamatergic projections from the cortex and thalamus as well as dopaminergic projections from the midbrain. The interaction between these two types of projections in the different regions of striatum, so critical to the generation and modification of adaptive behavior, is the focus of the present chapter.

Glutamate, released by cortical and thalamic projection neurons, is the major excitatory transmitter in the brain. Cortical and thalamic axons synapse onto the numerous dendritic spines of the striatal projection neurons known as "medium spiny neurons" (MSNs). These excitatory synapses are the first step in the transmission of signals from cortex and thalamus, the striatum being the "portal" of the entire basal ganglia (Nauta, 1979; Ding et al., 2008). All the projection neurons of the basal ganglia are GABAergic and presumably inhibitory. Without the excitatory drive to the striatum, the intrinsic oscillations in the output nuclei such as the substantia nigra pars reticulata and the internal globus pallidus result in tonic inhibition of their targets in the brainstem and thalamus (Chevalier et al., 1981; Chevalier and Deniau, 1990; Hikosaka et al., 2000).

Dopaminergic axons synapse on the neck of the spines of the MSNs, which form over 90% of the striatal neuronal population (Wilson, 2004). In slices and anesthetized animals, the subthreshold membrane potential of a given MSN can alternate between two states, a relatively depolarized UP state and a more hyperpolarized DOWN state (Wilson and Kawaguchi, 1996; Stern et al., 1998); for it to spike, it must be in the UP state when the excitatory input arrives. Because the activation of the N-methyl-D-aspartate (NMDA) glutamate receptors may be critical for the shift to the UP state, these receptors can play a permissive role in the spiking of MSNs (Kepecs and Raghavachari, 2007; Pomata et al., 2008). NMDA receptor activation is also thought to be critical for correlated firing among distributed groups of MSNs (Carrillo-Reid et al., 2008), and in spike bursting (Jin and Costa, 2010).

Bistability in the subthreshold membrane potential is not unique to striatal neurons, but it appears to be an important feature that is regulated by dopaminergic inputs. Since dopamine (DA) receptors are not directly coupled to ion channels, they do not carry any ionic current; modulation of membrane excitability and synaptic transmission are the chief means by which DA can affect behavior. Two major classes of DA receptors, D1- and D2-type receptors, are preferentially expressed in distinct groups of MSNs, with D1-expressing neurons giving rise to the striatonigral or direct pathway and D2-expressing neurons giving rise to the striatopallidal or indirect pathway. D1 and D2 receptors appear to exert opposing effects on membrane excitability (Kreitzer, 2009). D1 activation promotes hyperpolarization in the DOWN state, but promotes depolarization in the UP state (Hernandez-Lopez et al., 1997; Nisenbaum et al., 1998). The opposite is true of D2 activation (Hernandez-Lopez et al., 2000; Nicola et al., 2000). Likewise, D1 and D2 activation also exert different effects on glutamatergic transmission. D1 activation, for example, can potentiate NMDA currents, while D2 activation either has no effect or inhibits glutamatergic transmission (Morelli et al., 1992; Cepeda and Levine, 1998; Liu et al., 2006).

Thus, D1 activation may enhance the signal-to-noise ratio for the incoming excitatory drive from the thalamocortical system, whereas D2 activation may have the opposite effect. Consequently, a low level of synaptic DA, which presumably activates high-affinity D2 receptors, can prevent the selection of any particular circuit, whereas phasic DA, normally a result of bursting of DA neurons, can activate D1 as well as D2 receptors and promote the selection of specific cell groups. High DA may therefore promote a winner-take-all mechanism, whereas low DA could "level the playing field," so to speak, equalizing the activation of different cell groups. Thus, DA can play a major role in the online selection of active striatal cell groups. Since the firing of striatal neurons initiates the transmission of information in the basal ganglia circuitry, DA's modulatory influence can be critical for voluntary behavior. It can also have a role, to which we shall return, in controlling the long-term plasticity in the striatum, which preserves and amplifies the patterns created during online selection.

9.2   CONSEQUENCES OF DA DEPLETION

The functional significance of DA was recognized soon after the discovery of its role as a chemical transmitter in the brain. Carlsson found that the symptoms of Parkinson's disease are a consequence of DA depletion (Carlsson et al., 1957). A large literature has since accumulated on the effects of DA depletion on behavior.

As revealed by an established model of parkinsonism using 6-hydroxydopamine (6-OHDA) lesions of the nigrostriatal dopaminergic pathway, DA depletion results in a variety of symptoms, including slowness in initiating movements (bradykinesia), loss of exploratory behavior and curiosity, and trouble in feeding which can lead to death from starvation in the absence of help (Ungerstedt, 1971a). A clear demonstration of these classic effects is found in the recent work by Palmiter and colleagues, who developed DA-deficient (DD) mice with <1% of normal DA levels but with preserved norepinephrine (Palmiter, 2008). Surprisingly, no early developmental deficits were found with normal feeding and suckling behavior, suggesting that, immediately after birth, such behaviors do not require dopaminergic signaling. But starting around 10 days after birth, to stay alive the DD mice must be given daily injections of L-dopa, a DA precursor that can cross the blood–brain barrier. They become active within minutes after the first L-dopa injection. Like the classical depletion studies, Palmiter and colleagues were able to show that such dramatic effects on behavior were largely due to the nigrostriatal pathway, from the substantia nigra pars compacta to the dorsal striatum (Palmiter, 2008).

Following DA depletion, considerable compensatory changes, such as receptor supersensitivity, can take place (Ungerstedt, 1971b). They are at least partly responsible for the commonly observed recovery of function after depletion of DA by 6-OHDA. It is therefore unclear which impairments are due to the lack of DA *per se* and which are caused by the long-term adaptations that are a consequence of DA depletion. Thus long-term depletion or genetic deletion may not be the ideal way to reveal how DA modulates transmission and signaling online. An alternative approach is rapid and reversible DA depletion, using a combination of pharmacological and genetic manipulations (DD DAT-KO mice or DDD mice) (Sotnikova et al., 2005). DA, when released in the striatum (particularly in the lateral or sensorimotor striatum), is taken back into the presynaptic terminal via the DA transporter (DAT) (Jones et al., 1998; Gainetdinov and Caron, 2003). Lack of DA re-uptake in DAT knock-out (DAT-KO) mice results in a pronounced increase in striatal extracellular DA (Gainetdinov and Caron, 2003). In the striatum, this recycling of DA via DAT accounts for the majority of DA released from the terminal. Without DAT, the concentration of DA in the presynaptic terminal is diminished, and all the DA results from *de novo* synthesis (Sotnikova et al., 2005). Therefore, the administration of a tyrosine hydroxylase inhibitor (like AMPT, α-methyl-p-tyrosine) produces rapid DA depletion, because once all the available DA has been released, no more will be replenished. This depletion method, however, is reversible (Sotnikova et al., 2005). DDD mice become akinetic, developing extreme rigidity, body tremor, and ptosis very rapidly after treatment with AMPT, and these symptoms are readily recovered by L-dopa administration.

By common consensus, the functional significance of DA in behavior is largely due to its actions on target striatal neurons. The most popular model postulates that upon DA depletion there is increased activity in striatopallidal (indirect pathway) neurons, which express D2-type receptors, and decreased activity of striatonigral (direct pathway) neurons, which express D1-type receptors, ultimately resulting in inhibition of motor cortex activity (Albin et al., 1989; Alexander and Crutcher, 1990). Recent studies using optogenetic manipulation of direct and indirect pathway neurons support this view (Kravitz et al., 2010). But other studies failed to observe decreased overall firing rate in motor cortex after DA depletion, either in monkeys with dopaminergic lesions (Goldberg et al., 2004) or in DDD mice (Costa et al., 2006a), even with pronounced changes in striatal firing rate following reversible DA depletion (Costa et al., 2006a). According to an alternative view, the lack of DA results in abnormal oscillatory activity and increased synchrony in the basal ganglia. A variety of studies using pharmacological, pharmacogenetic, and lesion models of PD in rodents and primates, and studies of idiopathic PD patients, have shown the appearance and possible propagation of rhythmic and synchronous firing in basal ganglia structures (Bergman et al., 1994; Plenz and Kital, 1999; Levy et al., 2000; Raz et al., 2001; Brown et al., 2002, 2004; Brown, 2003; Costa et al., 2006a; Burkhardt et al., 2007). Also, local field potential (LFP) recordings show decreased power in the gamma frequencies (30–60 Hz) and increased power of lower frequency oscillations in the beta band (10–30 Hz), in the subthalamic nucleus, cortex, and dorsal striatum after DA depletion (Kuhn et al., 2005; Sharott et al., 2005; Androulidakis et al., 2008). These changes in oscillatory activity and synchrony in basal ganglia are a direct consequence of DA depletion in DDD mice (Costa et al., 2006a); they are also observed after acute DA receptor blockade (Burkhardt et al., 2009) and reversed by L-dopa administration (Costa et al., 2006a).

DA also modulates the spike timing of individual neurons in relation to the phase of the LFP oscillations. After DA depletion, more MSNs are entrained to the LFP oscillations, mostly during the trough of the oscillations, which correspond to the point of maximum intracellular depolarization (Lampl and Yarom, 1993). The changes in synchrony and entrainment could occur through several mechanisms. For example, DA depletion can reduce collateral GABAergic inhibition between MSNs, leading to increased synchrony (Taverna et al., 2008). This does not easily explain the decrease in firing rate and increase in entrainment observed after DA depletion (Costa et al., 2006a, Burkhardt et al., 2009). Another possibility is that DA depletion reduces neuronal excitability in MSNs, permitting neurons to fire only during depolarized phases of the LFP oscillations, which would result in increased synchrony; however, the neurons that decrease firing rate are not the ones showing changes in LFP entrainment (Burkhardt et al., 2009).

In short, DA depletion can rapidly abolish voluntary behavior. These effects have traditionally been attributed to rapid changes in firing rate of different basal ganglia circuits, namely to excessive activation of the indirect pathway, but DA can also modulate the temporal coordination between the activity of different basal ganglia neurons. These two views are not necessarily mutually exclusive, even if the effects of DA depletion basal ganglia firing rate and oscillatory activity seem to be somewhat dissociable (Burkhardt et al., 2009). In this respect, it remains to be determined if deep brain stimulation acts by changing the firing rate of basal ganglia output nuclei or their oscillatory activity (Gradinaru et al., 2009) (Figure 9.1).

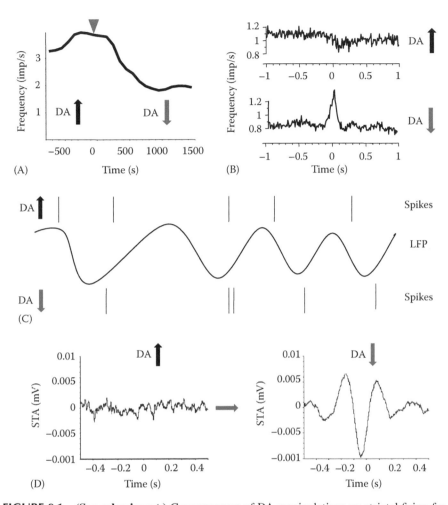

FIGURE 9.1 **(See color insert.)** Consequences of DA manipulations on striatal firing frequency and synchrony. (A) Acute DA depletion in DDD mice results in a rapid decrease in the average activity of striatal neurons. The population activity in the striatum of a DDD mouse before and after DA depletion is depicted (time 0). (B) Rapid DA depletion results in an increase in synchrony in the activity of striatal MSNs. Example of cross-correlation of firing between the same pair of striatal MSNs before (top) and after (bottom) DA depletion in DDD mice. (C) Scheme depicting increased entrainment of MSN firing to the striatal LFP oscillations after DA depletion. (D) Increased entrainment of a striatal MSN to the LFP oscillations after acute DA receptor blockade. (Adapted from Costa, et al., 2006a; Burkhardt, J.M., et al., *Front Integr. Neurosci.*, 3, 28, 2009.) The spike-triggered average (STA) of the LFP oscillations measures if spikes occur preferentially at a particular phase of the LFP oscillations. Significant fluctuations of the STA of the LFP around time zero indicates that spikes occur at a particular phase of the LFP oscillations while a flat STA of the LFP around time zero indicates no relation between the time of the spike and the LFP oscillations.

9.3 INCREASED DOPAMINERGIC SIGNALING AND EFFECTS ON BEHAVIOR

The behavioral consequences of too little DA, then, are clear. But what happens with too much DA? Although the importance of DA was first discovered using a depletion model, the complementary approach—increasing DA—has also contributed to our understanding of its function.

Various psychiatric disorders are thought to involve increased dopaminergic transmission, ranging from schizophrenia to Tourette's syndrome (Moore et al., 1999; Singer et al., 2002; Carlsson, 2006). Deletion of the DAT results in excessive DA tone and in a variety of symptoms including novelty-induced hyperactivity and increased behavioral stereotypy (Giros et al., 1996). Excessive DA tone, in contrast to DA depletion, decreases synchrony and entrainment to the LFP oscillations; for example, it can decrease beta oscillations and increase gamma oscillations (Costa et al., 2006a). Animal models of hyperdopaminergic conditions commonly use a pharmacological approach, for example, giving psychostimulants to boost dopaminergic signaling. By blocking DA transport and reuptake or enhancing release, psychostimulants can increase the level of synaptic DA, resulting in striking effects on behavior (Pierce and Kalivas, 1997). In humans, psychostimulants can cause psychosis, with symptoms ranging from excessive exploratory and searching behavior to paranoia. Rigid, repetitive, and uncontrollable movements are a common feature of disorders involving abnormalities in the basal ganglia: for example, repetitions in speech and tics in Tourette's syndrome and repetitive rituals of obsessive-compulsive disorder (Graybiel and Rauch, 2000). In rodents, psychostimulants have been shown to enhance exploratory behavior (Rebec and Bashore, 1984). At low doses, behavioral activation is more general, with increased locomotion mediated by the limbic corticostriatal circuit (Swanson, 2000). Yet with higher doses, a more specific and stereotyped pattern emerges including behaviors such as head bobbing, sniffing, and rearing behaviors that require the dorsal striatum (Joyce and Iversen, 1984). Mice with high levels of DA as a result of genetic knockdown of the DAT exhibit increased stereotypy of their grooming sequence—a complex action pattern that requires the dorsal striatum (Cromwell and Berridge, 1996; Aldridge and Berridge, 1998; Berridge et al., 2005).

The behavioral effects of psychostimulants are a result of abnormal dopaminergic modulation of the glutamatergic inputs to the striatum. For example, DAT-KO mice show increased levels of extracellular DA, but their novelty-induced hyperactivity does not seem to result from further increase in extracellular DA (Costa et al., 2006a), leading to the hypothesis that DA modulation of glutamatergic inputs is responsible for the behavioral abnormalities observed. Furthermore, the long-term consequences of psychostimulants can be seen in the significant structural plasticity in the striatal projection neurons, for example, dendritic dysmorphisms and increased spine density in the ventral striatum of rats (Robinson and Kolb, 1997; Lee et al., 2006).

DA is thought to enhance glutamatergic signaling via the activation of D1 receptors, increasing the gain of the information flow through the cortico-basal ganglia circuit by potentiating strong inputs and suppressing weaker inputs (Lorenz, 1973; Rolls et al., 1984; Nicola et al., 2004; Rebec, 2006). According to the available evidence, the development of stereotypy either in natural grooming sequences or in other behaviors requires D1 receptor activation, since they are enhanced by D1 agonists (Berridge and Aldridge, 2000a,b). The increased neuronal activity following administration of methamphetamine, as measured by the expression of immediate early genes like c-fos, is reversed by antagonists of D1 receptors (Gross and Marshall, 2009). Electrophysiological studies of striatal activity also found D1-dependent potentiation of glutamatergic transmission. In awake rats, iontophoretic application of glutamate caused a dose-dependent activation of striatal neurons, and DA applied simultaneously enhanced the excitatory actions of glutamate (Pierce and Rebec, 1995).

The psychostimulant-induced c-fos expression requires the activation of ionotropic glutamate receptors (Torres and Rivier, 1993). Antagonists of NMDA or α-amino-3-hydroxy-5-methyl-4-propionate (AMPA) receptors typically reduce the c-fos expression induced by psychostimulants (Gross and Marshall, 2009). Given the role of Gs-coupled D1 receptors in increasing adenylate cyclase and potentiating NMDA receptors (Cepeda and Levine, 1998), it is not surprising that D1 activation can generally enhance glutamatergic transmission onto MSNs. By contrast, the activation of D2 receptors, which are $G_{i/o}$ coupled, can dampen glutamatergic transmission onto MSNs via a variety of mechanisms, such as reducing glutamate release and inhibiting NMDA receptor activation (Bamford et al., 2004; Liu et al., 2006; Yin and Lovinger, 2006).

As mentioned already, because NMDA receptors play a critical role in shifting the membrane potential of MSNs to a more depolarized UP state, the activation of the low-affinity D1 receptors by phasic DA activity is expected to support this process in the direct pathway neurons. The activation of the D1-expressing striatonigral neurons by cortical and thalamic inputs can disinhibit the tonically active brainstem and thalamic neurons, thus leading to the initiation of specific central pattern generators for action sequences (Grillner et al., 2005; Hikosaka, 2007). The low affinity of D1 receptors also suggests that they should be preferentially activated by high levels of DA.

Other studies, however, have shown that the pattern of activation after psychostimulant administration is more complex. Whether the D1-expressing direct pathway or the D2-expressing indirect pathway is engaged depends in part on the behavioral context (Badiani et al., 2000; Uslaner et al., 2001; Ferguson and Robinson, 2004). The expression of immediate early genes in different neurochemical compartments in the striatum, like the striosome and matrix, is highly correlated with the degree of stereotypy (Canales and Graybiel, 2000). But concurrent activation of both D1 and D2 receptors after cocaine exposure is necessary for the striosome-dominant gene expression and behavioral stereotypy. Despite the considerable size of the literature on gene expression patterns following psychostimulant administration and the consequent behavioral adaptations, therefore, there is no clear consensus on the mechanisms involved.

9.4 DOPAMINE AND REWARD

The question of whether DA is crucial for "reward" has been debated for some three decades. The link between DA and reward was originally based on several key observations. First, effective sites for intracranial self-stimulation are close to the dopaminergic fibers, and self-stimulation can be blocked by DA antagonists (Wise, 1982, 1984). DA depletion, on the other hand, reduces feeding and other appetitive behaviors, and DA antagonists reduce the level of motivation for operant responding.

In addition, psychostimulants, like cocaine, which enhance DA levels also produce subjective states of euphoria. Hence the "anhedonia" hypothesis was formulated to explain such findings (Wise, 1982). According to this hypothesis, DA serves as a common currency for different types of rewards. After DA depletion, the animal can no longer experience the pleasure associated with the rewards and stimuli associated with rewards—hence the reduction in motivated behaviors.

Berridge and Robinson, however, observed that the actual consummatory responses, such as facial expressions indicating taste reactivity to sucrose and quinine, are not affected after DA depletion (Berridge and Robinson, 1998; Berridge, 2007). Rather, what is reduced by DA depletion and enhanced by psychostimulants is seeking behavior controlled by the desire for the specific goal, regardless of the actual hedonic response to it. Consequently, they suggested that the function of DA is better described as "incentive salience." Whereas the anhedonia hypothesis attributes the lack of motivated behavior to a lack of "liking," the incentive sensitization hypothesis attributes it to the lack of "wanting." That is, only the more preparatory aspects of appetitive behavior require DA.

The claim that DA is a pleasure signal does not predict the observation that an animal seeks repeatedly something that it does not actually like, though this is a common observation in studies of addictive behavior (Vanderschuren and Everitt, 2004). Moreover, the activity of DA neurons is often correlated with salient events in general, regardless of their hedonic valence (Horvitz, 2000). A recent study, in particular, has clearly demonstrated phasic activity of DA neurons in response to aversive air puffs (Matsumoto and Hikosaka, 2009). Finally, in a two-bottle preference test, DD mice still display a preference for "hedonically positive" sweet tastants over water (Cannon and Palmiter, 2003), and DAT-KO animals with elevated DA (500% compared to controls) display normal preference for sweet and bitter stimuli (Costa et al., 2006b).

9.5 FUNCTIONAL HETEROGENEITY OF BASAL GANGLIA CIRCUITS

In short, then, DA is more important for preparatory behaviors than for consummatory behaviors. The latter is often involuntary and organized at the level of the brain stem, involving relatively fixed action patterns that can be expressed even in decerebrate animals (Grill and Norgren, 1978). Any simple account of the "meaning" of DA in terms of either pleasure or salience fails to capture its actual role in behavior. A more useful approach is to relate its effects on specific neurons in well-defined neural circuits to the specific behaviors generated by such circuits. So far, most studies have not distinguished between different groups of DA neurons and their target structures, especially the different striatal regions. As a hub in the cerebral network, the striatum occupies a central place where a variety of inputs converge, yet it is also a large and highly heterogeneous structure. It can be divided into different functional domains based on the area of origin in the cortical, thalamic, and dopaminergic projections (McGeorge and Faull, 1989; Joel and Weiner, 2000). The projection from the primary sensorimotor cortices targets largely the so-called sensorimotor striatum. On the other hand, the limbic cortical regions, basolateral amygdala, hippocampus, orbitofrontal project to the limbic regions in the striatum.

Association cortices, for example, those in the prefrontal and parietal areas, project to the associative striatum, which receives a mixture of limbic, associative, and sensorimotor inputs (McGeorge and Faull, 1989). These diverse glutamatergic inputs are modulated by different groups of DA neurons. The different functional domains are preserved through the entire basal ganglia circuitry, also characterizing structures like the pallidum and thalamus (Joel and Weiner, 1994). There is therefore a rough medial to lateral gradient in the cortical, thalamic, striatal, pallidal, and dopaminergic cell groups (Joel and Weiner, 2000). The more medial ventral tegmental area projects largely to the ventral or limbic striatum, whereas the more lateral substantia nigra pars compacta projects to the dorsal striatum, with the medial portion projecting to medial striatum and the lateral portion projecting to lateral striatum. This gradient reflects a continuum from more limbic regions, which are critical for the processing of emotion, hedonic valence, and motivational salience, to more sensorimotor regions, which are critical for the organization and control of movements.

The cortico-basal ganglia networks are therefore organized as a series of iterating macro-circuits (Yin and Knowlton, 2006; Zahm, 2006; Yin et al., 2008). The basic circuit serves as an important motif in the anatomical organization of the brain, much like the alpha helix in biochemistry (Swanson, 2000). Previous work has established at least four such overlapping, iterating, and reentrant cortico-basal ganglia networks (Yin et al., 2008). As shown in Figure 9.2, the accumbens shell, core, dorsomedial or associative striatum, and dorsolateral or sensorimotor striatum are the corresponding striatal regions in these networks.

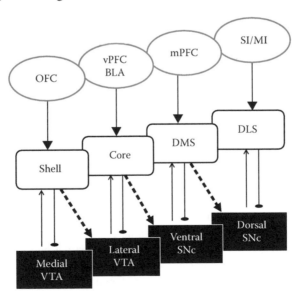

FIGURE 9.2 The cortico-basal ganglia network as a basic motif in cerebral anatomy. BLA, basolateral amygdala complex; core, nucleus accumbens core; DLS, dorsolateral striatum; DMS, dorsomedial striatum; mPFC, medial prefrontal cortex; OFC, orbitofrontal cortex; shell, nucleus accumbens shell; SI/MI, primary sensory and motor cortices; SNc, substantia nigra pars compacta; vPFC, ventral prefrontal cortex; VTA, ventral tegmental area. (Adapted from Yin, H.H. et al., *Eur. J. Neurosci.*, 28, 1437, 2008.)

Because glutamatergic inputs from corresponding cortical and thalamic regions as well as dopaminergic inputs from corresponding midbrain regions interact in each of the networks, primarily at the level of the striatum, in the analysis of such interactions it is critical to identify the networks involved. In the absence of excitatory drive, the striatum is relatively quiescent in awake animals, which results in lack of inhibition of the output nuclei that tonically inhibit the brain stem motor networks for various movements (Swanson et al., 1984; Hikosaka, 2007). The activation of striatal neurons in the direct pathway can disinhibit these brainstem control centers. For example, the center for locomotor activity is under the control of ventral pallidum (substantia innominata), the output nucleus of the limbic basal ganglia, so activation of areas like the nucleus accumbens core can lead to locomotor and other exploratory behaviors (Swerdlow et al., 1984; Swanson and Kalivas, 2000).

Given the convergent pattern of cortical projections, and the high threshold to spike observed in MSNs, only the activated (spiking) striatal neurons can influence downstream pallidal neurons. But whereas a particular cortical region might represent a particular body region, at the level of the striatum these glutamatergic inputs are mixed and recombined in the service of behavioral programs. DA, as already mentioned, has a major role in determining the threshold of striatal neurons, and therefore also in the initiation and selection of cell assemblies involved in specific behaviors.

It has been proposed that the cortico-basal ganglia networks are hierarchically organized (Yin and Knowlton, 2006), an idea in accord with recent work on the spiraling projections between the striatum and the substantia nigra (Haber, 2003). The limbic networks are important for primary hedonic reactions to rewards as well as exploratory and approach behaviors. The basolateral amygdaloid complex and other limbic cortical areas such as the orbitofrontal cortex provide the major inputs to the ventral striatum, which in turn projects to ventral pallidum (Swerdlow et al., 1984; Swanson, 2000). The associative network involves the association cortices in prefrontal and parietal regions, their striatal target in the dorsomedial or associative striatum, and downstream pallidal and thalamic components.

It is critical for the acquisition of instrumental action–outcome contingencies and keeping track of outcome values (Balleine and Dickinson, 1998; Yin et al., 2005b). The sensorimotor network is critical for the automatization, timing, and sequencing of behaviors (Jog et al., 1999; Yin et al., 2004; Yin and Knowlton, 2006; Lau and Glimcher, 2007, 2008; Jin and Costa 2010). In natural behaviors, all four networks are likely to be involved. Thus the effects of DA on different networks will partly depend on its modulation of striatal glutamatergic transmission in the respective striatal regions. These networks are implicated in distinct and experimentally dissociable types of behavioral control.

The major challenge in the analysis of voluntary behavior is a central ambiguity inherent in the observation of behavior. Behavior as commonly observed, for example, pressing a lever for food, is ambiguous: the "final common path" between motor neuron and muscle is shared by multiple higher systems. It is well established that the contraction of the muscles can be a result of spinal mechanisms alone, or involve additional circuits in the brainstem, midbrain, or diencephalon (Grillner et al., 2005). In other words, a press is not always a press, since direct observation does not readily reveal what the distinct central influences are. For that reason, special behavioral assays are needed to unmask the central neural systems involved in the control of a specific type of behavior. The current understanding of these networks is still crude, based on large scale perturbation of function using lesions and inactivation of specific neural structures. But studies have begun to reveal the role of different cortico-basal ganglia networks in these types of behavioral control. Later, we review what is currently known about the type of behavioral control implemented by each network.

9.6   VENTRAL STRIATUM AND PAVLOVIAN CONDITIONING

The ventral striatum, part of the limbic cortico-basal ganglia network, appears to be important for certain types of Pavlovian conditioning, for example, approaching some stimulus that predicts reward. As a result of the predictive relationship between stimulus and reward, the stimulus comes to elicit sign tracking behavior, when the animal approaches and "tracks" the stimulus that predicts reward. This type of behavior is independent of the action–outcome contingency (Williams and Williams, 1969; Schwartz and Gamzu, 1977). Sign tracking is reliably correlated with phasic activity of dopaminergic neurons projecting to the nucleus accumbens (Day et al., 2006, 2007).

In Pavlovian conditioning, a predictor of a biologically significant event such as a reward comes to elicit a number of behaviors that prepare the animal for the upcoming reward (Konorski, 1967). Such behaviors are highly specific to the outcome. Thus, the preparatory behavior might be different for water and for solid food. For example, salivation to dry foods might be elicited by predictors of food; anticipatory licking for liquid reward might be elicited by predictors of juice. Despite the great variety of behaviors under Pavlovian control, they are all independent of arbitrary behavior–outcome relationships, or acquired feedback functions. This is because the behaviors involved are usually "reflexive" and innately organized. They are based on relatively permanent features of the organism–environment relationship, that is, approaching predictors of food (sign tracking) usually brings one closer to food.

The behavioral significance associated with predictive learning, of which Pavlovian conditioning is one type, is clear. When some event is outside of one's control, that is, independent of the behavior of the organism, adaptive learning can nevertheless prepare the organism for the expected event. That is, a variety of innately organized anticipatory behaviors can alter and prepare the organism for the biologically significant event, be it danger or reward.

Most current ideas about the role of DA in learning are based on the finding that phasic DA activity can reflect the prediction error, a difference between actual and predicted reward. These findings have attracted widespread attention because popular models of Pavlovian conditioning use a prediction error between actual outcome and expected outcome as a teaching signal. Such a prediction error is found in the phasic activity of DA cells (Waelti et al., 2001). Phasic activity of DA neurons shifts in the course of learning from being responsive to the outcome to being responsive to the predictor. In learning models, a prediction error drives the change in associative strength between a stimulus (CS) and a reward (US), which directly enhances the ability of the CS to elicit a conditional response. This type of learning has been described with formal models such as the Rescorla–Wagner model and the related temporal difference model (Schultz et al., 1997; Schultz, 1998; Sutton and Barto, 1998; Waelti et al., 2001; Tobler et al., 2003). Thus the learning, or associative strength, is directly reflected in the performance of an elicited response. Presumably in appetitive learning, a stimulus (e.g., a picture) repeatedly predicts reward (juice), so that this increased prediction leads to the anticipatory licking response during the presentation of the picture.

The correlation between phasic activity of DA cells and a reward prediction error, however, does not suffice to demonstrate that DA is necessary for learning; for that one must show that eliminating phasic DA impairs learning and restoring or mimicking phasic DA promotes learning. But such experiments are more difficult to perform. Given DA's role in voluntary movement, motivation, and arousal, it is difficult to dissociate learning from performance with manipulations of DA signaling in the brain (Smith-Roe and Kelley, 2000). As a result, it has been difficult to implicate DA in a well-defined type of learning, including the type of appetitive Pavlovian conditioning studied in the *in vivo* electrophysiological experiments.

Recent studies taking advantage of the latest molecular genetic tools, however, have made some progress in showing a critical role of phasic DA in learning. Phasic DA is a result of a burst of action potentials in the DA neurons, whereas tonic DA is a result of much slower firing rate of the same neurons. As NMDA receptors are known to be critical for the bursting activity of dopaminergic neurons, recent studies have attempted to abolish phasic DA without affecting tonic DA by deleting NMDA receptors specifically in dopaminergic neurons through DAT-Cre-mediated excision of the *NMDAR1* gene. Results from these studies suggest that learning is impaired (Engblom et al., 2008; Zweifel et al., 2009). However, because NMDA receptors are often critical for synaptic plasticity, these results cannot dissociate the role of phasic DA from that of NMDA receptor-dependent plasticity at excitatory synapses onto dopaminergic neurons. Nonetheless, other studies using a different Cre line (tyrosine hydroxylase driven) that expresses preferentially in VTA DA neurons abolishes cue-evoked phasic firing of DA neurons and impairs Pavlovian learning (Cui and Costa, Annual Meeting of the Society for Neuroscience, 2009). Phasic activation of dopaminergic neurons using optogenetics seems to be sufficient for conditioned place preference, which is typically considered a type of contextual Pavlovian conditioning (Tsai et al., 2009).

9.7   INSTRUMENTAL LEARNING AND BEHAVIOR

Instrumental behavior refers to what is commonly called "voluntary" behavior, in contrast to more "reflexive" behaviors modified by Pavlovian conditioning. A variety of studies have implicated the dorsal striatum and associated cortico-basal ganglia networks in the learning and expression of instrumental behaviors, and DA in the dorsal striatum appears to be essential. For example, viral restorations of DA in the ventral striatum in DD mice is not sufficient to restore behavior, while dorsal striatal restoration, particularly in the dorsolateral striatum, is sufficient (Hnasko et al., 2005; Robinson et al., 2005, 2007; Palmiter, 2008; Darvas and Palmiter, 2009).

In instrumental learning, there is a feedback function from behavior to reward rate, whereas in Pavlovian conditioning this feedback function is deliberately removed. There is no unconditional reflexive response to the reward, but some relatively arbitrary behavioral pattern, also known as the operant, is selected and repeated by the animal (Skinner, 1938). The role of instrumental learning is to expand the existing behavioral repertoire of the organism by modifying behavioral policies and recombining behavioral elements to yield new behaviors. Instrumental actions, such as lever pressing, are by definition learned from experience, initially through trial and error. Learning such actions allows the animal to acquire control over desired outcomes (Konorski, 1967; Dickinson, 1994; Balleine and Dickinson, 1998).

Studies have shown that in the course of instrumental learning, say pressing a lever for food, the animal learns the causal relationship between its action and the consequence in order to select the appropriate action every time it desires the outcome. This relationship, called the action–outcome contingency, is a simple "if–then" statement: If I press the lever, then I get food. It selects a motor output that specifically controls the desired rate of reward. But under certain conditions, such as extensive training, stress, or uncertainty, an alternative system is engaged to control lever pressing (Dias-Ferreira et al., 2009; Derusso et al., 2010). The behavior controlled by the second system is called "habitual"; it is not controlled by the explicit anticipation of the goal.

Manipulations that change the current value of the outcome, for example, devaluation, will also reduce performance on a probe test if the behavior in question is truly goal directed. In contrast, if the behavior is habitual, it will be impervious to treatments that change the value of the outcome (Dickinson, 1989). Likewise, manipulations that change the causal relationship between action and outcome will change the performance of goal-directed actions, but not that of habitual responses. Thus, manipulations of current value of the reward and of the contingency between action and reward are the two tests that can dissociate goal-directed control and habitual control (Dickinson, 1985).

Using behavioral assays that specifically probe the control process engaged, experiments have dissociated these two systems both behaviorally and in terms of their neural substrates (Yin and Knowlton, 2006). Lesions of the dorsomedial striatum impair the goal directedness of behavior and render it habitual (Yin et al., 2005b). Local infusion of NMDA receptor antagonists into the dorsomedial striatum also impairs the acquisition of new action–outcome contingencies (Yin et al., 2005a). Lesions or inactivation of the dorsolateral striatum, by contrast, impairs habit formation, resulting in goal-directed control (Yin et al., 2004, 2006). Depletion of DA in this region appears to have a similar effect, rendering behavior goal directed even under training conditions that generated habitual behavior in control animals (Faure et al., 2005), while amphetamine sensitization (which increases dopaminergic tone in this area) favors a shift from goal directed to habitual behavior (Nelson and Killcross, 2006).

9.8 BEHAVIORAL SEQUENCES AND SKILLS

The dorsal striatum and its dopaminergic afferents have been implicated in the acquisition of action sequences. Recent studies showed that during the learning of particular behavioral sequence, dopaminergic neurons and striatal MSNs develop neural activity specifically signaling the initiation or termination of the newly acquired behavioral sequences (Jin and Costa, 2010). Furthermore, a striatum-specific mutation of the NMDA receptors selectively impaired this type of learning without affecting improvements in performance speed, indicating that glutamatergic inputs to the striatum are necessary to acquire and consolidate action sequences. The role of striatal circuits in sequence learning may be dissociable from other functions of these circuits such as improvement in motor performance or the ability to learn the expected value of actions. These findings are consistent with the deficits in the initiation and termination of movement sequences observed in disorders affecting the striatum and its dopaminergic inputs, for example, Parkinson's and Huntington's diseases. Thus, individual actions may be linked together as a sequence via NMDA receptor-dependent plasticity.

That DA plays a significant role in long-term synaptic plasticity at the glutamatergic inputs to the striatum is now widely appreciated (Chapters 4 and 5). Because synaptic plasticity is the primary hypothesis for a biological substrate of learning, DA has also been implicated in the learning process (Miller, 1981). Like glutamatergic transmission elsewhere in the brain, the cortico-striatal and thalamostriatal inputs to the striatum are plastic—they can change as a consequence of experience, in the form of long-term potentiation (LTP) and long-term depression (LTD). Both direct and indirect pathway MSNs exhibit Hebbian plasticity; D1 receptors, in particular, have been implicated in LTP, especially in the direct pathway (Shen et al., 2008). DA is also known to be important for short-term and long-term depression of excitatory transmission, primarily by its activation of D2 receptors (Wang et al., 2006; Yin and Lovinger, 2006). A stimulation protocol used for effective intracranial self-stimulation can induce LTP in the associative striatum *in vivo*; and this form of LTP appears to require the activation of D1 receptors (Kerr and Wickens, 2001; Reynolds et al., 2001).

Another form of instrumental learning is skill learning—the refining of arbitrarily organized movement patterns. Previous studies have shown changes in neural activity in the striatum during motor and procedural learning (Carelli et al., 1997; Ungerleider et al., 2002; Brasted and Wise, 2004; Barnes et al., 2005). Furthermore, different striatal circuits seem to be involved in the early versus late phases of skill learning (Miyachi et al., 1997, 2002; Costa et al., 2004; Yin et al., 2009). For example, the dorsomedial or associative striatum, which receives input primarily from association cortices such as the prefrontal cortex (McGeorge and Faull, 1989; Voorn et al., 2004), seems to be more involved in the initial stages of skill and procedural learning than in later stages (Miyachi et al., 1997, 2002; Yin et al., 2009). On the other hand, the dorsolateral or sensorimotor striatum, which receives inputs from sensorimotor cortex (McGeorge and Faull, 1989; Voorn et al., 2004), is critical for the more gradual acquisition of habitual and automatic behavior (Miyachi et al., 1997, 2002; Yin et al., 2009). These phases of learning are accompanied by synaptic plasticity at glutamatergic synapses onto MSNs, as *ex vivo* studies have shown that skill learning results in long-lasting changes in glutamatergic transmission. Changes in the associative striatum seem to be predominant early in training, while changes in the sensorimotor striatum evolved only after extensive training (Yin et al., 2009). Furthermore, the observed potentiation in glutamatergic transmission in sensorimotor striatum after a skill is consolidated results in a decreased threshold for spiking in sensorimotor striatum. Furthermore, this potentiation of glutamatergic transmission is not only limited to an increase in the AMPA component, but also involves an increase in the NMDA-mediated component, which may play an important role in temporal summation and thus the bursting firing of MSNs (Yin et al., 2009).

Not surprisingly, blockade of dopaminergic signaling during the early stages of training impairs the execution of the newly learned actions (Choi et al., 2005; Yin et al., 2009), and rats with diminished nigrostriatal DA release are impaired in learning a new skill in a rotarod (Akita et al., 2006). However, with extended training, the execution of these actions can become less DA dependent (Choi et al., 2005; Yin et al., 2009). In particular, phasic DA may no longer be necessary for execution, since blockade of the low-affinity D1 receptors no longer disrupts performance. The reduced role of DA signaling after extended training could be attributed to the glutamatergic plasticity in the sensorimotor striatum; presumably, after training, glutamatergic inputs can drive MSNs more easily without the need for DA modulation. This is consistent with studies showing that a striatum-specific deletion of the NMDA receptor, which impairs LTP at glutamatergic synapses, displays deficits in skill learning (Dang et al., 2006). However, although phasic DA does not seem necessary for the performance of a well-learned skill, it seems to be important for the continuous refining of the skill upon additional training (Beeler et al., 2010).

Interestingly, there is some evidence indicating that the potentiation in glutamatergic transmission observed in the sensorimotor striatum after the consolidation of rotarod learning was observed predominantly in striatopallidal MSNs (i.e., the indirect pathway) (Yin et al., 2009). This finding is consistent with recent work showing that A2A adenosine receptors, which are mainly expressed in striatopallidal neurons and critical for habit formation (Yu et al., 2009). A2A receptors have been shown to be critical for LTP in the indirect pathway, so the lack of A2A-dependent LTP may be responsible for impaired habit formation in mice with striatum-specific A2A deletion (Shen et al., 2008).

9.9   CONCLUSIONS

In the striatum, glutamatergic projections from the cortex and thalamus, modulated by the dopaminergic input from the midbrain, are critical for voluntary behavior. DA can change the degree to which the striatal projection neurons are entrained to the cortical inputs (Costa et al., 2006a; Costa, 2007). By such interactions, both firing rate and the coordination of the neural activity are profoundly influenced by DA.

At the cellular level, DA's effects on striatal MSNs appear to be fairly uniform, and vary only according to receptor subtypes, which show distinct expression patterns in different cell populations. The excitatory inputs serve to provide the basis for selection in the striatum, which is the input nucleus of the basal ganglia. As the striatum receives many inputs from the entire cerebral cortex, which of the inputs will combine to exert some effect downstream is determined at this stage, when glutamatergic and dopaminergic inputs interact at the first excitatory synapse in the cortico-basal ganglia networks. At a computational level, the role of DA is to select the type of excitatory inputs to the basal ganglia that can have an impact on the eventual behavior itself. This can be achieved through DA's short-term effects—, modulation of synaptic transmission and long-term effects, that is, modulation of synaptic plasticity, which result in consolidation and maintenance of the short-term effects.

Having made significant progress in our understanding of the effects of DA at the cellular and molecular levels in primarily *in vitro* preparations, the crucial next step will be to elucidate the functional role of DA at the level of neural circuits and behavior. Current research is focused on the acquisition and expression of behavioral control using quantitative assays based on operational definitions. Distinct functional domains in the striatum reflect distinct cortico-basal ganglia circuits, which differ in their representation of behavioral parameters and their control capability. Anatomically, they also show significant variation in the expression of receptors, which can potentially shed light on regional variations in the modulatory influence of DA on glutamatergic transmission and behavior. As the previous review suggests, the challenge in explaining the myriad effects of dopaminergic and glutamatergic manipulations on behavior stems partly from the complexity of the anatomical organization, and partly from the complexity of the behaviors generated. To meet such a challenge, we will need more refined analysis of behavior combined with new tools for mapping neural circuits and *in vivo* measure of neural activity.

ACKNOWLEDGMENTS

This work is supported by AA018018 and AA016991 to HHY and by ERC StG 243393 to RMC.

REFERENCES

Akita H, Ogata M, Jitsuki S, Ogura T, Oh-Nishi A, Hoka S, Saji M (2006) Nigral injection of antisense oligonucleotides to synaptotagmin I using HVJ-liposome vectors causes disruption of dopamine release in the striatum and impaired skill learning. *Brain Res* 1095:178–189.

Albin RL, Young AB, Penney JB (1989) The functional anatomy of basal ganglia disorders. *Trends Neurosci* 12:366–375.

Aldridge JW, Berridge KC (1998) Coding of serial order by neostriatal neurons: A "natural action" approach to movement sequence. *J Neurosci* 18:2777–2787.

Alexander GE, Crutcher MD (1990) Functional architecture of basal ganglia circuits: Neural substrates of parallel processing. *Trends Neurosci* 13:266–271.

Androulidakis AG, Brucke C, Kempf F, Kupsch A, Aziz T, Ashkan K, Kuhn AA, Brown P (2008) Amplitude modulation of oscillatory activity in the subthalamic nucleus during movement. *Eur J Neurosci* 27:1277–1284.

Badiani A, Oates MM, Fraioli S, Browman KE, Ostrander MM, Xue CJ, Wolf ME, Robinson TE (2000) Environmental modulation of the response to amphetamine: Dissociation between changes in behavior and changes in dopamine and glutamate overflow in the rat striatal complex. *Psychopharmacology (Berl)* 151:166–174.

Balleine BW, Dickinson A (1998) Goal-directed instrumental action: Contingency and incentive learning and their cortical substrates. *Neuropharmacology* 37:407–419.

Bamford NS, Zhang H, Schmitz Y, Wu NP, Cepeda C, Levine MS, Schmauss C, Zakharenko SS, Zablow L, Sulzer D (2004) Heterosynaptic dopamine neurotransmission selects sets of corticostriatal terminals. *Neuron* 42:653–663.

Barnes TD, Kubota Y, Hu D, Jin DZ, Graybiel AM (2005) Activity of striatal neurons reflects dynamic encoding and recoding of procedural memories. *Nature* 437:1158–1161.

Beeler JA, Cao ZF, Kheirbek MA, Ding Y, Koranda J, Murakami M, Kang UJ, Zhuang X (2010) Dopamine-dependent motor learning: Insight into levodopa's long-duration response. *Ann Neurol* 67:639–647.

Bergman H, Wichmann T, Karmon B, DeLong MR (1994) The primate subthalamic nucleus. II. Neuronal activity in the MPTP model of parkinsonism. *J Neurophysiol* 72:507–520.

Berridge KC (2007) The debate over dopamine's role in reward: The case for incentive salience. *Psychopharmacology (Berl)* 191:391–431.

Berridge KC, Aldridge JW (2000a) Super-stereotypy I: Enhancement of a complex movement sequence by systemic dopamine D1 agonists. *Synapse* 37:194–204.

Berridge KC, Aldridge JW (2000b) Super-stereotypy II: Enhancement of a complex movement sequence by intraventricular dopamine D1 agonists. *Synapse* 37:205–215.

Berridge KC, Aldridge JW, Houchard KR, Zhuang X (2005) Sequential super-stereotypy of an instinctive fixed action pattern in hyper-dopaminergic mutant mice: A model of obsessive compulsive disorder and Tourette's. *BMC Biol* 3:4.

Berridge KC, Robinson TE (1998) What is the role of dopamine in reward: Hedonic impact, reward learning, or incentive salience? *Brain Res Brain Res Rev* 28:309–369.

Brasted PJ, Wise SP (2004) Comparison of learning-related neuronal activity in the dorsal premotor cortex and striatum. *Eur J Neurosci* 19:721–740.

Brown P (2003) Oscillatory nature of human basal ganglia activity: Relationship to the pathophysiology of Parkinson's disease. *Mov Disord* 18:357–363.

Brown P, Kupsch A, Magill PJ, Sharott A, Harnack D, Meissner W (2002) Oscillatory local field potentials recorded from the subthalamic nucleus of the alert rat. *Exp Neurol* 177:581–585.

Brown P, Mazzone P, Oliviero A, Altibrandi MG, Pilato F, Tonali PA, Di Lazzaro V (2004) Effects of stimulation of the subthalamic area on oscillatory pallidal activity in Parkinson's disease. *Exp Neurol* 188:480–490.

Burkhardt JM, Constantinidis C, Anstrom KK, Roberts DC, Woodward DJ (2007) Synchronous oscillations and phase reorganization in the basal ganglia during akinesia induced by high-dose haloperidol. *Eur J Neurosci* 26:1912–1924.

Burkhardt JM, Jin X, Costa RM (2009) Dissociable effects of dopamine on neuronal firing rate and synchrony in the dorsal striatum. *Front Integr Neurosci* 3:28.

Canales JJ, Graybiel AM (2000) A measure of striatal function predicts motor stereotypy. *Nat Neurosci* 3:377–383.

Cannon CM, Palmiter RD (2003) Reward without dopamine. *J Neurosci* 23:10827–10831.

Carelli RM, Wolske M, West MO (1997) Loss of lever press-related neuronal firing of rat striatal forelimb neurons after repeated sessions in a lever pressing task. *J Neurosci* 17:1804–1814.

Carlsson A (2006) The neurochemical circuitry of schizophrenia. *Pharmacopsychiatry* 39 (Suppl 1):S10–S14.

Carlsson A, Lindqvist M, Magnusson T (1957) 3,4-Dihydroxyphenylalanine and 5-hydroxytryptophan as reserpine antagonists. *Nature* 180:1200.

Carrillo-Reid L, Tecuapetla F, Tapia D, Hernandez-Cruz A, Galarraga E, Drucker-Colin R, Bargas J (2008) Encoding network states by striatal cell assemblies. *J Neurophysiol* 99:1435–1450.

220 Dopamine–Glutamate Interactions in the Basal Ganglia

Cepeda C, Levine MS (1998) Dopamine and N-methyl-D-aspartate receptor interactions in the neostriatum. *Dev Neurosci* 20:1–18.

Chevalier G, Deniau JM (1990) Disinhibition as a basic process in the expression of striatal functions. *Trends Neurosci* 13:277–280.

Chevalier G, Deniau JM, Thierry AM, Feger J (1981) The nigro-tectal pathway. An electrophysiological reinvestigation in the rat. *Brain Res* 213:253–263.

Choi WY, Balsam PD, Horvitz JC (2005) Extended habit training reduces dopamine mediation of appetitive response expression. *J Neurosci* 25:6729–6733.

Costa RM (2007) Plastic corticostriatal circuits for action learning: What's dopamine got to do with it? *Ann N Y Acad Sci* 1104:172–191.

Costa RM, Gutierrez R, de Araujo IE, Coelho MR, Kloth AD, Gainetdinov RR, Caron MG, Nicolelis MA, Simon SA (2006b) Dopamine levels modulate the updating of tastant values. *Genes Brain Behav* 6(4):314–320.

Costa RM, Lin SC, Sotnikova TD, Cyr M, Gainetdinov RR, Caron MG, Nicolelis MA (2006a) Rapid alterations in corticostriatal ensemble coordination during acute dopamine-dependent motor dysfunction. *Neuron* 52:359–369.

Cromwell HC, Berridge KC (1996) Implementation of action sequences by a neostriatal site: A lesion mapping study of grooming syntax. *J Neurosci* 16:3444–3458.

Dang MT, Yokoi F, Yin HH, Lovinger DM, Wang Y, Li Y (2006) Disrupted motor learning and long-term synaptic plasticity in mice lacking NMDAR1 in the striatum. *Proc Natl Acad Sci USA* 103:15254–15259.

Darvas M, Palmiter RD (2009) Restriction of dopamine signaling to the dorsolateral striatum is sufficient for many cognitive behaviors. *Proc Natl Acad Sci USA* 106:14664–14669.

Day JJ, Roitman MF, Wightman RM, Carelli RM (2007) Associative learning mediates dynamic shifts in dopamine signaling in the nucleus accumbens. *Nat Neurosci* 10:1020–1028.

Day JJ, Wheeler RA, Roitman MF, Carelli RM (2006) Nucleus accumbens neurons encode Pavlovian approach behaviors: Evidence from an autoshaping paradigm. *Eur J Neurosci* 23:1341–1351.

Derusso AL, Fan D, Gupta J, Shelest O, Costa RM, Yin HH (2010) Instrumental uncertainty as a determinant of behavior under interval schedules of reinforcement. *Front Integr Neurosci* 4:17.

Dias-Ferreira E, Sousa JC, Melo I, Morgado P, Mesquita AR, Cerqueira JJ, Costa RM, Sousa N (2009) Chronic stress causes frontostriatal reorganization and affects decision-making. *Science* 325:621–625.

Dickinson A (1985) Actions and habits: The development of behavioural autonomy. *Philos Trans R Soc B* 308:67–78.

Dickinson A (1989) Expectancy theory in animal conditioning. In: *Contemporary Learning Theories* (Klein SB, Mowrer RR, eds.), pp. 279–308. Hillsdale, NJ: Lawrence Erlbaum Associates.

Dickinson A (1994) Instrumental conditioning. In: *Animal Learning and Cognition* (Mackintosh NJ, ed.), pp. 45–79. Orlando, FL: Academic.

Ding J, Peterson JD, Surmeier DJ (2008) Corticostriatal and thalamostriatal synapses have distinctive properties. *J Neurosci* 28:6483–6492.

Engblom D, Bilbao A, Sanchis-Segura C, Dahan L, Perreau-Lenz S, Balland B, Parkitna JR, Lujan R, Halbout B, Mameli M, Parlato R, Sprengel R, Luscher C, Schutz G, Spanagel R (2008) Glutamate receptors on dopamine neurons control the persistence of cocaine seeking. *Neuron* 59:497–508.

Faure A, Haberland U, Conde F, El Massioui N (2005) Lesion to the nigrostriatal dopamine system disrupts stimulus-response habit formation. *J Neurosci* 25:2771–2780.

Ferguson SM, Robinson TE (2004) Amphetamine-evoked gene expression in striatopallidal neurons: Regulation by corticostriatal afferents and the ERK/MAPK signaling cascade. *J Neurochem* 91:337–348.

Gainetdinov RR, Caron MG (2003) Monoamine transporters: From genes to behavior. *Annu Rev Pharmacol Toxicol* 43:261–284.

Giros B, Jaber M, Jones SR, Wightman RM, Caron MG (1996) Hyperlocomotion and indifference to cocaine and amphetamine in mice lacking the dopamine transporter. *Nature* 379:606–612.

Goldberg JA, Rokni U, Boraud T, Vaadia E, Bergman H (2004) Spike synchronization in the cortex/basal-ganglia networks of Parkinsonian primates reflects global dynamics of the local field potentials. *J Neurosci* 24:6003–6010.

Gradinaru V, Mogri M, Thompson KR, Henderson JM, Deisseroth K (2009) Optical deconstruction of parkinsonian neural circuitry. *Science* 324:354–359.

Graybiel AM, Rauch SL (2000) Toward a neurobiology of obsessive-compulsive disorder. *Neuron* 28:343–347.

Grill HJ, Norgren R (1978) The taste reactivity test. II. Mimetic responses to gustatory stimuli in chronic thalamic and chronic decerebrate rats. *Brain Res* 143:281–297.

Grillner S, Hellgren J, Menard A, Saitoh K, Wikstrom MA (2005) Mechanisms for selection of basic motor programs—Roles for the striatum and pallidum. *Trends Neurosci* 28(7): 364–370.

Gross NB, Marshall JF (2009) Striatal dopamine and glutamate receptors modulate methamphetamine-induced cortical Fos expression. *Neuroscience* 161:1114–1125.

Haber SN (2003) The primate basal ganglia: Parallel and integrative networks. *J Chem Neuroanat* 26:317–330.

Hernandez-Lopez S, Bargas J, Surmeier DJ, Reyes A, Galarraga E (1997) D1 receptor activation enhances evoked discharge in neostriatal medium spiny neurons by modulating an L-type Ca^{2+} conductance. *J Neurosci* 17:3334–3342.

Hernandez-Lopez S, Tkatch T, Perez-Garci E, Galarraga E, Bargas J, Hamm H, Surmeier DJ (2000) D2 dopamine receptors in striatal medium spiny neurons reduce L-type Ca^{2+} currents and excitability via a novel PLC[beta]1-IP3-calcineurin-signaling cascade. *J Neurosci* 20:8987–8995.

Hikosaka O (2007) GABAergic output of the basal ganglia. *Prog Brain Res* 160:209–226.

Hikosaka O, Takikawa Y, Kawagoe R (2000) Role of the basal ganglia in the control of purposive saccadic eye movements. *Physiol Rev* 80:953–978.

Hnasko TS, Sotak BN, Palmiter RD (2005) Morphine reward in dopamine-deficient mice. *Nature* 438:854–857.

Horvitz JC (2000) Mesolimbocortical and nigrostriatal dopamine responses to salient non-reward events. *Neuroscience* 96:651–656.

Jin X, Costa RM (2010) Start/stop signals emerge in nigrostriatal circuits during sequence learning. *Nature* 466:457–462.

Joel D, Weiner I (1994) The organization of the basal ganglia-thalamocortical circuits: Open interconnected rather than closed segregated. *Neuroscience* 63:363–379.

Joel D, Weiner I (2000) The connections of the dopaminergic system with the striatum in rats and primates: An analysis with respect to the functional and compartmental organization of the striatum. *Neuroscience* 96:451–474.

Jog MS, Kubota Y, Connolly CI, Hillegaart V, Graybiel AM (1999) Building neural representations of habits. *Science* 286:1745–1749.

Jones SR, Gainetdinov RR, Wightman RM, Caron MG (1998) Mechanisms of amphetamine action revealed in mice lacking the dopamine transporter. *J Neurosci* 18:1979–1986.

Joyce EM, Iversen SD (1984) Dissociable effects of 6-OHDA-induced lesions of neostriatum on anorexia, locomotor activity and stereotypy: The role of behavioural competition. *Psychopharmacology (Berl)* 83:363–366.

Kepecs A, Raghavachari S (2007) Gating information by two-state membrane potential fluctuations. *J Neurophysiol* 97:3015–3023.

Kerr JN, Wickens JR (2001) Dopamine D-1/D-5 receptor activation is required for long-term potentiation in the rat neostriatum in vitro. *J Neurophysiol* 85:117–124.

Konorski J (1967) *Integrative Activity of the Brain.* Chicago, IL: University of Chicago Press.

Kravitz AV, Freeze BS, Parker PR, Kay K, Thwin MT, Deisseroth K, Kreitzer AC (2010) Regulation of parkinsonian motor behaviours by optogenetic control of basal ganglia circuitry. *Nature* 466:622–626.

Kreitzer AC (2009) Physiology and pharmacology of striatal neurons. *Annu Rev Neurosci* 32:127–147.

Kuhn AA, Trottenberg T, Kivi A, Kupsch A, Schneider GH, Brown P (2005) The relationship between local field potential and neuronal discharge in the subthalamic nucleus of patients with Parkinson's disease. *Exp Neurol* 194:212–220.

Lampl I, Yarom Y (1993) Subthreshold oscillations of the membrane potential: A functional synchronizing and timing device. *J Neurophysiol* 70:2181–2186.

Lau B, Glimcher PW (2007) Action and outcome encoding in the primate caudate nucleus. *J Neurosci* 27:14502–14514.

Lau B, Glimcher PW (2008) Value representations in the primate striatum during matching behavior. *Neuron* 58:451–463.

Lee KW, Kim Y, Kim AM, Helmin K, Nairn AC, Greengard P (2006) Cocaine-induced dendritic spine formation in D1 and D2 dopamine receptor-containing medium spiny neurons in nucleus accumbens. *Proc Natl Acad Sci USA* 103:3399–3404.

Levy R, Hutchison WD, Lozano AM, Dostrovsky JO (2000) High-frequency synchronization of neuronal activity in the subthalamic nucleus of parkinsonian patients with limb tremor. *J Neurosci* 20:7766–7775.

Liu XY, Chu XP, Mao LM, Wang M, Lan HX, Li MH, Zhang GC, Parelkar NK, Fibuch EE, Haines M, Neve KA, Liu F, Xiong ZG, Wang JQ (2006) Modulation of D2R-NR2B interactions in response to cocaine. *Neuron* 52:897–909.

Lorenz K (1973) Autobiography. In: *Les Prix Nobel.* Stockholm, Sweden.

Matsumoto M, Hikosaka O (2009) Two types of dopamine neuron distinctly convey positive and negative motivational signals. *Nature* 459:837–841.

McGeorge AJ, Faull RL (1989) The organization of the projection from the cerebral cortex to the striatum in the rat. *Neuroscience* 29:503–537.

Miller R (1981) *Meaning and Purpose in the Intact Brain.* New York: Oxford University Press.

Miyachi S, Hikosaka O, Lu X (2002) Differential activation of monkey striatal neurons in the early and late stages of procedural learning. *Exp Brain Res* 146:122–126.

Miyachi S, Hikosaka O, Miyashita K, Karadi Z, Rand MK (1997) Differential roles of monkey striatum in learning of sequential hand movement. *Exp Brain Res* 115:1–5.

Moore H, West AR, Grace AA (1999) The regulation of forebrain dopamine transmission: Relevance to the pathophysiology and psychopathology of schizophrenia. *Biol Psychiatry* 46:40–55.

Morelli M, Fenu S, Pinna A, Di Chiara G (1992) Opposite effects of NMDA receptor blockade on dopaminergic D1- and D2-mediated behavior in the 6-hydroxydopamine model of turning: Relationship with c-fos expression. *J Pharmacol Exp Ther* 260:402–408.

Nauta HJ (1979) Projections of the pallidal complex: An autoradiographic study in the cat. *Neuroscience* 4:1853–1873.

Nelson A, Killcross S (2006) Amphetamine exposure enhances habit formation. *J Neurosci* 26:3805–3812.

Nicola SM, Surmeier J, Malenka RC (2000) Dopaminergic modulation of neuronal excitability in the striatum and nucleus accumbens. *Annu Rev Neurosci* 23:185–215.

Nicola SM, Woodward Hopf F, Hjelmstad GO (2004) Contrast enhancement: A physiological effect of striatal dopamine? *Cell Tissue Res* 318:93–106.

Nisenbaum ES, Mermelstein PG, Wilson CJ, Surmeier DJ (1998) Selective blockade of a slowly inactivating potassium current in striatal neurons by (+/−) 6-chloro-APB hydrobromide (SKF82958). *Synapse* 29:213–224.

Palmiter RD (2008) Dopamine signaling in the dorsal striatum is essential for motivated behaviors: Lessons from dopamine-deficient mice. *Ann NY Acad Sci* 1129:35–46.

Pierce RC, Kalivas PW (1997) A circuitry model of the expression of behavioral sensitization to amphetamine-like psychostimulants. *Brain Res Brain Res Rev* 25:192–216.

Pierce RC, Rebec GV (1995) Iontophoresis in the neostriatum of awake, unrestrained rats: Differential effects of dopamine, glutamate and ascorbate on motor- and nonmotor-related neurons. *Neuroscience* 67:313–324.

Plenz D, Kital ST (1999) A basal ganglia pacemaker formed by the subthalamic nucleus and external globus pallidus. *Nature* 400:677–682.

Pomata PE, Belluscio MA, Riquelme LA, Murer MG (2008) NMDA receptor gating of information flow through the striatum in vivo. *J Neurosci* 28:13384–13389.

Raz A, Frechter-Mazar V, Feingold A, Abeles M, Vaadia E, Bergman H (2001) Activity of pallidal and striatal tonically active neurons is correlated in mptp-treated monkeys but not in normal monkeys. *J Neurosci* 21:RC128.

Rebec GV (2006) Behavioral electrophysiology of psychostimulants. *Neuropsychopharmacology* 31:2341–2348.

Rebec GV, Bashore TR (1984) Critical issues in assessing the behavioral effects of amphetamine. *Neurosci Biobehav Rev* 8:153–159.

Reynolds JN, Hyland BI, Wickens JR (2001) A cellular mechanism of reward-related learning. *Nature* 413:67–70.

Robinson TE, Kolb B (1997) Persistent structural modifications in nucleus accumbens and prefrontal cortex neurons produced by previous experience with amphetamine. *J Neurosci* 17:8491–8497.

Robinson S, Rainwater AJ, Hnasko TS, Palmiter RD (2007) Viral restoration of dopamine signaling to the dorsal striatum restores instrumental conditioning to dopamine-deficient mice. *Psychopharmacology (Berl)* 191:567–578.

Robinson S, Sandstrom SM, Denenberg VH, Palmiter RD (2005) Distinguishing whether dopamine regulates liking, wanting, and/or learning about rewards. *Behav Neurosci* 119:5–15.

Rolls ET, Thorpe SJ, Boytim M, Szabo I, Perrett DI (1984) Responses of striatal neurons in the behaving monkey. 3. Effects of iontophoretically applied dopamine on normal responsiveness. *Neuroscience* 12:1201–1212.

Schultz W (1998) Predictive reward signal of dopamine neurons. J *Neurophysiol* 80:1–27.

Schultz W, Dayan P, Montague PR (1997) A neural substrate of prediction and reward. *Science* 275:1593–1599.

Schwartz B, Gamzu E (1977) Pavlovian control of operant behavior. In: *Handbook of Operant Behavior* (Honig W, Staddon J.E.R, eds.), pp. 53–97. Old Tappan, NJ: Prentice Hall.

Sharott A, Magill PJ, Harnack D, Kupsch A, Meissner W, Brown P (2005) Dopamine depletion increases the power and coherence of beta-oscillations in the cerebral cortex and subthalamic nucleus of the awake rat. *Eur J Neurosci* 21:1413–1422.

Shen W, Flajolet M, Greengard P, Surmeier DJ (2008) Dichotomous dopaminergic control of striatal synaptic plasticity. *Science* 321:848–851.

Singer HS, Szymanski S, Giuliano J, Yokoi F, Dogan AS, Brasic JR, Zhou Y, Grace AA, Wong DF (2002) Elevated intrasynaptic dopamine release in Tourette's syndrome measured by PET. *Am J Psychiatry* 159:1329–1336.

Skinner B (1938) *The Behavior of Organisms*. New York: Appleton-Century-Crofts.

Smith-Roe SL, Kelley AE (2000) Coincident activation of NMDA and dopamine D1 receptors within the nucleus accumbens core is required for appetitive instrumental learning. *J Neurosci* 20:7737–7742.

Sotnikova TD, Beaulieu JM, Barak LS, Wetsel WC, Caron MG, Gainetdinov RR (2005) Dopamine-independent locomotor actions of amphetamines in a novel acute mouse model of Parkinson disease. *PLoS Biol* 3:e271.

Stern EA, Jaeger D, Wilson CJ (1998) Membrane potential synchrony of simultaneously recorded striatal spiny neurons in vivo. *Nature* 394:475–478.

Sutton RS, Barto AG (1998) *Reinforcement Learning*. Cambridge, MA: MIT Press.

Swanson LW (2000) Cerebral hemisphere regulation of motivated behavior. *Brain Res* 886:113–164.

Swanson CJ, Kalivas PW (2000) Regulation of locomotor activity by metabotropic glutamate receptors in the nucleus accumbens and ventral tegmental area. *J Pharmacol Exp Ther* 292:406–414.

Swanson LW, Mogenson GJ, Gerfen CR, Robinson P (1984) Evidence for a projection from the lateral preoptic area and substantia innominata to the 'mesencephalic locomotor region' in the rat. *Brain Res* 295:161–178.

Swerdlow NR, Swanson LW, Koob GF (1984) Substantia innominata: Critical link in the behavioral expression of mesolimbic dopamine stimulation in the rat. *Neurosci Lett* 50:19–24.

Taverna S, Ilijic E, Surmeier DJ (2008) Recurrent collateral connections of striatal medium spiny neurons are disrupted in models of Parkinson's disease. *J Neurosci* 28:5504–5512.

Tobler PN, Dickinson A, Schultz W (2003) Coding of predicted reward omission by dopamine neurons in a conditioned inhibition paradigm. *J Neurosci* 23:10402–10410.

Torres G, Rivier C (1993) Cocaine-induced expression of striatal c-fos in the rat is inhibited by NMDA receptor antagonists. *Brain Res Bull* 30:173–176.

Tsai HC, Zhang F, Adamantidis A, Stuber GD, Bonci A, de Lecea L, Deisseroth K (2009) Phasic firing in dopaminergic neurons is sufficient for behavioral conditioning. *Science* 324:1080–1084.

Ungerleider LG, Doyon J, Karni A (2002) Imaging brain plasticity during motor skill learning. *Neurobiol Learn Mem* 78:553–564.

Ungerstedt U (1971a) Adipsia and aphagia after 6-hydroxydopamine induced degeneration of the nigro-striatal dopamine system. *Acta Physiol Scand Suppl* 367:95–122.

Ungerstedt U (1971b) Postsynaptic supersensitivity after 6-hydroxy-dopamine induced degeneration of the nigro-striatal dopamine system. *Acta Physiol Scand* Suppl 367:69–93.

Uslaner J, Badiani A, Day HE, Watson SJ, Akil H, Robinson TE (2001) Environmental context modulates the ability of cocaine and amphetamine to induce c-fos mRNA expression in the neocortex, caudate nucleus, and nucleus accumbens. *Brain Res* 920:106–116.

Vanderschuren LJ, Everitt BJ (2004) Drug seeking becomes compulsive after prolonged cocaine self-administration. *Science* 305:1017–1019.

Voorn P, Vanderschuren LJ, Groenewegen HJ, Robbins TW, Pennartz CM (2004) Putting a spin on the dorsal-ventral divide of the striatum. *Trends Neurosci* 27:468–474.

Waelti P, Dickinson A, Schultz W (2001) Dopamine responses comply with basic assumptions of formal learning theory. *Nature* 412:43–48.

Wang Z, Kai L, Day M, Ronesi J, Yin HH, Ding J, Tkatch T, Lovinger DM, Surmeier DJ (2006) Dopaminergic control of corticostriatal long-term synaptic depression in medium spiny neurons is mediated by cholinergic interneurons. *Neuron* 50:443–452.

Williams DR, Williams H (1969) Automaintenance in the pigeon: Sustained pecking despite contingent non-reinforcement. *J Exp Anal Behav* 12:511–520.

Wilson CJ (2004) Basal ganglia. In: *The Synaptic Organization of the Brain*, 5th edn., (Shephard GM, ed.). New York: Oxford University Press.

Wilson CJ, Kawaguchi Y (1996) The origins of two-state spontaneous membrane potential fluctuations of neostriatal spiny neurons. *J Neurosci* 16:2397–2410.

Wise RA (1982) Neuroleptics and operant behavior: The anhedonia hypothesis. *Behav Brain Sci* 5:39–87.

Wise RA (1984) Neural mechanisms of the reinforcing action of cocaine. *NIDA Res Monogr* 50:15–33.

Yin HH, Knowlton BJ (2006) The role of the basal ganglia in habit formation. *Nat Rev Neurosci* 7:464–476.

Yin HH, Knowlton BJ, Balleine BW (2004) Lesions of dorsolateral striatum preserve outcome expectancy but disrupt habit formation in instrumental learning. *Eur J Neurosci* 19:181–189.

Yin HH, Knowlton BJ, Balleine BW (2005a) Blockade of NMDA receptors in the dorsomedial striatum prevents action-outcome learning in instrumental conditioning. *Eur J Neurosci* 22:505–512.

Yin HH, Knowlton BJ, Balleine BW (2006) Inactivation of dorsolateral striatum enhances sensitivity to changes in the action-outcome contingency in instrumental conditioning. *Behav Brain Res* 166:189–196.

Yin HH, Lovinger DM (2006) Frequency-specific and D2 receptor-mediated inhibition of glutamate release by retrograde endocannabinoid signaling. *Proc Natl Acad Sci USA* 103:8251–8256.

Yin HH, Mulcare SP, Hilario MR, Clouse E, Holloway T, Davis MI, Hansson AC, Lovinger DM, Costa RM (2009) Dynamic reorganization of striatal circuits during the acquisition and consolidation of a skill. *Nat Neurosci* 12:333–341.

Yin HH, Ostlund SB, Balleine BW (2008) Reward-guided learning beyond dopamine in the nucleus accumbens: The integrative functions of cortico-basal ganglia networks. *Eur J Neurosci* 28:1437–1448.

Yin HH, Ostlund SB, Knowlton BJ, Balleine BW (2005b) The role of the dorsomedial striatum in instrumental conditioning. *Eur J Neurosci* 22:513–523.

Yu C, Gupta J, Chen JF, Yin HH (2009) Genetic deletion of A2A adenosine receptors in the striatum selectively impairs habit formation. *J Neurosci* 29:15100–15103.

Zahm DS (2006) The evolving theory of basal forebrain functional-anatomical 'macrosystems.' *Neurosci Biobehav Rev* 30:148–172.

Zweifel LS, Parker JG, Lobb CJ, Rainwater A, Wall VZ, Fadok JP, Darvas M, Kim MJ, Mizumori SJ, Paladini CA, Phillips PE, Palmiter RD (2009) Disruption of NMDAR-dependent burst firing by dopamine neurons provides selective assessment of phasic dopamine-dependent behavior. *Proc Natl Acad Sci USA* 106:7281–7288.

10 Impaired Dopamine–Glutamate Receptor Interactions in Some Neurological Disorders

Miriam A. Hickey and Carlos Cepeda

CONTENTS

10.1 INTRODUCTION

The role of dopamine (DA) in neurological and psychiatric disorders is well established. In Parkinson's disease (PD), reduced levels of DA cause some of the cardinal motor symptoms (Riederer et al., 2001), and in schizophrenia, the DA hypothesis has occupied a prominent role for explaining positive symptoms. However, recently it has been recognized that glutamate (Glu) and its various receptors also play an important role in the etiology of schizophrenia. Similarly, based on the excitotoxic hypothesis of striatal neuronal death in Huntington's disease (HD), the role of Glu receptors is beyond doubt (DiFiglia, 1990). However, the role of DA

227

neurotransmission also has been recognized. Reduction in DA receptor expression is an early sign of neuropathology (Cha et al., 1999), and decreases in DA content have been reported in genetic mouse models (Hickey et al., 2002; Johnson et al., 2006). In fact, the late stages of HD bear resemblance to PD (Thompson et al., 1988). Finally, although PD is caused by reduced DA, PD symptoms and the dyskinesias caused by L-dopa are also modulated by Glu receptors (Brotchie, 2005).

These observations have led to the idea that in most neurological and psychiatric diseases, a single neurotransmitter probably cannot explain the full gamut of symptoms. Indeed, it is the way different neurotransmitter systems interact that leads to the variety of disease manifestations. Thus, the study of DA and Glu receptor interactions is becoming essential in order to understand neurological and psychiatric diseases. It must be emphasized that DA does not behave like a classical neurotransmitter as by itself it has no clear effects. DA acts like a modulator, increasing or decreasing voltage- and ligand-gated currents (Surmeier et al., 2007). In the striatum, DA alters the effects of Glu and γ-aminobutyric acid (GABA) receptor activation. Our laboratory pioneered the study of DA–Glu receptor interactions, and 10 years ago, we proposed a paradigm to predict the possible outcomes of these interactions (Cepeda and Levine, 1998). As it would be impossible to provide a deep and extensive review of the literature on impaired DA–Glu receptor interactions in the multitude of neurologic and psychiatric diseases where DA plays a role, in this chapter we are going to concentrate only on the role of DA–Glu receptor interactions in HD, PD, dyskinesias, and dystonia. The main focus is the corticostriatal synapse, a prime site of DA–Glu receptor interactions (Chapters 4 and 5).

10.2 BASIS FOR DA–GLU RECEPTOR INTERACTIONS IN THE STRIATUM

Medium-sized spiny neurons (MSNs), which constitute 90%–95% of all striatal neurons, utilize GABA as their principal neurotransmitter (Kita and Kitai, 1988) and also co-localize different peptides (Gerfen, 1992). The dorsal striatum receives Glu inputs from almost all neocortical areas (Wilson, 1990) and these inputs synapse onto spines of MSNs (Kemp and Powell, 1970). The striatum also receives DA projections from the substantia nigra pars compacta (Graybiel, 1990). These elements constitute the structural basis for DA–Glu receptor interactions in the striatum (Smith and Bolam, 1990).

The effects of DA are mediated by at least five identified receptor subtypes (Civelli et al., 1991; Sibley and Monsma, 1992). These subtypes have been classified into two pharmacological families, D1 (which includes D1 and D5 subtypes) and D2 (which includes D2, D3, and D4 subtypes). D2 receptors also are localized on corticostriatal endings (Wang and Pickel, 2002) where they regulate Glu release (Mercuri et al., 1985; Kornhuber and Kornhuber, 1986; Cepeda et al., 2001b; Bamford et al., 2004b). Similarly, Glu receptors are classified in two principal groups; ionotropic and metabotropic (Chapters 1 and 2). Ionotropic [α-amino-3-hydroxy-5-methyl-4-propionate (AMPA), kainate (KA), and NMDA] Glu receptors are ligand-gated cation channels, whereas metabotropic Glu receptors are coupled to various signal transduction systems (Monaghan et al., 1989; Hollmann and Heinemann, 1994;

Nakanishi, 1994). NMDA receptors are unique in that their activation is governed by a strong voltage dependence due to receptor-channel blockade by Mg^{2+} at hyperpolarized membrane potentials (Nowak et al., 1984). Mg^{2+} block gives NMDA receptors their characteristic negative slope conductance.

10.3 ELECTROPHYSIOLOGICAL PROPERTIES OF STRIATAL D1- AND D2-RECEPTOR–EXPRESSING MSNs

The striatal output is largely segregated into two populations of GABAergic MSNs with distinct projections, although some overlap exists (Shuen et al., 2008). The direct pathway consists of MSNs that predominantly express D1 DA receptors (Gerfen et al., 1990) and substance P (Haber and Nauta, 1983) and project to the substantia nigra pars reticulata and the internal segment of the globus pallidus (Albin et al., 1989; Gerfen et al., 1990). The indirect pathway is comprised of MSNs that express predominantly D2 receptors (Gerfen et al., 1990) and met-enkephalin (Haber and Nauta, 1983) and project to the external segment of the globus pallidus (Albin et al., 1989; Kawaguchi et al., 1990).

Anatomical evidence also has demonstrated differential excitatory inputs onto D1- and D2-receptor-containing MSNs (Reiner et al., 2003). Two types of pyramidal corticostriatal projections have been identified; one is ipsilateral and arises from collaterals of the pyramidal tract (PT-type), and the other is bilateral and projects only intratelencephalically to the cortex and striatum (IT-type). Terminals making asymmetric axospinous contact with striatonigral or D1-receptor-containing neurons are significantly smaller than those making contact with striatopallidal or D2-receptor-containing MSNs (Lei et al., 2004). The direct pathway neurons preferentially receive inputs from IT-type cortical neurons, whereas indirect pathway neurons receive greater inputs from PT-type cortical neurons, suggesting that D2 cells are subject to increased Glu release from corticostriatal terminals (Reiner et al., 2003).

It was believed that MSNs originating the direct and indirect pathways were morphologically and electrophysiologically identical. This assumption is changing due to the recent generation of mice expressing enhanced green fluorescent protein (EGFP) in the promoter region of D1 and D2 receptors that allows visualization of specific subpopulations of MSNs for recording. Studies *in vitro* demonstrate that cell input resistance, capacitance, time constant, and resting membrane potential are similar in D1- and D2-receptor-containing MSNs (Kreitzer and Malenka, 2007; Cepeda et al., 2008), although one recent study found that D2 cells have higher input resistances and are slightly but significantly more depolarized than D1 cells (Gertler et al., 2008). More consistent are studies of D1 and D2 MSNs, demonstrating that D2 cells are more excitable than D1 cells as, at similar current intensities, D2 cells fire more action potentials (Kreitzer and Malenka, 2007) or they have a lower threshold for action potential generation (Cepeda et al., 2008). Rheobase estimates indicate that the current required to induce firing is significantly reduced in D2 compared to D1 cells (Gertler et al., 2008). One factor contributing to this difference between subpopulations of MSNs appears to be that D2 cells have significantly smaller dendritic surface areas, due to fewer primary dendrites, which makes them more compact and consequently more excitable (Gertler et al., 2008).

Differences in synaptic inputs between subpopulations of MSNs also have been reported. Although initial studies examining spontaneous excitatory postsynaptic currents (EPSCs) in D1 and D2 MSNs did not find differences in frequency (Day et al., 2006), other studies have demonstrated significant differences. For example, in standard cerebrospinal fluid, the frequency of spontaneous EPSCs is higher in D2 compared to D1 cells and large-amplitude events (>100 pA) only occur in D2 cells (Cepeda et al., 2008). After the addition of the sodium channel blocker tetrodotoxin to isolate miniature EPSCs, the difference in frequency of EPSCs between D1 and D2 cells is reduced but the cumulative inter-event interval distributions are still significantly different (Kreitzer and Malenka, 2007; Cepeda et al., 2008). Further, after addition of the $GABA_A$ receptor blockers bicuculline or picrotoxin, which induce epileptiform activity in cortical pyramidal neurons, D2 cells display large membrane depolarizations rarely seen in D1 cells (Cepeda et al., 2008). The preferential propagation of epileptiform activity onto D2 cells supports previous data demonstrating that enkephalin-positive neurons that also express D2 receptors are selectively activated by cortical stimulation (Uhl et al., 1988; Berretta et al., 1997).

These results imply that D2-receptor-containing MSNs reflect ongoing cortical activity, particularly the activity generated by PT-type neurons, more faithfully than D1 cells, that is, inputs from PT-type neurons may provide D2 MSNs with a copy of the cortical motor signal (Reiner et al., 2003). This signal could be crucial for motor coordination. Another implication of increased intrinsic excitability and tighter corticostriatal synaptic coupling of D2 MSNs is that cells of the indirect pathway are more readily available for activation than cells of the direct pathway. In contrast to the tight synaptic coupling between cortical PT terminals and D2 MSNs, the smaller size and diffuse nature of the projections of the IT-type terminals will produce less activation of D1 neurons.

MSNs also receive important inputs from several thalamic nuclei. The nature of these inputs is beginning to be unraveled thanks to the use of a slice preparation that preserves both cortico- and thalamostriatal inputs (Smeal et al., 2007; Ding et al., 2008). In rats, at least 50% of striatal neurons receive convergent signals from the cerebral cortex and thalamic nuclei. In addition, at similar stimulation intensities, responses evoked by thalamic stimulation were smaller and revealed significant paired-pulse facilitation relative to cortical stimulation, which produced larger responses that showed insignificant changes in paired-pulse ratios (Smeal et al., 2007). In a more recent study, this group reported that the synapses mediating thalamostriatal input to a given MSN have a greater NMDA/AMPA ratio than do the synapses mediating cortical input to the same MSN (Smeal et al., 2008). As rats were used in these studies, it was not possible to separate D1 and D2 MSNs. This was done recently in EGFP mice (Ding et al., 2008). Interestingly, the results were almost exactly opposite from those reported in rats. For example, corticostriatal synapses in mice displayed paired-pulse facilitation attributable to a low-release probability at presynaptic terminals, whereas thalamostriatal synapses displayed a prominent paired-pulse depression attributable to a high-release probability. However, these synaptic features were indistinguishable in D1 and D2 MSNs (Ding et al., 2008). Further, in mice, the NMDA/AMPA ratio was larger at corticostriatal synapses than at thalamostriatal synapses in both D1 and D2 MSNs (Ding et al., 2008).

The basis for these differences is unclear, but it could be related to procedural differences including the species examined, recording electrode position in the striatum, and the age of the animals (Smeal et al., 2008).

10.4 DA MODULATION OF GLUTAMATE RECEPTOR–MEDIATED RESPONSES IN STRIATUM

DA modulation of responses mediated by the activation of Glu receptors is critical for striatal function. Our studies demonstrated that the outcome of DA modulation of Glu inputs depends not only on the type of DA receptor preferentially activated but also on the Glu receptor subtype activated. Thus, DA *via* D1 receptors enhances NMDA receptor–mediated responses whereas *via* D2 receptors it reduces AMPA receptor–mediated responses (Cepeda et al., 1993; Levine et al., 1996; Hernandez-Echeagaray et al., 2004). Although the other two interactions are less predictable, activation of D1 receptors generally enhances AMPA responses (Yan et al., 1999) and activation of D2 receptors decreases NMDA responses (Levine et al., 1996). The mechanisms by which DA produces differential effects on Glu transmission are complex and involve modulation of voltage-gated currents as well as a number of intracellular signaling pathways (Cepeda and Levine, 2006).

10.4.1 POSTSYNAPTIC MODULATION

Enhancement of NMDA responses by DA and D1 receptors can be mediated by a number of redundant and cooperative signaling cascades in the striatum (Cepeda and Levine, 1998; Seamans and Yang, 2004). The most prominent involve PKA and the DA- and adenosine $3',5'$-monophosphate (cAMP)-regulated phosphoprotein of 32 kDa (DARPP-32) (Colwell and Levine, 1995; Blank et al., 1997; Flores-Hernandez et al., 2002), phosphorylation of NMDA receptor NR1 subunits (Snyder et al., 1998), and activation of voltage-gated Ca^{2+} channels, particularly L-type channels (Cepeda et al., 1998; Tseng and O'Donnell, 2004). Activation of D1 receptors can also alter the surface distribution of NMDA receptors (Snyder et al., 2000; Hallett et al., 2006). For example, D1 receptor activation produces an increase in NR1, NR2A, and NR2B proteins in the synaptosomal membrane fraction (Dunah and Standaert, 2001) that is dependent on Fyn protein tyrosine kinase but not DARPP-32 (Dunah et al., 2004). Based on the fact that NMDA and D1 receptors partially overlap in dendritic spines, protein–protein interactions might direct the trafficking of D1 and NMDA receptors to the same subcellular domain. Evidence suggests that D1 and NMDA receptors are assembled as oligomeric units in the endoplasmic reticulum and transported to the cell surface as a preformed complex (Fiorentini et al., 2003). Compared to D1–NMDA receptor interactions, much less is known about the mechanisms by which D2 receptor activation leads to reduction of NMDA currents. Decreased cAMP production and PKA activity are certainly potential mechanisms. D2 receptors also can modulate neuronal excitability by activating the $PLC–IP_3–Ca^{2+}$ cascade (Hernandez-Lopez et al., 2000).

D1 receptors can also increase AMPA receptor–mediated responses. Activation of D1 receptors in cultured striatal neurons also promoted phosphorylation of AMPA receptors by PKA as well as potentiation of current amplitude (Price et al., 1999).

Similarly, in acutely isolated MSNs, activation of D1 receptors stabilized AMPA currents by preventing the rundown that is observed during repeated applications of the agonist (Yan et al., 1999). This effect was explained by increased phosphorylation and decreased dephosphorylation of AMPA channels that required inhibition of PP-1 activity by phosphorylated DARPP-32 (Yan et al., 1999). In contrast, DA and D2 receptors decrease AMPA receptor–mediated responses postsynaptically (Hernandez-Echeagaray et al., 2004), and this effect could involve diverse mechanisms. In MSNs, there is evidence that D2 receptor activation affects intracellular Ca^{2+} concentrations (Hernandez-Lopez et al., 2000) and therefore kinase and phosphatase activity. Any disruption in PKA function which favors phosphatase activity may reduce AMPA currents. D2 receptor activation reduces cAMP production through a G protein–mediated mechanism that also reduces the phosphorylation of DARPP-32 (Nishi et al., 1997; Lindgren et al., 2003). D2 receptor activation also could increase the activity of the protein phosphatase calcineurin, which dephosphorylates DARPP-32. Both signaling cascades lead to a decrease in PKA activity (Nishi et al., 1997) and GluR1 phosphorylation which, in turn, could reduce AMPA currents.

10.4.2 PRESYNAPTIC MODULATION OF STRIATAL GLUTAMATERGIC INPUTS BY DA

DA also alters Glu inputs onto MSNs by presynaptic mechanisms. DA receptors occur on presynaptic terminals where they can modulate neurotransmitter release (Wang and Pickel, 2002). In the dorsal striatum, D2 receptors decrease Glu release by presynaptic mechanisms (Flores-Hernandez et al., 1997; Cepeda et al., 2001b). Studies in mice lacking D2 receptors provide evidence of increased Glu release, indicating that such receptors function as gatekeepers, that is, primarily preventing excessive excitation in the striatum (Cepeda et al., 2001b). Further, optical techniques visualizing neurotransmitter release *via* destaining of terminals after incorporation of FM1-43, a styryl dye, have provided definite confirmation that D2 receptor activation alters Glu release at corticostriatal synapses (Bamford et al., 2004b). Inhibition of Glu release is frequency dependent, and it can act as a low-pass filter selective for terminals with low probability of release (Dani and Zhou, 2004). In this way, DA released by salient stimuli can directly regulate striatal neurotransmission by selecting specific sets of corticostriatal projections (Bamford et al., 2004b).

10.5 DA–GLU INTERACTIONS IN HUNTINGTON'S DISEASE

As mentioned earlier, it is becoming increasingly recognized that alterations in DA or Glu systems in the basal ganglia do not occur in isolation. Alterations in one system will necessarily affect the other. This has been recognized in schizophrenia as well as in other psychiatric and neurological disorders such as HD and PD.

HD is a genetic, progressive neurological disorder that is inherited in an autosomal dominant fashion. The symptoms include abnormal dance-like movements (chorea), cognitive disturbances, and disorders of mood (Harper, 1996). The HD gene is located on the short arm of chromosome 4 and contains an expansion in the normal number of CAG (glutamine) repeats, generally >40 (The Huntington's Disease Collaborative Research Group, 1993). HD is typically a late onset disease, although juvenile variants occur, usually when more CAG repeats are present. In young children with HD, the symptoms almost invariably include epileptic seizures (Rasmussen et al., 2000). Neuropathologically, HD is primarily characterized by neuronal loss in striatum and cortex (Vonsattel and DiFiglia, 1998). In the striatum, MSNs are most affected (Vonsattel et al., 1985). Although it has been generally believed that the progression of symptoms in the disorder is due to neurodegeneration of MSNs, it has become apparent that severe neuronal dysfunction precedes degeneration and is probably the major cause of many symptoms (Levine et al., 2004). Of equal importance, the understanding of mechanisms causing neuronal dysfunction will provide new targets for therapeutics that can be useful before degeneration has occurred.

The protein coded by the HD gene (*huntingtin*) is a large protein that is highly conserved and expressed ubiquitously throughout the body (Strong et al., 1993). Huntingtin is a cytoplasmic protein closely associated with vesicle membranes and microtubules, suggesting it may have a role in vesicle trafficking, exocytosis, and endocytosis (DiFiglia et al., 1995). In addition, its distribution is very similar to that of synaptophysin (Wood et al., 1996), and it has been shown to associate with various proteins involved in synaptic function. Thus, there is considerable evidence that mutant huntingtin can cause abnormal synaptic transmission in HD (Smith et al., 2005).

10.5.1 HUMAN STUDIES

The DA system has proven extremely useful for tracking neural dysfunction in HD patients. Several researchers have shown loss of D1 and D2 receptors in HD, both in postmortem tissue and through imaging techniques in patients. Asymptomatic HD patient striatal tissue has been shown to display reduced D1 and/or D2 receptors (Antonini et al., 1996; Weeks et al., 1996), while Grade 0–1* tissue shows reduced binding of D1 and D2 (Glass et al., 2000), indicative of their important role in HD pathogenesis. Imaging studies reveal loss in raclopride (D2) receptor binding in both manifest and presymptomatic HD patients, which researchers found correlated with increased microglial activation (Pavese et al., 2003; Tai et al., 2007) and also with reduced planning ability (Pavese et al., 2003). In addition to these striatal changes, reduced D2 binding and increased microglial activation are now also noted in hypothalamus, served by the tuberoinfundibular DA pathway (Politis et al., 2008). Importantly, imaging loss of receptors, in addition to measures of atrophy, could provide insight into the ongoing neural dysfunction, concomitant to neuronal degeneration in HD patients. There is even profound loss of the relatively newly discovered intrinsic DA neurons in postmortem grade 3/4 HD striatum (Huot and Parent, 2007; Huot et al., 2007). These profound changes in DA and DA signaling may underlie symptoms in HD patients, and indeed correlations are beginning to emerge. In humans, bradykinesia is an underlying manifestation of HD and becomes more apparent in late stages of the disease when choreic movements begin to subside (Thompson et al., 1988), reminiscent of the motor dysfunction in PD, a disease associated with profound loss of DA signaling.

---

* HD patient brain tissue is classified into five grades, from 0 to 4. The grading is based on the extent of striatal neuropathology, with no discernible degeneration observed in grade 0, whereas severe atrophy, neuronal loss, and astrocytosis are observed in grade 4 (Vonsattel et al., 1985).

Drugs that modulate the DA and Glu systems have a long history of use in HD. For example, tetrabenazine, a drug that reduces DA levels, is now the first drug approved by the US Food and Drug Administration for the treatment of chorea in HD. Ongoing trials continue to examine other agents to modulate the DA (and serotonin) system in HD (Bonelli et al., 2003; Brusa et al., 2009). In addition, modulators of Glu action also have been utilized. One particularly promising agent is memantine, a noncompetitive NMDA receptor antagonist. In two small HD clinical trials, it showed some promise and improved motor dysfunction (Ondo et al., 2007) while it retarded disease progression in another (Beister et al., 2004).

10.5.2 Animal Studies

Genetic mouse models of HD permit examination of the progression of the disease with great detail. Like HD in humans, most animal models also undergo changes in motor activity from hyper- to hypokinesia (Levine et al., 2004). Mouse models include transgenic (with a fragment or full-length human mutant gene), knock-in, and conditional models. None of the rodent models recapitulates the human disorder in its entirety nor displays the degree of neurodegeneration seen in human HD, but each has relevance to some of the symptoms (Levine et al., 2004).

The R6 line of transgenic mice (Mangiarini et al., 1996) is one of the most widely used models, not only because it was the first model generated but also because it offers many advantages. In particular, R6/2 mice (~150 CAG repeats) manifest a very aggressive, rapidly progressing form of HD. Transgenic R6/2 animals display overt behavioral symptoms as early as 4–5 weeks of age (Hickey et al., 2005) and die of unknown causes at about 15 weeks (Mangiarini et al., 1996). Alterations include the formation of neuronal intranuclear inclusions (Davies et al., 1997), changes in neurotransmitter receptor expression (Cha et al., 1998; Ariano et al., 2002), and altered components of signaling cascades (Bibb et al., 2000; Luthi-Carter et al., 2000; Menalled et al., 2000). Many of these alterations are correlated with ages at which first motor (Carter et al., 1999) and learning deficits (Lione et al., 1999; Murphy et al., 2000) occur.

The most widely used full-length models use yeast artificial chromosomes (YAC) expressing normal (YAC18) and mutant (YAC46 and YAC72, and YAC128) huntingtin (Hodgson et al., 1999; Slow et al., 2003). YAC72 mice display behavioral changes at around 7 months, as well as selective degeneration of MSNs in the lateral striatum by 12 months. Neurodegeneration can be present in the absence of aggregates in YAC mice, indicating that aggregates may not be essential to initiate neuronal death (Hodgson et al., 1999). YAC128 mice display similar but more severe alterations which occur earlier than in YAC72 mice (Slow et al., 2003); exhibiting increased open field activity at about 3 months, followed by rotarod abnormalities at 6 months. By 12 months, open field activity is diminished significantly compared to controls. In addition, modest (~10%) striatal atrophy and neuronal loss occur in the striatum and cortex of YAC128 mice (Van Raamsdonk et al., 2005).

Knock-in models also have emerged as major contributors to our understanding of HD. Several models that differ mainly in the number of CAG repeats (from 48 to 150) have been generated (White et al., 1997; Levine et al., 1999; Shelbourne et al., 1999;

Wheeler et al., 2000; Lin et al., 2001). Relative to the R6/2 mice, overt behavioral changes in knock-in mice are very late developing (2 years); however, more sensitive and careful testing demonstrates behavioral abnormalities as early as 1–2 months of age (Menalled et al., 2002, 2003; Hickey et al., 2008). Although delayed in onset in other models, nuclear staining and microaggregates are present in striatum at 2–6 months in the CAG140 and Q111 KI lines, which is relatively early in the course of the disease (Wheeler et al., 2002; Menalled et al., 2003). Nuclear inclusions are observed by 4 months of age in CAG140 mice but not until much later in other knock-in models (Menalled et al., 2002; Wheeler et al., 2002; Tallaksen-Greene et al., 2005), while loss of striatal neurons occurs at a late stage (2 years) (Heng et al., 2007; Hickey et al., 2008).

In R6/2 and YAC128 HD mice, alterations in glutamatergic function along the corticostriatal pathway are not uniform, but change dynamically in a biphasic manner. Although a progressive reduction in spontaneous and evoked glutamatergic synaptic activity, coinciding with the appearance of overt behavioral alterations, is the most noticeable change in R6/2 mice (Klapstein et al., 2001; Cepeda et al., 2003), dysregulation of glutamatergic input occurs early and is manifested by the presence of large-amplitude and complex synaptic events that peak around 5–7 weeks of age (Cepeda et al., 2003). We attributed these events to increased cortical excitability and possibly a reduction in presynaptic receptor function, including D2, $mGluR_{2-3}$, and CB_1 receptors (Cha et al., 1998; Luthi-Carter et al., 2000; Ariano et al., 2002). Predicted hyperexcitability in cortical networks has been confirmed in mouse models. Decreased inhibition in cortical pyramidal neurons, manifested by a reduction in spontaneous IPSCs, was also observed in the BAC HD model at 6 months, when motor dysfunction occurs (Spampanato et al., 2008).

Recently, we examined alterations in Glu release in the corticostriatal pathway of YAC128 mice at different stages of disease progression (1, 7, and 12 months), using combined optical and electrophysiological methods. Similar to results from R6/2 mice, the results in YAC128 mice demonstrated biphasic age-dependent changes in corticostriatal function. At 1 month, before the behavioral phenotype develops, AMPA receptor–mediated synaptic currents and Glu release evoked by cortical stimulation were increased. At 7 and 12 months, after the development of the behavioral phenotype, Glu release and AMPA synaptic currents were significantly reduced (Joshi et al., 2009). These effects were due to combined pre- and postsynaptic alterations.

Multiple alterations in corticostriatal synaptic function also occur depending on the pathogenicity of mutant huntingtin (Milnerwood and Raymond, 2007). Presynaptic dysfunction and a propensity toward synaptic depression in YAC72 and YAC128 compared to YAC18 mice occur at 1 month. In YAC128 mice, reduced AMPA responses evoked by intrastriatal stimulation also were observed. In contrast, when normalized to evoked AMPA currents, postsynaptic NMDA currents were enhanced in all three pathologic HD YAC variants (Milnerwood and Raymond, 2007). Increased NMDA receptor function was observed also in a subpopulation of MSNs in symptomatic R6/2 mice (Cepeda et al., 2001a). In a neurochemical animal model of HD, the NMDA receptor antagonist memantine reduced lesion size and apoptosis (Lee et al., 2006) and *in vitro*, it protected mutant (YAC128) striatal MSNs from Glu-induced cell death (Wu et al., 2006).

With regard to the DA system, profound and very early changes in several aspects of DA signaling are described in HD models, including loss of DA receptors (Cha et al., 1998) and DARPP-32 (Bibb et al., 2000; Hickey et al., 2008), reduced DA and DA metabolite levels (Abercrombie and Russo, 2002; Hickey et al., 2002), and reduced DA release (Johnson et al., 2006). Reductions in DA receptor function in R6/1 mice also could be responsible for deficits in cortical LTD as these can be reversed by quinpirole, a D2 receptor agonist (Cummings et al., 2006). In addition, in YAC128 mice, it was found that quinpirole produced a relative increase in corticostriatal inhibition in mutant tissue at 1 month but it became much less effective at 12 months compared to wild type (WT) mice (Joshi et al., 2009). Interestingly, in the CAG140 KI model, aggregates first develop in areas of DA innervation (Menalled et al., 2003) and enhancing DA levels in another knock-in model increases aggregates in these areas (Cyr et al., 2006). Increasing DA using L-dopa impairs fine motor performance of mice on rotarod (Hickey et al., 2002; Tang et al., 2007). However, subtle short-term benefits may also arise from increased DA in open field in mice (Hickey et al., 2002), and using other DA modulators in patients (Tedroff et al., 1999). On the other hand, tetrabenazine ameliorated motor dysfunction in YAC128 mice (Tang et al., 2007).

D2 receptor activation mediates toxicity in both *in vivo* and *in vitro* genetic models of HD (Charvin et al., 2005, 2008; Benchoua et al., 2008), although D1 receptor activation may also be involved (Robinson et al., 2008). However, a wealth of data also show that activation of D2 receptors in particular may be protective in several models of neurochemical striatal neurodegeneration (Bozzi and Borrelli, 2006). Imaging loss of receptors is a strategy also used in preclinical trials in mice, where D2 receptor binding was found to increase following cystamine treatment in R6/2 mice (Wang et al., 2005). Finally, both lesions of the DA nigrostriatal pathway and of the cortex are protective in HD models (Stack et al., 2007), again demonstrating the importance of both dopaminergic and glutamatergic pathways to HD pathogenesis.

10.6 DA–GLU INTERACTIONS IN PARKINSON'S DISEASE

PD is a degenerative disorder of the central nervous system that produces impairments in motor skills, language, and cognition (Jankovic, 2008). In humans, PD is characterized by muscle rigidity, tremor, bradykinesia, or, in extreme cases, akinesia. The primary symptoms are the result of decreased activation of the motor cortex by the basal ganglia, normally caused by insufficient DA in the striatum. PD, as well as other neurodegenerative disorders that affect motor function, is associated with abnormal neurotransmission along the corticostriatal pathway. Alterations in glutamatergic neurotransmission along this pathway could be sufficient to produce some of the cardinal symptoms of PD. Amantadine, a weak NMDA receptor antagonist, has long been used in PD and does show some symptom relief (Hallett and Standaert, 2004). In animal models of DA depletion, NMDA receptor blockade also produces antiparkinsonian effects (Blanchet et al., 1997; Chase et al., 2000). Additionally, NMDA and AMPA receptor blockade, can potentiate L-dopa antiparkinsonian effects, while also reducing dyskinetic movements (Blanchet et al., 1997; Chase et al., 2000; Bibbiani et al., 2005). NR2B-containing NMDA receptors appear to be important for this action and in reducing L-dopa-induced dyskinesias

(Steece-Collier et al., 2000; Hallett and Standaert, 2004; Nutt et al., 2008), and research continues to develop more selective agents for NR2B-containing NMDA receptors, with less cognitive side effects.

10.6.1 DOPAMINE DEPLETION MODELS OF PD

With the introduction of genetic mouse models of PD and electrophysiological characterization of alterations in the corticostriatal pathway, it has become evident that the mechanisms of motor dysfunction are different in genetic compared to neurotoxic models. Initial data from DA-depleted rats and cats indicated that the firing rate of striatal neurons increased in the ipsilateral side of the lesion (Hull et al., 1974; Schultz and Ungerstedt, 1978). Furthermore, after DA depletion, there was an increase in the frequency of spontaneous synaptic membrane depolarizations in MSNs (Galarraga et al., 1987; Cepeda et al., 1989; Calabresi et al., 1993), along with increased gap junctional communication (Cepeda et al., 1989), suggesting that presynaptic filtering by D2 receptors on corticostriatal terminals was reduced in PD. DA depletion also increased cell firing in striatopallidal neurons (Mallet et al., 2006), decreased the threshold to evoke cortical responses (Florio et al., 1993), and facilitated the occurrence of cortically generated membrane oscillations in a subpopulation of striatal neurons (Tseng et al., 2001). It is thus likely that the facilitation of corticostriatal input is selective to MSNs of the indirect pathway. In fact, the cells of the direct pathway appear to have reduced firing frequency probably as a consequence of decreased cortical activity (Mallet et al., 2006).

DA-deficient (DD) states sensitize presynaptic D2 receptor responses (Schultz, 1982; Calabresi et al., 1993; Kim et al., 2000) and likely influence cortical function by modifying cortical-basal ganglia circuits. Optical recordings have determined the effect of DA deficiency and replenishment at single cortical synaptic terminals in mouse models of acute and chronic DA depletion (Bamford et al., 2004a). Using reserpine-treated (Bamford et al., 2004a) and DD (Zhou and Palmiter, 1995) mice, these investigations demonstrated that DA depletion produced sensitized presynaptic D2 receptor responses and altered DA-mediated responses from subsets of corticostriatal terminals (Bamford et al., 2004a). Sensitized D2 receptors more broadly inhibited corticostriatal release and promoted further inhibition at the slow-releasing terminals, thus impairing presynaptic filtering. Similar alterations in DA receptor sensitivity (Zhou and Palmiter, 1995) associated with DA deficiency in humans would likely lead to bradykinesia under DA-depleted conditions or dyskinesias following DA replenishment (Bamford et al., 2004a).

Loss of spines in models of PD selectively affects D2 EGFP-positive MSNs (Day et al., 2006). Although strong evidence was presented that this effect could be attributed to dysregulation of postsynaptic L-type Ca^{2+} channels, a presynaptic contribution also is possible. As indicated earlier, DA-depleting lesions increase spontaneous Glu-mediated synaptic events. As this increase selectively affects D2-receptor-containing MSNs, it is possible that excessive Glu release can become neurotoxic and induce spine elimination, similar to effects that may occur in HD. Membrane loss can potentially induce increases in input resistance (Galarraga et al., 1987) making these cells even more electrotonically compact and excitable, thereby increasing sensitivity to Glu. Supporting experimental evidence for this idea was obtained recently by the demonstration that dendritic remodeling of MSNs seen in models of PD occurs secondary only to increases in corticostriatal glutamatergic drive (Neely et al., 2007).

10.6.2 GENETIC MODELS OF PD

Based on data from genetic models, it is becoming increasingly clear that changes in the corticostriatal pathway are different and sometimes opposite to those found in acute or chronic toxic models of DA depletion. Thus, the initial perturbation in genetic models appears to be increased DA tissue concentration that leads to reduced D2 receptor function. For example, α-synuclein-overexpressing mice (ASO) exhibit a significantly lower frequency of spontaneous EPSCs in MSNs compared to age-matched WT littermates (Wu et al., 2010). In addition, whereas application of amphetamine reduces spontaneous EPSC frequency in control mice, it had little or no effect in ASO mice and DA D2 receptor agonists or antagonists produced contrasting effects in cells from ASO compared to WT mice. Together, these observations suggest that abnormal accumulation of α-synuclein alters corticostriatal synaptic function and DA modulation and contributes to some of the behavioral abnormalities in ASO mice (Wu et al., 2007). Alterations in corticostriatal synaptic plasticity in the corticostriatal pathway of transgenic ASO mice also are observed (Watson et al., 2009). Whereas striatal long-term depression (LTD) occurred in ASO striatum, it did not occur in WTs. In contrast, corticostriatal LTD was absent in DJ-1 knock-out mice, while it was present in WT mice, and LTD was rescued in the mutants by treating with quinpirole (Goldberg et al., 2005). Mutations causing loss of DJ-1 protein underlie some familial forms of PD, and it is thought to play a role in the response of the cell to oxidative stress (Fleming and Chesselet, 2006). Loss of DJ-1 protein is also associated with increased susceptibility to oxidative stress, in the form of MPTP-induced SNc degeneration (Kim et al., 2005), whereas overexpression of DJ-1 protects against DA-induced toxicity and oxidative stress (Lev et al., 2009). Interestingly, paired-pulse facilitation increased after induction of LTD, suggesting that α-synuclein may impact long-term forms of synaptic plasticity by preferentially reducing presynaptic Glu release from corticostriatal terminals (Watson et al., 2009). In addition, the altered electrophysiology of the DA system could explain why these mice do not show stereotypies when challenged with amphetamine (Fleming et al., 2006).

The results in ASO mice resemble those found in DA transporter (DAT) knock-down mice (Wu et al., 2007), which display increased DA in the extracellular space due to reduced clearance. Whereas in control mice amphetamine reduced the frequency of spontaneous EPSCs by activation of D2 receptors, in DAT knock-down mice either no changes or small increases in frequency occur, suggesting altered sensitivity of D2 receptors (Wu et al., 2007). Increased extracellular DA in striatum appears to be a common finding in several genetic animal models of PD (Goldberg et al., 2003), including ASO mice (Maidment et al., 2006) and DJ-1 knock-out mice [(Chen et al., 2005), however see (Yamaguchi and Shen, 2007)], suggesting that increased extracellular DA may contribute to substantia nigra DA neuron stress and eventual degeneration, possibly by a retrograde mechanism. This is reminiscent of the retrograde degeneration of DA neurons seen after 6-hydroxydopamine injections in the striatum (Sauer and Oertel, 1994).

10.7 DA AND GLU IN DYSTONIA

Dystonia is a neurological condition in which patients exhibit sustained muscle contractions of opposing muscles, which leads to twisting or abnormal postures (Albanese et al., 2006). It is a highly disabling disorder and symptom manifestations are quite diverse. Generalized dystonia affects several parts of the body, whereas focal dystonia affects a specific body area, for example, cervical dystonia (Albanese et al., 2006). Primary dystonia, by definition, must occur in the absence of other neurological disorders, with no evidence of brain injury, and most cases of dystonia are idiopathic; however, some are inherited or induced by other diseases or treatments (Jankovic, 2006). For example, antipsychotic treatments can induce dystonia, but careful modulation of therapeutics can help prevent the dystonic movements while retaining antipsychotic capability (Gourzis et al., 2005; Ciranni et al., 2009). Interestingly, fluorodopa uptake is increased in striatum in idiopathic dystonia (Otsuka et al., 1992), and during dystonic episodes, striatal DA is elevated in the dtsz hamster model of paroxysmal dystonia (Hamann and Richter, 2004).

Using imaging in patients, several studies have shown presynaptic and postsynaptic alterations involving the striatal DA system in idiopathic dystonia patients (Otsuka et al., 1992; Perlmutter et al., 1997; Naumann et al., 1998). Polymorphisms within the coding sequence for the DA D5 receptor have been linked to cervical dystonia (Placzek et al., 2001; Brancati et al., 2003) and in blepharospasm, where patients exhibit involuntary closing of the eyelids (Misbahuddin et al., 2002). However, more research is required to understand how these polymorphisms cause dystonia. Additional evidence supports a role for loss of striatal D2 receptors in patients with dystonia (Perlmutter et al., 1997; Naumann et al., 1998). Remarkably, patients with writer's cramp, a type of focal dystonia, show loss of D2 receptor binding in striatum which is corrected following biofeedback-based sensorimotor training (Horstink et al., 1997; Berger et al., 2007). These latter studies suggest that neuronal dysfunction, rather than neuronal loss, underlies dystonia. Modulation of DA signaling through the use of risperidone (5HT2 and D2 antagonist) may also be beneficial in treating dystonia (Grassi et al., 2000).

Although rare, several forms of dystonia are inherited, and they also show striatal DA system involvement. Several mutations are in receptors and other components of the DA-signaling cascade. In particular, a small subset of dystonia patients can be treated successfully, and symptoms may even resolve, with low-dose L-dopa (Albanese et al., 2006). These juvenile patients have L-dopa-responsive dystonia, which is caused by mutations in genes involved in DA synthesis, including tyrosine hydroxylase, GTP cyclohydrolase 1, or sepiapterin reductase (Furukawa et al., 1999; de Carvalho Aguiar and Ozelius, 2002; Segawa et al., 2003; Jankovic, 2006). GTP cyclohydrolase 1 and sepiopterin reductase are involved in the production of cofactors for TH, and TH is the rate-limiting enzyme in the generation of DA from norepinephrine (Furukawa et al., 1999; de Carvalho Aguiar and Ozelius, 2002;

Albanese et al., 2006). Several other mutations in the DA-signaling pathway also lead to the development of dystonia including D2 receptors (DYT11), L-aromatic acid decarboxylase deficiency, or biopterin deficiencies, highlighting the importance of the DA system in these inherited dystonias (Furukawa et al., 1999; de Carvalho Aguiar and Ozelius, 2002; Segawa et al., 2003; Thony and Blau, 2006; Beukers et al., 2009). Rodent models carrying some of these mutations also develop DA-signaling striatal dysfunction (Zeng et al., 2004; Balcioglu et al., 2007; Sato et al., 2008; Zhao et al., 2008). A single-residue deletion in the torsinA protein results in DYT1 dystonia, the most common inherited dystonia, and a mouse overexpressing this mutant protein shows disinhibition of striatal GABAergic synaptic activity, related in part to D2 receptor dysfunction (Sciamanna et al., 2009).

Perhaps unsurprisingly given the data described earlier on other diseases affecting movement, a central role for Glu signaling and its modulation in dystonia is emerging. In the dtsz hamster model of paroxysmal dystonia, systemic MK-801 exerts antidystonic effects, as well as potentiating the antidystonic effects of the atypical neuroleptic clozapine (Richter et al., 1991; Richter and Loscher, 1997), while extracellular striatal levels of Glu are normal (Hamann et al., 2008). Subsequent research showed that striatal AMPA receptors, at least in part, may play an important role for modulating dystonia in this model (Sander and Richter, 2007). Dystonia patients show reduced temporal discrimination of sensory stimuli (Tinazzi et al., 2003), and sensorimotor cortical and network overactivity is observed in patients with several types of dystonia which is also associated with reduced D2 receptor binding in striatum (Carbon et al., 2010a,b; Simonyan and Ludlow, 2010). Although much has been learned, more research is required in order to develop a clearer model for the subtle interplay between DA and glut in striatum in dystonia. This research may have the added benefit of being applicable to other diseases, in which dystonic movements manifest, such as HD and PD (Harper, 1996; Albanese et al., 2006).

10.8   CONCLUSIONS

Neurologic and psychiatric diseases that in the past were attributed only to a single neurotransmitter system can no longer be viewed as such. In particular, changes in DA and Glu neurotransmission reciprocally affect each other and regulatory processes occur in order to compensate for down- or upregulation of the other neurotransmitter and its receptors. As shown in this chapter, symptoms attributed to one neurotransmitter can be modified by activating receptors of other neurotransmitters or neuromodulators. This fact opens a wide array of therapeutic alternatives and better ways to reduce potential side effects. Much has still to be learned; however, increased knowledge of this interplay may aid in reducing treatment side effects and improving therapeutics, in addition to better understanding of the pathophysiology underlying these devastating disorders.

ACKNOWLEDGMENTS

This work was supported by USPHS grants NS33538 and NS41574 and the Cure HD Initiative.

REFERENCES

Abercrombie E, Russo M (2002) *Neurochemistry in the R6/2 Transgenic Mouse Model of Huntington's Disease*. Society for Neuroscience, Orlando, FL.

Albanese A, Barnes MP, Bhatia KP, Fernandez-Alvarez E, Filippini G, Gasser T, Krauss JK, Newton A, Rektor I, Savoiardo M, Valls-Sole J (2006) A systematic review on the diagnosis and treatment of primary (idiopathic) dystonia and dystonia plus syndromes: Report of an EFNS/MDS-ES Task Force. *Eur J Neurol* 13:433–444.

Albin RL, Young AB, Penney JB (1989) The functional anatomy of basal ganglia disorders. *Trends Neurosci* 12:366–375.

Antonini A, Leenders KL, Spiegel R, Meier D, Vontobel P, Weigell-Weber M, Sanchez-Pernaute R, de Yebenez JG, Boesiger P, Weindl A, Maguire RP (1996) Striatal glucose metabolism and dopamine D2 receptor binding in asymptomatic gene carriers and patients with Huntington's disease. *Brain* 119(Pt 6):2085–2095.

Ariano MA, Aronin N, Difiglia M, Tagle DA, Sibley DR, Leavitt BR, Hayden MR, Levine MS (2002) Striatal neurochemical changes in transgenic models of Huntington's disease. *J Neurosci Res* 68:716–729.

Balcioglu A, Kim MO, Sharma N, Cha JH, Breakefield XO, Standaert DG (2007) Dopamine release is impaired in a mouse model of DYT1 dystonia. *J Neurochem* 102:783–788.

Bamford NS, Robinson S, Joyce JA, Palmiter RD (2004a) Dopamine-deficient mice demonstrate hypersensitive corticostriatal D2 receptors. *Ann Neurol* 56:S85.

Bamford NS, Zhang H, Schmitz Y, Wu NP, Cepeda C, Levine MS, Schmauss C, Zakharenko SS, Zablow L, Sulzer D (2004b) Heterosynaptic dopamine neurotransmission selects sets of corticostriatal terminals. *Neuron* 42:653–663.

Beister A, Kraus P, Kuhn W, Dose M, Weindl A, Gerlach M (2004) The N-methyl-D-aspartate antagonist memantine retards progression of Huntington's disease. *J Neural Transm* 68(Suppl):117–122.

Benchoua A, Trioulier Y, Diguet E, Malgorn C, Gaillard MC, Dufour N, Elalouf JM, Krajewski S, Hantraye P, Deglon N, Brouillet E (2008) Dopamine determines the vulnerability of striatal neurons to the N-terminal fragment of mutant huntingtin through the regulation of mitochondrial complex II. *Hum Mol Genet* 17:1446–1456.

Berger HJ, van der Werf SP, Horstink CA, Cools AR, Oyen WJ, Horstink MW (2007) Writer's cramp: Restoration of striatal D2-binding after successful biofeedback-based sensorimotor training. *Parkinsonism Relat Disord* 13:170–173.

Berretta S, Parthasarathy HB, Graybiel AM (1997) Local release of GABAergic inhibition in the motor cortex induces immediate-early gene expression in indirect pathway neurons of the striatum. *J Neurosci* 17:4752–4763.

Beukers RJ, Booij J, Weisscher N, Zijlstra F, van Amelsvoort TA, Tijssen MA (2009) Reduced striatal D2 receptor binding in myoclonus-dystonia. *Eur J Nucl Med Mol Imaging* 36:269–274.

Bibb JA, Yan Z, Svenningsson P, Snyder GL, Pieribone VA, Horiuchi A, Nairn AC, Messer A, Greengard P (2000) Severe deficiencies in dopamine signaling in presymptomatic Huntington's disease mice. *Proc Natl Acad Sci USA* 97:6809–6814.

Bibbiani F, Oh JD, Kielaite A, Collins MA, Smith C, Chase TN (2005) Combined blockade of AMPA and NMDA glutamate receptors reduces levodopa-induced motor complications in animal models of PD. *Exp Neurol* 196:422–429.

Blanchet PJ, Papa SM, Metman LV, Mouradian MM, Chase TN (1997) Modulation of levodopa-induced motor response complications by NMDA antagonists in Parkinson's disease. *Neurosci Biobehav Rev* 21:447–453.

Blank T, Nijholt I, Teichert U, Kugler H, Behrsing H, Fienberg A, Greengard P, Spiess J (1997) The phosphoprotein DARPP-32 mediates cAMP-dependent potentiation of striatal N-methyl-D-aspartate responses. *Proc Natl Acad Sci USA* 94:14859–14864.

Bonelli R, Mayr B, Niederwieser G, Reisecker F, Kapfhammer H-P (2003) Ziprasidone in Huntington's disease: The first case reports. *J Psychopharmacol* 17:459–460.

Bozzi Y, Borrelli E (2006) Dopamine in neurotoxicity and neuroprotection: What do D2 receptors have to do with it? *Trends Neurosci* 29:167–174.

Brancati F, Valente EM, Castori M, Vanacore N, Sessa M, Galardi G, Berardelli A, Bentivoglio AR, Defazio G, Girlanda P, Abbruzzese G, Albanese A, Dallapiccola B (2003) Role of the dopamine D5 receptor (DRD5) as a susceptibility gene for cervical dystonia. *J Neurol Neurosurg Psychiatry* 74:665–666.

Brotchie JM (2005) Nondopaminergic mechanisms in levodopa-induced dyskinesia. *Mov Disord* 20:919–931.

Brusa L, Orlacchio A, Moschella V, Iani C, Bernardi G, Mercuri N (2009) Treatment of the symptoms of Huntington's disease: Preliminary results comparing aripiprazole and tetrabenazine. *Mov Disord* 24:126–129.

Calabresi P, Mercuri NB, Sancesario G, Bernardi G (1993) Electrophysiology of dopamine-denervated striatal neurons. Implications for Parkinson's disease. *Brain* 116:433–452.

Carbon M, Argyelan M, Eidelberg D (2010a) Functional imaging in hereditary dystonia. *Eur J Neurol* 17 (Suppl 1):58–64.

Carbon M, Argyelan M, Habeck C, Ghilardi MF, Fitzpatrick T, Dhawan V, Pourfar M, Bressman SB, Eidelberg D (2010b) Increased sensorimotor network activity in DYT1 dystonia: A functional imaging study. *Brain* 133:690–700.

Carter RJ, Lione LA, Humby T, Mangiarini L, Mahal A, Bates GP, Dunnett SB, Morton AJ (1999) Characterization of progressive motor deficits in mice transgenic for the human Huntington's disease mutation. *J Neurosci* 19:3248–3257.

de Carvalho Aguiar PM, Ozelius LJ (2002) Classification and genetics of dystonia. *Lancet Neurol* 1:316–325.

Cepeda C, Andre VM, Yamazaki I, Wu N, Kleiman-Weiner M, Levine MS (2008) Differential electrophysiological properties of dopamine D1 and D2 receptor-containing striatal medium-sized spiny neurons. *Eur J Neurosci* 27:671–682.

Cepeda C, Ariano MA, Calvert CR, Flores-Hernandez J, Chandler SH, Leavitt BR, Hayden MR, Levine MS (2001a) NMDA receptor function in mouse models of Huntington disease. *J Neurosci Res* 66:525–539.

Cepeda C, Buchwald NA, Levine MS (1993) Neuromodulatory actions of dopamine in the neostriatum are dependent upon the excitatory amino acid receptor subtypes activated. *Proc Natl Acad Sci USA* 90:9576–9580.

Cepeda C, Colwell CS, Itri JN, Chandler SH, Levine MS (1998) Dopaminergic modulation of NMDA-induced whole cell currents in neostriatal neurons in slices: Contribution of calcium conductances. *J Neurophysiol* 79:82–94.

Cepeda C, Hurst RS, Altemus KL, Flores-Hernandez J, Calvert CR, Jokel ES, Grandy DK, Low MJ, Rubinstein M, Ariano MA, Levine MS (2001b) Facilitated glutamatergic transmission in the striatum of D2 dopamine receptor-deficient mice. *J Neurophysiol* 85:659–670.

Cepeda C, Hurst RS, Calvert CR, Hernandez-Echeagaray E, Nguyen OK, Jocoy E, Christian LJ, Ariano MA, Levine MS (2003) Transient and progressive electrophysiological alterations in the corticostriatal pathway in a mouse model of Huntington's disease. *J Neurosci* 23:961–969.

Cepeda C, Levine MS (1998) Dopamine and N-methyl-D-aspartate receptor interactions in the neostriatum. *Dev Neurosci* 20:1–18.

Cepeda C, Levine MS (2006) Where do you think you are going? The NMDA-D1 receptor trap. *Sci STKE* 2006:pe20.

Cepeda C, Walsh JP, Hull CD, Howard SG, Buchwald NA, Levine MS (1989) Dye-coupling in the neostriatum of the rat: I. Modulation by dopamine-depleting lesions. *Synapse* 4:229–237.

Cha JH, Frey AS, Alsdorf SA, Kerner JA, Kosinski CM, Mangiarini L, Penney JB, Jr., Davies SW, Bates GP, Young AB (1999) Altered neurotransmitter receptor expression in transgenic mouse models of Huntington's disease. *Philos Trans R Soc Lond B Biol Sci* 354:981–989.

Cha JH, Kosinski CM, Kerner JA, Alsdorf SA, Mangiarini L, Davies SW, Penney JB, Bates GP, Young AB (1998) Altered brain neurotransmitter receptors in transgenic mice expressing a portion of an abnormal human Huntington disease gene. *Proc Natl Acad Sci USA* 95:6480–6485.

Charvin D, Roze E, Perrin V, Deyts C, Betuing S, Pages C, Regulier E, Luthi-Carter R, Brouillet E, Deglon N, Caboche J (2008) Haloperidol protects striatal neurons from dysfunction induced by mutated huntingtin in vivo. *Neurobiol Dis* 29:22–29.

Charvin D, Vanhoutte P, Pages C, Borrelli E, Caboche J (2005) Unraveling a role for dopamine in Huntington's disease: The dual role of reactive oxygen species and D2 receptor stimulation. *Proc Natl Acad Sci USA* 102:12218–12223.

Chase TN, Oh JD, Konitsiotis S (2000) Antiparkinsonian and antidyskinetic activity of drugs targeting central glutamatergic mechanisms. *J Neurol* 247 2(Suppl):II36–II42.

Chen L, Cagniard B, Mathews T, Jones S, Koh HC, Ding Y, Carvey PM, Ling Z, Kang UJ, Zhuang X (2005) Age-dependent motor deficits and dopaminergic dysfunction in DJ-1 null mice. *J Biol Chem* 280:21418–21426.

Ciranni MA, Kearney TE, Olson KR (2009) Comparing acute toxicity of first- and second-generation antipsychotic drugs: A 10-year, retrospective cohort study. *J Clin Psychiatry* 70:122–129.

Civelli O, Bunzow JR, Grandy DK, Zhou QY, Van Tol HH (1991) Molecular biology of the dopamine receptors. *Eur J Pharmacol* 207:277–286.

Colwell CS, Levine MS (1995) Excitatory synaptic transmission in neostriatal neurons: Regulation by cyclic AMP-dependent mechanisms. *J Neurosci* 15:1704–1713.

Cummings DM, Milnerwood AJ, Dallerac GM, Waights V, Brown JY, Vatsavayai SC, Hirst MC, Murphy KP (2006) Aberrant cortical synaptic plasticity and dopaminergic dysfunction in a mouse model of Huntington's disease. *Hum Mol Genet* 15:2856–2868.

Cyr M, Sotnikova TD, Gainetdinov RR, Caron MG (2006) Dopamine enhances motor and neuropathological consequences of polyglutamine expanded huntingtin. *Faseb J* 20:2541–2543.

Dani JA, Zhou FM (2004) Selective dopamine filter of glutamate striatal afferents. *Neuron* 42:522–524.

Davies SW, Turmaine M, Cozens BA, DiFiglia M, Sharp AH, Ross CA, Scherzinger E, Wanker EE, Mangiarini L, Bates GP (1997) Formation of neuronal intranuclear inclusions underlies the neurological dysfunction in mice transgenic for the HD mutation. *Cell* 90:537–548.

Day M, Wang Z, Ding J, An X, Ingham CA, Shering AF, Wokosin D, Ilijic E, Sun Z, Sampson AR, Mugnaini E, Deutch AY, Sesack SR, Arbuthnott GW, Surmeier DJ (2006) Selective elimination of glutamatergic synapses on striatopallidal neurons in Parkinson disease models. *Nat Neurosci* 9:251–259.

DiFiglia M (1990) Excitotoxic injury of the neostriatum: A model for Huntington's disease. *Trends Neurosci* 13:286–289.

DiFiglia M, Sapp E, Chase K, Schwarz C, Meloni A, Young C, Martin E, Vonsattel JP, Carraway R, Reeves SA, et al. (1995) Huntingtin is a cytoplasmic protein associated with vesicles in human and rat brain neurons. *Neuron* 14:1075–1081.

Ding J, Peterson JD, Surmeier DJ (2008) Corticostriatal and thalamostriatal synapses have distinctive properties. *J Neurosci* 28:6483–6492.

Dunah AW, Sirianni AC, Fienberg AA, Bastia E, Schwarzschild MA, Standaert DG (2004) Dopamine D1-dependent trafficking of striatal N-methyl-D-aspartate glutamate receptors requires Fyn protein tyrosine kinase but not DARPP-32. *Mol Pharmacol* 65:121–129.

Dunah AW, Standaert DG (2001) Dopamine D1 receptor-dependent trafficking of striatal NMDA glutamate receptors to the postsynaptic membrane. *J Neurosci* 21:5546–5558.

Fiorentini C, Gardoni F, Spano P, Di Luca M, Missale C (2003) Regulation of dopamine D1 receptor trafficking and desensitization by oligomerization with glutamate N-methyl-D-aspartate receptors. *J Biol Chem* 278:20196–20202.

Fleming SM, Chesselet MF (2006) Behavioral phenotypes and pharmacology in genetic mouse models of Parkinsonism. *Behav Pharmacol* 17:383–391.

Fleming SM, Salcedo J, Hutson CB, Rockenstein E, Masliah E, Levine MS, Chesselet MF (2006) Behavioral effects of dopaminergic agonists in transgenic mice overexpressing human wildtype alpha-synuclein. *Neuroscience* 142:1245–1253.

Flores-Hernandez J, Cepeda C, Hernandez-Echeagaray E, Calvert CR, Jokel ES, Fienberg AA, Greengard P, Levine MS (2002) Dopamine enhancement of NMDA currents in dissociated medium-sized striatal neurons: Role of D1 receptors and DARPP-32. *J Neurophysiol* 88:3010–3020.

Flores-Hernandez J, Galarraga E, Bargas J (1997) Dopamine selects glutamatergic inputs to neostriatal neurons. *Synapse* 25:185–195.

Florio T, Di Loreto S, Cerrito F, Scarnati E (1993) Influence of prelimbic and sensorimotor cortices on striatal neurons in the rat: Electrophysiological evidence for converging inputs and the effects of 6-OHDA-induced degeneration of the substantia nigra. *Brain Res* 619:180–188.

Furukawa Y, Nygaard TG, Gutlich M, Rajput AH, Pifl C, DiStefano L, Chang LJ, Price K, Shimadzu M, Hornykiewicz O, Haycock JW, Kish SJ (1999) Striatal biopterin and tyrosine hydroxylase protein reduction in dopa-responsive dystonia. *Neurology* 53:1032–1041.

Galarraga E, Bargas J, Martinez-Fong D, Aceves J (1987) Spontaneous synaptic potentials in dopamine-denervated neostriatal neurons. *Neurosci Lett* 81:351–355.

Gerfen CR (1992) The neostriatal mosaic: Multiple levels of compartmental organization in the basal ganglia. *Annu Rev Neurosci* 15:285–320.

Gerfen CR, Engber TM, Mahan LC, Susel Z, Chase TN, Monsma FJ, Jr., Sibley DR (1990) D1 and D2 dopamine receptor-regulated gene expression of striatonigral and striatopallidal neurons. *Science* 250:1429–1432.

Gertler TS, Chan CS, Surmeier DJ (2008) Dichotomous anatomical properties of adult striatal medium spiny neurons. *J Neurosci* 28:10814–10824.

Glass M, Dragunow M, Faull RL (2000) The pattern of neurodegeneration in Huntington's disease: A comparative study of cannabinoid, dopamine, adenosine and GABA(A) receptor alterations in the human basal ganglia in Huntington's disease. *Neuroscience* 97:505–519.

Goldberg MS, Fleming SM, Palacino JJ, Cepeda C, Lam HA, Bhatnagar A, Meloni EG, Wu N, Ackerson LC, Klapstein GJ, Gajendiran M, Roth BL, Chesselet MF, Maidment NT, Levine MS, Shen J (2003) Parkin-deficient mice exhibit nigrostriatal deficits but not loss of dopaminergic neurons. *J Biol Chem* 278:43628–43635.

Goldberg MS, Pisani A, Haburcak M, Vortherms TA, Kitada T, Costa C, Tong Y, Martella G, Tscherter A, Martins A, Bernardi G, Roth BL, Pothos EN, Calabresi P, Shen J (2005) Nigrostriatal dopaminergic deficits and hypokinesia caused by inactivation of the familial Parkinsonism-linked gene DJ-1. *Neuron* 45:489–496.

Gourzis P, Polychronopoulos P, Papapetropoulos S, Assimakopoulos K, Argyriou AA, Beratis S (2005) Quetiapine in the treatment of focal tardive dystonia induced by other atypical antipsychotics: A report of 2 cases. *Clin Neuropharmacol* 28:195–196.

Grassi E, Latorraca S, Piacentini S, Marini P, Sorbi S (2000) Risperidone in idiopathic and symptomatic dystonia: Preliminary experience. *Neurol Sci* 21:121–123.

Graybiel AM (1990) Neurotransmitters and neuromodulators in the basal ganglia. *Trends Neurosci* 13:244–254.

Haber SN, Nauta WJ (1983) Ramifications of the globus pallidus in the rat as indicated by patterns of immunohistochemistry. *Neuroscience* 9:245–260.

Hallett PJ, Spoelgen R, Hyman BT, Standaert DG, Dunah AW (2006) Dopamine D1 activation potentiates striatal NMDA receptors by tyrosine phosphorylation-dependent subunit trafficking. *J Neurosci* 26:4690–4700.

Hallett PJ, Standaert DG (2004) Rationale for and use of NMDA receptor antagonists in Parkinson's disease. *Pharmacol Ther* 102:155–174.

Hamann M, Richter A (2004) Striatal increase of extracellular dopamine levels during dystonic episodes in a genetic model of paroxysmal dyskinesia. *Neurobiol Dis* 16:78–84.

Hamann M, Sohr R, Morgenstern R, Richter A (2008) Extracellular amino acid levels in the striatum of the dt(sz) mutant, a model of paroxysmal dystonia. *Neuroscience* 157:188–195.

Harper PS (1996) *Huntington's Disease*, 2nd edn., London, U.K.: Saunders.

Heng MY, Tallaksen-Greene SJ, Detloff PJ, Albin RL (2007) Longitudinal evaluation of the Hdh(CAG)150 knock-in murine model of Huntington's disease. *J Neurosci* 27:8989–8998.

Hernandez-Echeagaray E, Starling AJ, Cepeda C, Levine MS (2004) Modulation of AMPA currents by D2 dopamine receptors in striatal medium-sized spiny neurons: Are dendrites necessary? *Eur J Neurosci* 19:2455–2463.

Hernandez-Lopez S, Tkatch T, Perez-Garci E, Galarraga E, Bargas J, Hamm H, Surmeier DJ (2000) D2 dopamine receptors in striatal medium spiny neurons reduce L-type Ca^{2+} currents and excitability via a novel PLC[beta]1-IP3-calcineurin-signaling cascade. *J Neurosci* 20:8987–8995.

Hickey MA, Gallant K, Gross GG, Levine MS, Chesselet MF (2005) Early behavioral deficits in R6/2 mice suitable for use in preclinical drug testing. *Neurobiol Dis* 20:1–11.

Hickey MA, Kosmalska A, Enayati J, Cohen R, Zeitlin S, Levine MS, Chesselet MF (2008) Extensive early motor and non-motor behavioral deficits are followed by striatal neuronal loss in knock-in Huntington's disease mice. *Neuroscience* 157:280–295.

Hickey MA, Reynolds GP, Morton AJ (2002) The role of dopamine in motor symptoms in the R6/2 transgenic mouse model of Huntington's disease. *J Neurochem* 81:46–59.

Hodgson JG, Agopyan N, Gutekunst CA, Leavitt BR, LePiane F, Singaraja R, Smith DJ, Bissada N, McCutcheon K, Nasir J, Jamot L, Li XJ, Stevens ME, Rosemond E, Roder JC, Phillips AG, Rubin EM, Hersch SM, Hayden MR (1999) A YAC mouse model for Huntington's disease with full-length mutant huntingtin, cytoplasmic toxicity, and selective striatal neurodegeneration. *Neuron* 23:181–192.

Hollmann M, Heinemann S (1994) Cloned glutamate receptors. *Annu Rev Neurosci* 17:31–108.

Horstink CA, Praamstra P, Horstink MW, Berger HJ, Booij J, Van Royen EA (1997) Low striatal D2 receptor binding as assessed by [123I]IBZM SPECT in patients with writer's cramp. *J Neurol Neurosurg Psychiatry* 62:672–673.

Hull CD, Levine MS, Buchwald NA, Heller A, Browning RA (1974) The spontaneous firing pattern of forebrain neurons. I. The effects of dopamine and non-dopamine depleting lesions on caudate unit firing patterns. *Brain Res* 73:241–262.

Huot P, Levesque M, Parent A (2007) The fate of striatal dopaminergic neurons in Parkinson's disease and Huntington's chorea. *Brain* 130:222–232.

Huot P, Parent A (2007) Dopaminergic neurons intrinsic to the striatum. *J Neurochem* 101:1441–1447.

Jankovic J (2006) Treatment of dystonia. *Lancet Neurol* 5:864–872.

Jankovic J (2008) Parkinson's disease: Clinical features and diagnosis. *J Neurol Neurosurg Psychiatry* 79:368–376.

Johnson MA, Rajan V, Miller CE, Wightman RM (2006) Dopamine release is severely compromised in the R6/2 mouse model of Huntington's disease. *J Neurochem* 97:737–746.

Joshi PR, Wu NP, Andre VM, Cummings DM, Cepeda C, Joyce JA, Carroll JB, Leavitt BR, Hayden MR, Levine MS, Bamford NS (2009) Age-dependent alterations of corticostriatal activity in the YAC128 mouse model of Huntington disease. *J Neurosci* 29:2414–2427.

Kawaguchi Y, Wilson CJ, Emson PC (1990) Projection subtypes of rat neostriatal matrix cells revealed by intracellular injection of biocytin. *J Neurosci* 10:3421–3438.

Kemp JM, Powell TP (1970) The cortico-striate projection in the monkey. *Brain* 93:525–546.

Kim RH, Smith PD, Aleyasin H, Hayley S, Mount MP, Pownall S, Wakeham A, You-Ten AJ, Kalia SK, Horne P, Westaway D, Lozano AM, Anisman H, Park DS, Mak TW (2005) Hypersensitivity of DJ-1-deficient mice to 1-methyl-4-phenyl-1,2,3,6-tetrahydropyridine (MPTP) and oxidative stress. *Proc Natl Acad Sci USA* 102:5215–5220.

Kim DS, Szczypka MS, Palmiter RD (2000) Dopamine-deficient mice are hypersensitive to dopamine receptor agonists. *J Neurosci* 20:4405–4413.

Kita H, Kitai ST (1988) Glutamate decarboxylase immunoreactive neurons in rat neostriatum: Their morphological types and populations. *Brain Res* 447:346–352.

Klapstein GJ, Fisher RS, Zanjani H, Cepeda C, Jokel ES, Chesselet MF, Levine MS (2001) Electrophysiological and morphological changes in striatal spiny neurons in R6/2 Huntington's disease transgenic mice. *J Neurophysiol* 86:2667–2677.

Kornhuber J, Kornhuber ME (1986) Presynaptic dopaminergic modulation of cortical input to the striatum. *Life Sci* 39:699–674.

Kreitzer AC, Malenka RC (2007) Endocannabinoid-mediated rescue of striatal LTD and motor deficits in Parkinson's disease models. *Nature* 445:643–647.

Lee ST, Chu K, Jung KH, Kim J, Kim EH, Kim SJ, Sinn DI, Ko SY, Kim M, Roh JK (2006) Memantine reduces hematoma expansion in experimental intracerebral hemorrhage, resulting in functional improvement. *J Cereb Blood Flow Metab* 26:536–544.

Lei W, Jiao Y, Del Mar N, Reiner A (2004) Evidence for differential cortical input to direct pathway versus indirect pathway striatal projection neurons in rats. *J Neurosci* 24:8289–8299.

Lev N, Ickowicz D, Barhum Y, Lev S, Melamed E, Offen D (2009) DJ-1 protects against dopamine toxicity. *J Neural Transm* 116:151–160.

Levine MS, Altemus KL, Cepeda C, Cromwell HC, Crawford C, Ariano MA, Drago J, Sibley DR, Westphal H (1996) Modulatory actions of dopamine on NMDA receptor-mediated responses are reduced in D1A-deficient mutant mice. *J Neurosci* 16:5870–5882.

Levine MS, Cepeda C, Hickey MA, Fleming SM, Chesselet MF (2004) Genetic mouse models of Huntington's and Parkinson's diseases: Illuminating but imperfect. *Trends Neurosci* 27:691–697.

Levine MS, Klapstein GJ, Koppel A, Gruen E, Cepeda C, Vargas ME, Jokel ES, Carpenter EM, Zanjani H, Hurst RS, Efstratiadis A, Zeitlin S, Chesselet MF (1999) Enhanced sensitivity to N-methyl-D-aspartate receptor activation in transgenic and knockin mouse models of Huntington's disease. *J Neurosci Res* 58:515–532.

Lin CH, Tallaksen-Greene S, Chien WM, Cearley JA, Jackson WS, Crouse AB, Ren S, Li XJ, Albin RL, Detloff PJ (2001) Neurological abnormalities in a knock-in mouse model of Huntington's disease. *Hum Mol Genet* 10:137–144.

Lindgren N, Usiello A, Goiny M, Haycock J, Erbs E, Greengard P, Hokfelt T, Borrelli E, Fisone G (2003) Distinct roles of dopamine D2L and D2S receptor isoforms in the regulation of protein phosphorylation at presynaptic and postsynaptic sites. *Proc Natl Acad Sci USA* 100:4305–4309.

Lione LA, Carter RJ, Hunt MJ, Bates GP, Morton AJ, Dunnett SB (1999) Selective discrimination learning impairments in mice expressing the human Huntington's disease mutation. *J Neurosci* 19:10428–10437.

Luthi-Carter R, Strand A, Peters NL, Solano SM, Hollingsworth ZR, Menon AS, Frey AS, Spektor BS, Penney EB, Schilling G, Ross CA, Borchelt DR, Tapscott SJ, Young AB, Cha JH, Olson JM (2000) Decreased expression of striatal signaling genes in a mouse model of Huntington's disease. *Hum Mol Genet* 9:1259–1271.

Maidment NT, Lam HA, Ackerson LC, Rockenstein E, Masliah E (2006) Dysregulation of dopamine transmission in mice overexpressing human wildtype alpha-synuclein. Society for Neuroscience Abstracts: Program No. 378.4.

Mallet N, Ballion B, Le Moine C, Gonon F (2006) Cortical inputs and GABA interneurons imbalance projection neurons in the striatum of parkinsonian rats. *J Neurosci* 26:3875–3884.

Mangiarini L, Sathasivam K, Seller M, Cozens B, Harper A, Hetherington C, Lawton M, Trottier Y, Lehrach H, Davies SW, Bates GP (1996) Exon 1 of the HD gene with an expanded CAG repeat is sufficient to cause a progressive neurological phenotype in transgenic mice. *Cell* 87:493–506.

Menalled LB, Sison JD, Dragatsis I, Zeitlin S, Chesselet MF (2003) Time course of early motor and neuropathological anomalies in a knock-in mouse model of Huntington's disease with 140 CAG repeats. *J Comp Neurol* 465:11–26.

Menalled LB, Sison JD, Wu Y, Olivieri M, Li XJ, Li H, Zeitlin S, Chesselet MF (2002) Early motor dysfunction and striosomal distribution of huntingtin microaggregates in Huntington's disease knock-in mice. *J Neurosci* 22:8266–8276.

Menalled L, Zanjani H, MacKenzie L, Koppel A, Carpenter E, Zeitlin S, Chesselet MF (2000) Decrease in striatal enkephalin mRNA in mouse models of Huntington's disease. *Exp Neurol* 162:328–342.

Mercuri N, Bernardi G, Calabresi P, Cotugno A, Levi G, Stanzione P (1985) Dopamine decreases cell excitability in rat striatal neurons by pre- and postsynaptic mechanisms. *Brain Res* 358:110–121.

Milnerwood AJ, Raymond LA (2007) Corticostriatal synaptic function in mouse models of Huntington's disease: Early effects of huntingtin repeat length and protein load. *J Physiol* 585:817–831.

Misbahuddin A, Placzek MR, Chaudhuri KR, Wood NW, Bhatia KP, Warner TT (2002) A polymorphism in the dopamine receptor DRD5 is associated with blepharospasm. *Neurology* 58:124–126.

Monaghan DT, Bridges RJ, Cotman CW (1989) The excitatory amino acid receptors: Their classes, pharmacology, and distinct properties in the function of the central nervous system. *Annu Rev Pharmacol Toxicol* 29:365–402.

Murphy KP, Carter RJ, Lione LA, Mangiarini L, Mahal A, Bates GP, Dunnett SB, Morton AJ (2000) Abnormal synaptic plasticity and impaired spatial cognition in mice transgenic for exon 1 of the human Huntington's disease mutation. *J Neurosci* 20:5115–5123.

Nakanishi S (1994) Metabotropic glutamate receptors: Synaptic transmission, modulation, and plasticity. *Neuron* 13:1031–1037.

Naumann M, Pirker W, Reiners K, Lange KW, Becker G, Brucke T (1998) Imaging the pre- and postsynaptic side of striatal dopaminergic synapses in idiopathic cervical dystonia: A SPECT study using [123I] epidepride and [123I] beta-CIT. *Mov Disord* 13:319–323.

Neely MD, Schmidt DE, Deutch AY (2007) Cortical regulation of dopamine depletion-induced dendritic spine loss in striatal medium spiny neurons. *Neuroscience* 149:457–464.

Nishi A, Snyder GL, Greengard P (1997) Bidirectional regulation of DARPP-32 phosphorylation by dopamine. *J Neurosci* 17:8147–8155.

Nowak L, Bregestovski P, Ascher P, Herbet A, Prochiantz A (1984) Magnesium gates glutamate-activated channels in mouse central neurones. *Nature* 307:462–465.

Nutt JG, Gunzler SA, Kirchhoff T, Hogarth P, Weaver JL, Krams M, Jamerson B, Menniti FS, Landen JW (2008) Effects of a NR2B selective NMDA glutamate antagonist, CP-101,606, on dyskinesia and Parkinsonism. *Mov Disord* 23:1860–1866.

Ondo WG, Mejia NI, Hunter CB (2007) A pilot study of the clinical efficacy and safety of memantine for Huntington's disease. *Parkinsonism Relat Disord* 13:453–454.

Otsuka M, Ichiya Y, Shima F, Kuwabara Y, Sasaki M, Fukumura T, Kato M, Masuda K, Goto I (1992) Increased striatal 18F-dopa uptake and normal glucose metabolism in idiopathic dystonia syndrome. *J Neurol Sci* 111:195–199.

Pavese N, Andrews TC, Brooks DJ, Ho AK, Rosser AE, Barker RA, Robbins TW, Sahakian BJ, Dunnett SB, Piccini P (2003) Progressive striatal and cortical dopamine receptor dysfunction in Huntington's disease: A PET study. *Brain* 126:1127–1135.

Perlmutter JS, Stambuk MK, Markham J, Black KJ, McGee-Minnich L, Jankovic J, Moerlein SM (1997) Decreased [18F]spiperone binding in putamen in idiopathic focal dystonia. *J Neurosci* 17:843–850.

Placzek MR, Misbahuddin A, Chaudhuri KR, Wood NW, Bhatia KP, Warner TT (2001) Cervical dystonia is associated with a polymorphism in the dopamine (D5) receptor gene. *J Neurol Neurosurg Psychiatry* 71:262–264.

Politis M, Pavese N, Tai YF, Tabrizi SJ, Barker RA, Piccini P (2008) Hypothalamic involvement in Huntington's disease: An in vivo PET study. *Brain* 131:2860–2869.

Price CJ, Kim P, Raymond LA (1999) D1 dopamine receptor-induced cyclic AMP-dependent protein kinase phosphorylation and potentiation of striatal glutamate receptors. *J Neurochem* 73:2441–2446.

Rasmussen A, Macias R, Yescas P, Ochoa A, Davila G, Alonso E (2000) Huntington disease in children: Genotype-phenotype correlation. *Neuropediatrics* 31:190–194.

Reiner A, Jiao Y, Del Mar N, Laverghetta AV, Lei WL (2003) Differential morphology of pyramidal tract-type and intratelencephalically projecting-type corticostriatal neurons and their intrastriatal terminals in rats. *J Comp Neurol* 457:420–440.

Richter A, Fredow G, Loscher W (1991) Antidystonic effects of the NMDA receptor antagonists memantine, MK-801 and CGP 37849 in a mutant hamster model of paroxysmal dystonia. *Neurosci Lett* 133:57–60.

Richter A, Loscher W (1997) MK-801 potentiates antidystonic effects of clozapine but not of haloperidol in mutant dystonic hamsters. *Brain Res* 769:296–302.

Riederer P, Reichmann H, Janetzky B, Sian J, Lesch KP, Lange KW, Double KL, Nagatsu T, Gerlach M (2001) Neural degeneration in Parkinson's disease. *Adv Neurol* 86:125–136.

Robinson P, Lebel M, Cyr M (2008) Dopamine D1 receptor-mediated aggregation of N-terminal fragments of mutant huntingtin and cell death in a neuroblastoma cell line. *Neuroscience* 153:762–772.

Sander SE, Richter A (2007) Effects of intrastriatal injections of glutamate receptor antagonists on the severity of paroxysmal dystonia in the dtsz mutant. *Eur J Pharmacol* 563:102–108.

Sato K, Sumi-Ichinose C, Kaji R, Ikemoto K, Nomura T, Nagatsu I, Ichinose H, Ito M, Sako W, Nagahiro S, Graybiel AM, Goto S (2008) Differential involvement of striosome and matrix dopamine systems in a transgenic model of dopa-responsive dystonia. *Proc Natl Acad Sci USA* 105:12551–12556.

Sauer H, Oertel WH (1994) Progressive degeneration of nigrostriatal dopamine neurons following intrastriatal terminal lesions with 6-hydroxydopamine: A combined retrograde tracing and immunocytochemical study in the rat. *Neuroscience* 59:401–415.

Schultz W (1982) Depletion of dopamine in the striatum as an experimental model of Parkinsonism: Direct effects and adaptive mechanisms. *Prog Neurobiol* 18:121–166.

Schultz W, Ungerstedt U (1978) Short-term increase and long-term reversion of striatal cell activity after degeneration of the nigrostriatal dopamine system. *Exp Brain Res* 33:159–171.

Sciamanna G, Bonsi P, Tassone A, Cuomo D, Tscherter A, Viscomi MT, Martella G, Sharma N, Bernardi G, Standaert DG, Pisani A (2009) Impaired striatal D2 receptor function leads to enhanced GABA transmission in a mouse model of DYT1 dystonia. *Neurobiol Dis* 34:133–145.

Seamans JK, Yang CR (2004) The principal features and mechanisms of dopamine modulation in the prefrontal cortex. *Prog Neurobiol* 74:1–58.

Segawa M, Nomura Y, Nishiyama N (2003) Autosomal dominant guanosine triphosphate cyclohydrolase I deficiency (Segawa disease). *Ann Neurol* 54(Suppl 6):S32–S45.

Shelbourne PF, Killeen N, Hevner RF, Johnston HM, Tecott L, Lewandoski M, Ennis M, Ramirez L, Li Z, Iannicola C, Littman DR, Myers RM (1999) A Huntington's disease CAG expansion at the murine Hdh locus is unstable and associated with behavioural abnormalities in mice. *Hum Mol Genet* 8:763–774.

Shuen JA, Chen M, Gloss B, Calakos N (2008) Drd1a-tdTomato BAC transgenic mice for simultaneous visualization of medium spiny neurons in the direct and indirect pathways of the basal ganglia. *J Neurosci* 28:2681–2685.

Sibley DR, Monsma FJ, Jr. (1992) Molecular biology of dopamine receptors. *Trends Pharmacol Sci* 13:61–69.

Simonyan K, Ludlow CL (2010) Abnormal activation of the primary somatosensory cortex in spasmodic dysphonia: An fMRI study. *Cereb Cortex,* 20(11):2749–2759.

Slow EJ, Van Raamsdonk J, Rogers D, Coleman SH, Graham RK, Deng Y, Oh R, Bissada N, Hossain SM, Yang YZ, Li XJ, Simpson EM, Gutekunst CA, Leavitt BR, Hayden MR (2003) Selective striatal neuronal loss in a YAC128 mouse model of Huntington disease. *Hum Mol Genet* 12:1555–1567.

Smeal RM, Gaspar RC, Keefe KA, Wilcox KS (2007) A rat brain slice preparation for characterizing both thalamostriatal and corticostriatal afferents. *J Neurosci Methods* 159:224–235.

Smeal RM, Keefe KA, Wilcox KS (2008) Differences in excitatory transmission between thalamic and cortical afferents to single spiny efferent neurons of rat dorsal striatum. *Eur J Neurosci* 28:2041–2052.

Smith AD, Bolam JP (1990) The neural network of the basal ganglia as revealed by the study of synaptic connections of identified neurones. *Trends Neurosci* 13:259–265.

Smith R, Brundin P, Li JY (2005) Synaptic dysfunction in Huntington's disease: A new perspective. *Cell Mol Life Sci* 62:1901–1912.

Snyder GL, Allen PB, Fienberg AA, Valle CG, Huganir RL, Nairn AC, Greengard P (2000) Regulation of phosphorylation of the GluR1 AMPA receptor in the neostriatum by dopamine and psychostimulants in vivo. *J Neurosci* 20:4480–4488.

Snyder GL, Fienberg AA, Huganir RL, Greengard P (1998) A dopamine/D1 receptor/protein kinase A/dopamine- and cAMP-regulated phosphoprotein (Mr 32 kDa)/protein phosphatase-1 pathway regulates dephosphorylation of the NMDA receptor. *J Neurosci* 18:10297–10303.

Spampanato J, Gu X, Yang XW, Mody I (2008) Progressive synaptic pathology of motor cortical neurons in a BAC transgenic mouse model of Huntington's disease. *Neuroscience* 157:606–620.

Stack EC, Dedeoglu A, Smith KM, Cormier K, Kubilus JK, Bogdanov M, Matson WR, Yang L, Jenkins BG, Luthi-Carter R, Kowall NW, Hersch SM, Beal MF, Ferrante RJ (2007) Neuroprotective effects of synaptic modulation in Huntington's disease R6/2 mice. *J Neurosci* 27:12908–12915.

Steece-Collier K, Chambers LK, Jaw-Tsai SS, Menniti FS, Greenamyre JT (2000) Antiparkinsonian actions of CP-101,606, an antagonist of NR2B subunit-containing N-methyl-d-aspartate receptors. *Exp Neurol* 163:239–243.

Strong TV, Tagle DA, Valdes JM, Elmer LW, Boehm K, Swaroop M, Kaatz KW, Collins FS, Albin RL (1993) Widespread expression of the human and rat Huntington's disease gene in brain and nonneural tissues. *Nat Genet* 5:259–265.

Surmeier DJ, Ding J, Day M, Wang Z, Shen W (2007) D1 and D2 dopamine-receptor modulation of striatal glutamatergic signaling in striatal medium spiny neurons. *Trends Neurosci* 30:228–235.

Tai YF, Pavese N, Gerhard A, Tabrizi SJ, Barker RA, Brooks DJ, Piccini P (2007) Microglial activation in presymptomatic Huntington's disease gene carriers. *Brain* 130:1759–1766.

Tallaksen-Greene SJ, Crouse AB, Hunter JM, Detloff PJ, Albin RL (2005) Neuronal intranuclear inclusions and neuropil aggregates in HdhCAG(150) knockin mice. *Neuroscience* 131:843–852.

Tang TS, Chen X, Liu J, Bezprozvanny I (2007) Dopaminergic signaling and striatal neurodegeneration in Huntington's disease. *J Neurosci* 27:7899–7910.

Tedroff J, Ekesbo A, Sonesson C, Waters N, Carlsson A (1999) Long-lasting improvement following (-)-OSU6162 in a patient with Huntington's disease. *Neurology* 53:1605–1606.

The Huntington's Disease Collaborative Research Group (1993) A novel gene containing a trinucleotide repeat that is expanded and unstable on Huntington's disease chromosomes. *Cell* 72:971–983.

Thompson PD, Berardelli A, Rothwell JC, Day BL, Dick JP, Benecke R, Marsden CD (1988) The coexistence of bradykinesia and chorea in Huntington's disease and its implications for theories of basal ganglia control of movement. *Brain* 111(Pt 2):223–244.

Thony B, Blau N (2006) Mutations in the BH4-metabolizing genes GTP cyclohydrolase I, 6-pyruvoyl-tetrahydropterin synthase, sepiapterin reductase, carbinolamine-4a-dehydratase, and dihydropteridine reductase. *Hum Mutat* 27:870–878.

Tinazzi M, Rosso T, Fiaschi A (2003) Role of the somatosensory system in primary dystonia. *Mov Disord* 18:605–622.

Tseng KY, Kasanetz F, Kargieman L, Riquelme LA, Murer MG (2001) Cortical slow oscillatory activity is reflected in the membrane potential and spike trains of striatal neurons in rats with chronic nigrostriatal lesions. *J Neurosci* 21:6430–6439.

Tseng KY, O'Donnell P (2004) Dopamine-glutamate interactions controlling prefrontal cortical pyramidal cell excitability involve multiple signaling mechanisms. *J Neurosci* 24:5131–5139.

Uhl GR, Navia B, Douglas J (1988) Differential expression of preproenkephalin and preprodynorphin mRNAs in striatal neurons: High levels of preproenkephalin expression depend on cerebral cortical afferents. *J Neurosci* 8:4755–4764.

Van Raamsdonk JM, Pearson J, Slow EJ, Hossain SM, Leavitt BR, Hayden MR (2005) Cognitive dysfunction precedes neuropathology and motor abnormalities in the YAC128 mouse model of Huntington's disease. *J Neurosci* 25:4169–4180.

Vonsattel JP, DiFiglia M (1998) Huntington disease. *J Neuropathol Exp Neurol* 57:369–384.

Vonsattel JP, Myers RH, Stevens TJ, Ferrante RJ, Bird ED, Richardson EP, Jr. (1985) Neuropathological classification of Huntington's disease. *J Neuropathol Exp Neurol* 44:559–577.

Wang H, Pickel VM (2002) Dopamine D2 receptors are present in prefrontal cortical afferents and their targets in patches of the rat caudate-putamen nucleus. *J Comp Neurol* 442:392–404.

Wang X, Sarkar A, Cicchetti F, Yu M, Zhu A, Jokivarsi K, Saint-Pierre M, Brownell AL (2005) Cerebral PET imaging and histological evidence of transglutaminase inhibitor cystamine induced neuroprotection in transgenic R6/2 mouse model of Huntington's disease. *J Neurol Sci* 231:57–66.

Watson JB, Hatami A, David H, Masliah E, Roberts K, Evans CE, Levine MS (2009) Alterations in corticostriatal synaptic plasticity in mice overexpressing human alpha-synuclein. *Neuroscience* 159:501–513.

Weeks RA, Piccini P, Harding AE, Brooks DJ (1996) Striatal D1 and D2 dopamine receptor loss in asymptomatic mutation carriers of Huntington's disease. *Ann Neurol* 40:49–54.

Wheeler VC, Gutekunst CA, Vrbanac V, Lebel LA, Schilling G, Hersch S, Friedlander RM, Gusella JF, Vonsattel JP, Borchelt DR, MacDonald ME (2002) Early phenotypes that presage late-onset neurodegenerative disease allow testing of modifiers in Hdh CAG knock-in mice. *Hum Mol Genet* 11:633–640.

Wheeler VC, White JK, Gutekunst CA, Vrbanac V, Weaver M, Li XJ, Li SH, Yi H, Vonsattel JP, Gusella JF, Hersch S, Auerbach W, Joyner AL, MacDonald ME (2000) Long glutamine tracts cause nuclear localization of a novel form of huntingtin in medium spiny striatal neurons in HdhQ92 and HdhQ111 knock-in mice. *Hum Mol Genet* 9:503–513.

White JK, Auerbach W, Duyao MP, Vonsattel J-P, Gusella JF, Joyner AL, MacDonald ME (1997) Huntingtin is required for neurogenesis and is not impaired by the Huntington's disease CAG expansion. *Nat Genet* 17(4):404–410.

Wilson CJ (1990) Basal ganglia. In: *The Synaptic Organization of the Brain* (Shepherd GW, ed.), pp. 279–316. Oxford, U.K.: Oxford University Press.

Wood PT, MacMillan JC, Harper PS, Lowenstein PR, Jones AL (1996) Partial characterisation of murine huntingtin and apparent variations in the subcellular localisation of huntingtin in human, mouse and rat brain. *Hum Mol Genet* 5:481–487.

Wu N, Cepeda C, Zhuang X, Levine MS (2007) Altered corticostriatal neurotransmission and modulation in dopamine transporter knock-down mice. *J Neurophysiol* 98:423–432.

Wu N, Joshi PR, Cepeda C, Masliah E, Levine MS (2010) Alpha-synuclein overexpression in mice alters synaptic communication in the corticostriatal pathway. *J Neurosci Res* 88:1764–1776.

Wu J, Tang T, Bezprozvanny I (2006) Evaluation of clinically relevant glutamate pathway inhibitors in in vitro model of Huntington's disease. *Neurosci Lett* 407:219–223.

Yamaguchi H, Shen J (2007) Absence of dopaminergic neuronal degeneration and oxidative damage in aged DJ-1-deficient mice. *Mol Neurodegener* 2:10.

Yan Z, Hsieh-Wilson L, Feng J, Tomizawa K, Allen PB, Fienberg AA, Nairn AC, Greengard P (1999) Protein phosphatase 1 modulation of neostriatal AMPA channels: Regulation by DARPP-32 and spinophilin. *Nat Neurosci* 2:13–17.

Zeng BY, Heales SJ, Cancvari L, Rose S, Jenner P (2004) Alterations in expression of dopamine receptors and neuropeptides in the striatum of GTP cyclohydrolase-deficient mice. *Exp Neurol* 190:515–524.

Zhao Y, DeCuypere M, LeDoux MS (2008) Abnormal motor function and dopamine neurotransmission in DYT1 DeltaGAG transgenic mice. *Exp Neurol* 210:719–730.

Zhou QY, Palmiter RD (1995) Dopamine-deficient mice are severely hypoactive, adipsic, and aphagic. *Cell* 83:1197–1209.

Index

A

α-Amino-3-hydroxy-5-methyl-4-
       isoxazolepropionic acid (AMPA)
   Ca^{2+} channels, 38
   CaMKII-dependent trafficking, 82
   D1 and D2 receptor activation, 231–232
   dystonia, 240
   GluA2 subunit, 80
   glutamate synapses, 77–78
   linear current–voltage (I–V)
       relationships, 32–33
   midbrain dopamine neurons, 44–45
   potentiation, 79
   SNc and striatum, 40, 41
   SNr connectivity, 178–179
   subunits, neostriatum, 35
   thalamostriatal synapses, 230
   YAC128 mice, 235
AMPA, *see* α-Amino-3-hydroxy-5-methyl-4-
       isoxazolepropionic acid (AMPA)
Anhedonia hypothesis, 210
α-synuclein-overexpressing (ASO) mice
   DAT knock-down mice, 238–239
   D2 receptor, 238

B

Basal ganglia (BG)
   description, 2
   dopaminergic neurons, SNc, 4
   information flow, 2
   localization, metabotropic glutamate
       receptors, 2, 3
   neurotransmission by group III mGluRs
      agonists in animal models, PD, 19–20
      anatomical studies, 17
      chronic l-DOPA treatment, 17
      detection, 16
      dopaminergic modulation, 18
      electrophysiological studies, 17–18
      immunohistochemical studies, 17
      modulation, dopamine, 18–19
      protection against nigrostriatal
        degeneration, 20
   neurotransmission by group II mGluRs
      GPe and STN, 13
      SNr, 13–14
      striatum, 11–12

neurotransmission by group I mGluRs
      dopaminergic modulation, 7
      globus pallidus (external), 7–8
      striatum, 5–7
      substantia nigra pars reticulata, 8–9
      subthalamic nucleus, 8
   SNr and GPi, 2
   striatal neuron population, 2
BG, *see* Basal ganglia
BG circuits
   afferents, 72–73
   definition, 72
   functional heterogeneity
      analysis, voluntary behavior, 212–213
      association cortices, 211
      cortico-BG network, 211
      glutamatergic and dopaminergic
        inputs, 212
      limbic and sensorimotor network, 212
      locomotor activity, 212
      medial to lateral gradient, 211
      preparatory behaviors, 210
   MSN, 73–74
   SN network, 172
   subcortical structures, 72

C

CAMKII and plasticity
   AMPAR trafficking, 82
   LTP maintenance, 81
   phosphorylate AMPAR GluR1 subunits, 82
   signaling pathways, 80–81, 90
   threonine 286 phosphorylation, 81
Corticostriatal (CS) neurons network, *in vivo*
   anatomical and electrophysiological
      features, 142, 143
   antidromic activation, 141
   diverse groups, 141
   as electrophysiological unit
      antidromic activation, 142
      conduction velocity, 142–143
      corticostriatal inputs, 144
      evoked synaptic potentials, 145
      firing rate, 145
      intrinsic membrane properties, 143
      timing-dependent unbalance, 143
   integration, cortical information
      basic mechanisms, 146–148

T - #0396 - 071024 - C3 - 234/156/13 - PB - 9780367381974 - Gloss Lamination